100/N

SEROTONIN

FROM CELL BIOLOGY TO PHARMACOLOGY AND THERAPEUTICS

Medical Science Symposia Series

Volume 5

The proceedings of the first International Symposium on Serotonin from Cell Biology to Pharmacology and Therapeutics, edited by R. Paoletti, P.M. Vanhoutte, N. Brunello and F.M. Maggi, were published by Kluwer Academic Publishers in 1990. ISBN 0-7923-0531-0

The titles published in this series are listed at the end of this volume.

Serotonin

From Cell Biology to Pharmacology and Therapeutics

Edited by

P. M. Vanhoutte
Baylor College of Medicine, Houston, Texas, U.S.A.

P. R. Saxena
Department of Pharmacology, Erasmus University, Rotterdam, The Netherlands

R. Paoletti
Institute of Pharmacological Sciences, University of Milan, Milan, Italy

N. Brunello
Department of Pharmaceutical Sciences, University of Modena, Modena, Italy

and

A. S. Jackson (Assistant Editor)
Giovanni Lorenzini Medical Foundation, Houston, Texas, U.S.A.

KLUWER ACADEMIC PUBLISHERS
DORDRECHT / BOSTON / LONDON

Fondazione Giovanni Lorenzini, Milan, Italy
Giovanni Lorenzini Medical Foundation, Houston, U.S.A.

Library of Congress Cataloging-in-Publication Data

```
Serotonin : from cell biology to pharmacology and therapeutics /
   edited by P.M. Vanhoutte ... [et al.].
      p.   cm. -- (Medical science symposia series ; v. 5)
   "Proceedings of the invited lectures of the Second International
Symposium on Serotonin from Cell Biology to Pharmacology and
Therapeutics, held in Houston, Texas, September 15-18, 1992."
   Includes bibliographical references and index.
   ISBN 0-7923-2518-4
   1. Serotonin--Physiological effect--Congresses.  2. Serotonin-
-Receptors--Congresses.  3. Serotoninergic mechanisms--Congresses.
I. Vanhoutte, Paul M.  II. International Symposium on "Serotonin:
from Cell Biology to Pharmacology and Therapeutics" (2nd : 1992 :
Houston, Tex.)  III. Series.
QP801.S4S473  1993
615'.7--dc20                                                  93-20892
```

ISBN 0-7923-2518-4

Published by Kluwer Academic Publishers,
P.O. Box 17, 3300 AA Dordrecht, The Netherlands.

Kluwer Academic Publishers incorporates
the publishing programmes of
D. Reidel, Martinus Nijhoff, Dr W. Junk and MTP Press.

Sold and distributed in the U.S.A. and Canada
by Kluwer Academic Publishers,
101 Philip Drive, Norwell, MA 02061, U.S.A.

In all other countries, sold and distributed
by Kluwer Academic Publishers Group,
P.O. Box 322, 3300 AH Dordrecht, The Netherlands.

Printed on acid-free paper

All Rights Reserved
© 1993 Kluwer Academic Publishers and Fondazione Giovanni Lorenzini
No part of the material protected by this copyright notice may be reproduced or
utilized in any form or by any means, electronic or mechanical,
including photocopying, recording or by any information storage and
retrieval system, without written permission from the copyright owner.

Printed in the Netherlands

CONTENTS

Preface — xi

I. MOLECULAR AND CELL BIOLOGY

1. Different Organization of Human Monoamine Oxidase (MAO) A and B Promoters
   J.C. Shih, Q.-S. Zhu, J. Grimsby, K. Chen, and J. Shih — 1

2. Molecular Biology of Serotonin Uptake Sites
   B.J. Hoffman — 9

3. Molecular Biology of 5-HT1 Receptors
   P. Hartig, N. Adham, J. Bard, A. Hou-Yu, H.-T. Kao, M. Macchi, L. Schechter, D. Urquhart, J. Zgombick, M. Durkin, R. Weinshank, and T. Branchek — 21

4. 5-HT$_{B/D}$-Receptor Agonists Block C-fos-LI within Trigeminal Nucleus Caudalis in Response to Noxious Meningeal Stimulation
   M.A. Moskowitz and K. Nozaki — 33

II. PHARMACOLOGY OF SEROTONIN RECEPTORS

5. A Reappraisal of 5-HT Receptor Classification
   P.P.A. Humphrey, P. Hartig, and D. Hoyer — 41

6. Classification of Agents Used in Serotonin Research
   R.A. Glennon — 49

7. Electrophysiology of Central Serotonin Neurotransmission
   P. Blier, G. Piñeyro, T. Dennis, and C. de Montigny — 55

8. Immunohistochemical Visualization of 5-HT1A Receptors in the Rat Brain
   H. Gozlan, K. Kia, J. Gingrich, M. Riad, E. Doucet, C. Gerard, D. Verge, and M. Hamon — 65

9. In Vitro Functional Correlates of 5-HT1-Like Receptors
   D.N. Middlemiss, M.S. Beer, and L.O. Wilkinson — 73

10. 5-HT1D Receptor Subtypes
    D. Hoyer, A.T. Bruinvels, and J.M. Palacios — 85

11. 5-HT3 Receptors
    G.J. Kilpatrick and H. Rogers — 99

12.	5-HT4 Receptor: Current Status D.E. Clarke and J. Bockaert	107
13.	Serotonergic Receptors: From Ligands to Sequence P.A.J. Janssen and H. Moereels	119

III. CENTRAL CARDIOVASCULAR EFFECTS AND MIGRAINE

14.	Possible Mechanisms of Action of Drugs Effective Against Acute Migraine Attacks P. Saxena	127
15.	SB 203186, A Potent 5-HT4 Antagonist for Cardiac 5-HT4 Receptors A.J. Kaumann, A.D. Medhurst, P. Boyland, M. Vimal, and R.C. Young	135
16.	Role of the Lateral Tegmental Field in the Sympatholytic Action of 5-HT1A Agonists R.B. McCall and M.E. Clement	141
17.	Urapidil: The Role of 5-HT1A and Ó-Adrenergic Receptors in Blood Pressure Reduction N. Kolassa, K.D. Beller, R. Boer, H. Boss, and K.H. Sanders	151

IV. PREVENTION OF EMESIS

18.	Clinical Aspects of Antiemesis: Role of the 5-HT3 Antagonists S.M. Grunberg	159
19.	The Antiemetic Properties of 5-HT3 Receptor Antagonists B. Costall, R.J. Naylor, and J.A. Rudd	165
20.	Mechanisms of the Emetic Response to Chemotherapy and of the Antiemetic Action of 5-HT3 Receptor Antagonists: Clinical Studies L.X. Cubeddu and I.S. Hoffmann	171
21.	Serotonin and Serotonergic Drugs in Emesis U. Wells, M. Ravenscroft, P. Bhandari, and P.L.R. Andrews	179
22.	Clinical Effects of 5-HT3 Antagonists in Treating Nausea and Vomiting M.E. Butcher	187

V. ANXIETY, DEPRESSION, AND MOOD DISORDERS

23.	Inhibitors of Serotonin Uptake S.Z. Langer and D. Graham	197

24.	Preclinical Evidence for the Role of Serotonin Receptors in Anxiety B. Olivier, J. Mos, E. Molewijk, T. Zethof, and G. van der Poel	205
25.	The Effects of Tryptophan Depletion in Depression R.M. Salomon, H.L. Miller, P.L. Delgado, and D.S. Charney	211
26.	Anxiolytic Effects of Drugs Acting on 5-HT Receptor Subtypes D.L. Murphy, A. Broocks, C. Aulakh, and T.A. Pigott	223
27.	Obsessive Compulsive Disorders: A Neurobiological Hypothesis B.L. Jacobs	231
28.	Role of the Serotonergic System in Pituitary Hormone Secretion: The Pharmacologic Challenge Paradigm in Man H.Y. Meltzer	239
29.	Serotonin in Anxiety and Related Disorders H.G.M. Westenberg and J.A. Den Boer	249
30.	Alcoholic Violent Offenders: Behavioral, Endocrine, Diurnal Rhythm, and Genetic Correlates of Low CSF 5-HIAA Concentrations M. Linnoila, M. Virkkunen, R. Tokola, D. Nielsen, and D. Goldman	255
VI.	**SCHIZOPHRENIA**	
31.	Modulation of Dopaminergic Neurotransmission by 5-HT2 Antagonism T.H. Svensson, G.G. Nomikos, and J.L. Andersson	263
32.	Combined Serotonin 5-HT2 and Dopamine DA2 Antagonist Treatments in Chronic Schizophrenia R.L. Borison and B.I. Diamond	271
33.	Is Serotonin Involved in the Pathogenesis of Schizophrenia? H.M. van Praag	277
34.	Serotonin Antagonists as Antipsychotics B. L. Kinkead, M.J. Owens, and C.B. Nemeroff	289
VII.	**VASCULAR FUNCTION**	
35.	Pharmacological Analysis of 5-HT Receptors in Human Small Coronary Arteries J.A. Angus, T.M. Cocks, and M. Ross-Smith	297
36.	Platelets, Platelet Mediators, and Unstable Coronary Artery Disease Syndromes J.T. Willerson	307

37.	Abnormal Coronary Vasoconstrictor Responses to Serotonin in Patients with Ischemic Heart Disease C.J. Vrints, J. Bosmans, H. Bult, A.G. Herman, and J.P. Snoeck	321

VIII. REGULATION OF THE IMMUNE SYSTEM

38.	5-HT and the Immune System G. Fillion, M.-P. Fillion, I. Cloez-Tayarani, H. Sarhan, F. Haour, and F. Bolanos	329
39.	The Stimulatory Effects of d-Fenfluramine (d-FEN) on Blood and Splenic Immune Functions in the Fischer 344 Rat are Age and Sex Dependent S.A. Lorens, L. Petrovic, G. Hejna, X.W. Dong, and J. Clancy, Jr.	337
40.	Serotonergic Regulation of Natural Killer Cells: A Minireview K. Hellstrand, C. Dahlgren, and S. Hermodsson	345
41.	Interactions between Serotonin and the Immune System: An Overview J. Clancy, Jr., G. Fillion, K. Hellstrand, and S.A. Lorens	353

IX. 5-HT MECHANISMS IN DRUG ABUSE

42.	The Role of Serotonin in the Cellular Physiological Effects of Cocaine J.M. Lakoski and H. Zheng	359
43.	Serotonin Mechanisms in Ethanol Abuse H. Lal, S.M. Rezazadeh, and C.J. Wallis	367
44.	Serotonin Regulation of Alcohol Drinking W.J. McBride, J.M. Murphy, L. Lumeng, and T.-K. Li	375

X. FEEDING BEHAVIOR

45.	Hypothalamic Serotonin in Relation to Appetite for Macronutrients and Eating Disorders S.R. Leibowitz	383
46.	Serotonergic Pharmacology of Appetite G. Curzon, E.L. Gibson, A.J. Kennedy, and A.O. Oluyomi	393

XI. NEUROPROPTECTION, SLEEP, AND MEMORY

47.	Serotonin $(5-HT)_{1A}$ Receptor Agonists and Neuroprotection J. Traber, K.M. Bode-Greuel, and E. Horváth	401

48.	New Antidementia Molecules which Selectively Influence Serotonin Receptor Subtypes B. Costall, A.M. Domeney, M.E. Kelly, and R.J. Naylor	407
49.	Pharmacological Modulation of the Serotonergic System: An Overview of the Effects on Normal and Pathological Sleep C. Dugovic and G.H.C. Clincke	417
Index		425

PREFACE

This volume contains the Proceedings of the invited lectures of the Second International Symposium on **SEROTONIN from Cell Biology to Pharmacology and Therapeutics** held in Houston, Texas September 15-18, 1992. The meeting was held under the co-sponsorship of the Serotonin Club, the Giovanni Lorenzini Medical Foundation, and the Fondazione Giovanni Lorenzini. This volume discusses the major exploration in knowledge that has occurred recently of the complex role that 5- hydroxytryptamine (serotonin) plays in health and disease. In particular, these Proceedings highlight major breakthroughs in molecular biology and classification of receptor subtypes that are responsible for the many actions of the monoamine. The ever-increasing importance of serotonin in central regulation, whether autonomic or behavioral is represented by a large number of chapters prepared by world experts. Additionally, the role of serotonin in peripheral organs is also discussed. Hence, this volume provides the reader with a unique, up-to-date review of this exciting and novel area of science.

These Proceedings obviously are of great interest, not only to the researchers directly engaged in the quest for the understanding and unraveling of the actions of the interactions with serotonin as a major neurohumoral mediator, but also to all scholars and clinicians who wish to acquire a better understanding of the functioning of the brain and of peripheral organs.

Since this volume was constructed as a compilation of invited lectures, the scientific content and the opinions expressed in the chapters are the sole responsibility of the authors.

The editors wish to thank Dr. Marjorie Horning and Dr. Chantal Boulanger for their assistance and support, Mrs. Sherri Curry for her typing skills and her patience, and the staff at Kluwer for the prompt and efficient handling of the manuscripts.

ACKNOWLEDGEMENT

"We gratefully acknowledge the educational grant from CERENEX, a division of Glaxo Inc. which made this publication possible."

DIFFERENT ORGANIZATION OF HUMAN MONOAMINE OXIDASE (MAO) A AND B PROMOTERS

JEAN C. SHIH, QIN-SHI ZHU, JOSEPH GRIMSBY, KEVIN CHEN, and JACK SHIH
University of Southern California School of Pharmacy
Department of Molecular Pharmacology and Toxicology
1985 Zonal Avenue
Los Angeles, California 90033
USA

ABSTRACT. The promoter of human monoamine oxidase (MAO) A and B genes have been identified. The MAO A promoter activity is located in a 0.24 kb PvuII/DraII fragment containing two 90 bp repeats, each of which contains two Sp1 elements and lacks a TATA box. The highest MAO B promoter activity is detected in a 0.15 kb PstI/NaeI fragment which contains a Sp1-CACCC-Sp1-TATA structure. The different organization of the MAO A and B promoters may underlie their different cell and tissue specific expression.

1. Introduction

Monoamine oxidase A and B (MAO A and B; flavin-containing deaminating amine:oxygen oxidoreductase, EC 1.4.3.4) catalyze the oxidative deamination of a number of neurotransmitters, dietary amines and xenobiotics including the Parkinsonism-producing neurotoxin 1-methyl-4-phenyl-1,2,3,6-tetrahydropyridine[1]. Both forms are located in the outer mitochondrial membrane and are distinguished by their different substrate preference and sensitivity to inhibitors. Cloning of the cDNAs for MAO A and B demonstrates that these two forms of the enzyme are coded by different genes [2-5] which were derived from the same ancestral gene [6]. Both genes are closely linked and located on the X chromosome at Xp11.23 to Xp22.1 and are deleted in some patients with Norrie's disease [7, 8].

MAO A and B transcripts are coexpressed in most human tissues examined [9]. However, they do show different tissue and cell distribution, and they are regulated differently during development. In addition, abnormal MAO activity may be associated with mental disorders and MAO inhibitors have been used for the treatment of Parkinsonism and mental depression. In order to understand the mechanisms controlling the expression of these two forms of MAO, it is essential to characterize their promoters (figure 1). This report demonstrates that the immediate 5' flanking sequences of both MAO A and B genes contain cis-elements needed for active transcription, however, the organization of these elements are different.

2. Materials and Methods

The 5' flanking sequences of MAO A and B genes are isolated from corresponding genomic clones and sequenced after being subcloned into M13. On the basis of sequence analysis, various DNA fragments from the presumed promoter regions are isolated and measured for promoter activity by inserting them into the promoterless expression vector pOGH which contains human growth hormone as the reporter gene. These constructs are transfected into SHSY-5Y (human neuroblastoma) and NIH3T3 (mouse fibroblast) cells. The human growth hormone synthesized is secreted into the medium and is measured with a Kit from Nichols Institute Diagnostics. MAO catalytic activities are assayed as previously described [10], using serotonin and phenylethylamine as substrate for MAO A and B, respectively.

3. Results and Discussion

Sequence analysis of the 5' flanking region of MAO A and B exon 1 shows that the first 200 bp 5' of the cDNA start site for MAO A and B are comprised of 68% and 82% GC residues respectively. These two regions, later shown to contain the core region of promoter activity, also share the highest homology (61%, figure 2), suggesting that these two promoters may be derived from a common ancestral gene. However, none of the transcription factor binding sequences are conserved at their corresponding positions suggesting that these two promoters have functionally diverged during evolution and thus their expression are differently regulated [11]. The potential binding sites for transcription factors are also depicted in figures 3 and 4 in which they are numbered sequentially from 3' to 5'.

Promoter activity measurements with DNA fragments obtained from both 5' and 3' deletions show that the MAO A promoter activity is located in the two 90 bp direct repeat region (figure 3, each repeat is represented with an arrow). These two 90 bp repeats share 83% sequence identity and each contains two Sp1 elements in reversed orientation (figure 3, site 1-4). Deletion of the 3' 90 bp repeat from A0.24, a 0.24 kb PvuII/DraII fragment, which contains both 90 bp repeat and an upstream TAATAA sequence, decreases promoter activity by approximately 50% (compare A0.24 [53%] and A0.12 [25%] in figure 3). However, deletion from the 5' end which removes the TAATAA sequence and the Sp1 site 4 results in the most active fragment (A0.14) which contains Sp1 sites 1-3 and its activity is taken as 100%. A0.09 and A0.12 containing Sp1 sites 2-3 and 3-4 are capable of transcription activation although with only 20-25% the activity of A0.14. This means that at least three Sp1 sites or yet undetermined elements located within this region work cooperatively to activate transcription.

Figure 1. The way a promoter works

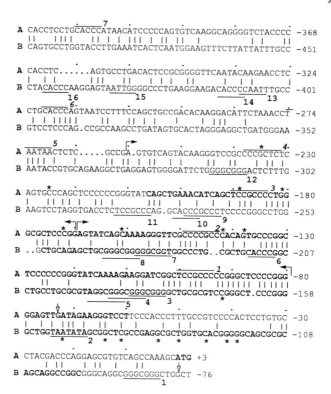

Figure 2. Sequence comparison of the core promoter regions of MAO A (A) and B (B) genes. The position of the two promoter regions are aligned to yield the highest degree of homology as determined by computer analysis. The potential cis-elements 1-7 of MAO A and 1-16 of MAO B are marked. The sequences of A0.14 and B0.15 fragments are in bold type. The two 90 bp repeats of MAO A are marked by solid arrows. The cDNA start site is indicated by an open arrow. Transcription initiation sites as determined by primer extension analysis (figure not shown) are marked with asterisks.

DNA fragments containing only one Sp1 site (site 4, see A0.37 and A0.25) display very low promoter activity. Lack of a Sp1 site (A0.21) has no detectable promoter activity. The decreased promoter activity by upstream sequences that contain possible binding sites for transcription factors (compare A1.4, A0.52, A0.24 and A0.14), (A0.41, A0.29 and A0.12) and (A0.37 and A0.25) may result from down regulating transcription factors or competition of transcription factors for the activating sites on the polymerase complex.

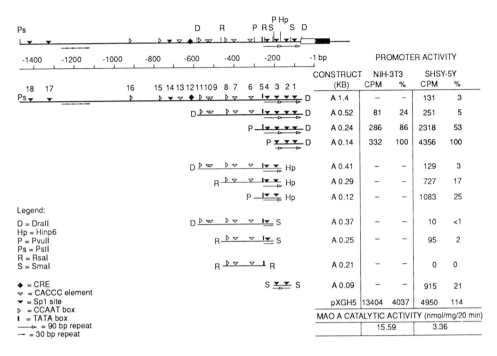

Figure 3. Restriction map of the MAO A promoter region and promoter activity measurements. At the top is the restriction enzyme map, where only restriction enzyme sites used for subcloning for promoter activity measurement are marked. The open box on the top line represents untranslated region and the closed box for the coding sequence of MAO exon 1. The potential transcription factor binding sites are numbered sequentially from 3' to 5'. The A of the codon ATG is defined as +1 bp. The pOGH constructs containing DNA fragments to be tested are named according to the size of the inserts (kb). "A" denotes DNA fragments of MAO A gene. The activity of the A0.14 construct is taken as 100% for each cell line. pXGH5 represents the plasmid containing mouse metallothionein promoter instead of the MAO promoter fragments, which was used to monitor the transfection efficiency. The last line shows MAO A enzymatic activity measured in each cell line.

Figure 4 shows that the 0.15 kb PstI/ NaeI fragment exhibits the highest MAO B promoter activity. This fragment contains a TAATATA box (site 2), three overlapping Sp1 elements (site 3,4 and 5), a CACCC element (site 6) and two overlapping Sp1 elements (site 7 and 8) therefore has a structure:

5' -(Sp1)$_2$-CACCC-(Sp1)$_3$-TAATATA-3'

Deletion from the 5' end of the B0.15 kb fragment, resulting in a stepwise loss of 5' Sp1 sites (site 7, 8: B0.13), the CACCC element (site 6: B0.12) and the 3' Sp1 sites (site 3, 4 and 5: B0.07) resulted in 38%, 28%, and 2% of the promoter activity, respectively.

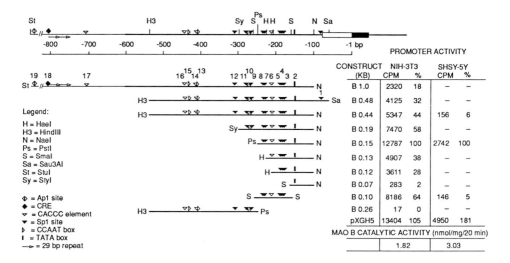

Figure 4. Restriction map of the MAO B promoter region and promoter activity measurements. The symbols used in this figure are the same as in figure 2. The potential transcription factor binding sites are numbered sequentially from 3' to 5'. "B" denotes DNA fragment of MAO B gene. The last line shows MAO B enzymatic activity measured in each cell line.

Deletion of the TATA box (B0.10) results in 64% activity thus demonstrating the positive role of the TATA box. Like the MAO A promoter, inclusion of further upstream sequences results in a stepwise decrease in the maximal promoter activity observed with B0.15 in NIH3T3 cells (B1.0: 18%, B0.48: 32%, B0.44: 44% and B0.19: 58%). B0.26 has no promoter activity despite the presence of potential cis-elements.

The fact that the core region of both MAO A and B promoters are GC rich, contain potential Sp1 binding sites and can activate transcription initiation in the absence of a TATA box or a CCAAT box demonstrates these two promoters are housekeeping like. On the other hand, the differences in their cis-element organization and their behavior in human (SHSY-5Y) and mouse (NIH3T3) cells indicate that the mechanisms regulating MAO A and B expression are quite different.

In the human cell line SHSY-5Y the core MAO A and B promoter activities (4356 cpm for A0.14 and 2742 cpm for B0.15) correlate well with catalytic activities (3.36 nmol/mg/20 min for MAO A and 3.03nmol/mg/20 min for MAO B). The unusually high MAO B promoter activity in NIH3T3 cells (12787 cpm) compared with moderate mouse MAO B enzymatic activity (1.82 nmol/mg/20 min) suggests that the mouse fibroblasts contain sufficient activating factors or less inhibiting factors for human MAO B constructs. The poor correlation observed in mouse (NIH3T3) cells between measured human MAO promoter activities (332 cpm for A0.14 and 12787 cpm for B0.15) and mouse catalytic activities (15.59 nmol/mg/29 min for MAO A and 1.82 nmol/mg/20 min for MAO B) suggests that mouse MAO promoters are different from humans.

Abnormal levels of MAO activity have been reported in a number of mental disorders and may result from promoter sequence changes. It will be interesting to investigate whether the varied MAO levels in disease states is caused by promoter modification (cis-element changes), or by changes of transcription factors required for MAO expression. The knowledge of the mechanism for MAO expression regulation can also be used to design new generation of MAO inhibitors based on altering gene transcription. This work opens up a new area of research concerning the molecular basis of MAO gene expression in disease states.

4. Acknowledgement

This work was supported by grant R01 MH37020, R37 MH39085 (Merit Award), and Research Scientist Award K05 MH00796, from the National Institute of Mental Health. Support from the Boyd and Elsie Welin Professorship is also appreciated.

5. References

1. Chiba, K., Trevor, A., and Castagnoli, N. (1984) 'Metabolism of the neurotoxic tertiary amine, MPTP by brain monoamine oxidase', Biochem. Biophys. Res. Commun. 120, 574-578.
2. Bach, A.W.J., Lan, N.C., Johnson, D.L., Abell, C.W., Bembenek, M.E., Kwan, S.-W., Seeberg, P.H., and Shih, J.C. (1988) 'cDNA cloning of human monoamine oxidase A and B: molecular basis of differences in enzymatic properties', Proc. Natl. Acad. Sci. USA 85, 4934-4938.
3. Hsu, Y.P., Weyler, W., Chen, S., Sims, K.B., Rinehart, W.B., Utterback, M., Powell, J.F., and Breakefield, X.O. (1988) 'Structural features of human monoamine oxidase A elucidated from cDNA and peptide sequences', J. Neurochem. 51, 1321-1324.
4. Lan, N.C., Chen, C.H., and Shih, J.C. (1989) 'Expression of functional human monoamine oxidase (MAO) A and B cDNA in mammalian cells',

J. Neurochem. 52, 1652-1654.

5. Shih, J.C. (1990) 'Molecular basis of human MAO A and B' Neuropsychopharmacology 4, 1-7.

6. Grimsby, J., Chen, K., Wang, L.J., Lan, N.C. and Shih, J.C. (1991) 'Human monoamine oxidase A and B genes exhibit identical exon-intron organization', Proc. Natl. Acad. Sci. USA 88, 3637-3641.

7. Ozelius, L., Hsu, Y.P., Bruns, G., Powell, J.F., Chen, S., Weyler, W., Utterback, M., Zucker, D., Haines, J., Trofalter, J.A., Conneally, P.M., Gusella, J.F., and Breakefield, X.O. (1988) 'Human monoamine oxidase gene (MAOA): Chromosome position (Xp21-p11) and DNA polymorphism', Genomics 3, 53-58.

8. Lan, N.C., Heinzmann, C., Gal, A., Klisak, I., Orth, U., Lai, E., Grimsby, J., Sparkes, R.S., Mohandas, T., and Shih, J.C. (1989) 'Human monoamine oxidase A and B genes map to Xp11.23 and are deleted in a patient with Norris Desease', Genomics 4, 552-55.

9. Grimsby, J., Lan, C., Neve, R., Chen, K., and Shih, J.C. (1990) 'Tissue distribution of human MAO A and B mRNA', J. Neurochem. 55, 1166-1169.

10. Chen, S., Shih, J.C., and Xu, Q.P. (1984) 'Interaction of N-(2-nitro-4-azidophenyl) serotonin with two types of monoamine oxidase in rat brain', J. Neurochem. 43, 1680-1687.

11. Zhu, Q.S., Grimsby, J., Chen, K., and Shih, J.C. (1992) 'Promoter organization and activity of human monoamine oxidase (MAO) A and B genes', J. Neuroscience, in press.

MOLECULAR BIOLOGY OF SEROTONIN UPTAKE SITES

BETH J. HOFFMAN
Laboratory of Cell Biology
National Institute of Mental Health
Bethesda, Maryland 20892

ABSTRACT. Serotonin (5-HT) is synthesized and packaged into secretory vesicles by the vesicular monoamine transporter prior ro release via exocytosis. After its release and action on receptors, 5-HT is inactivated by reuptake via the plasma membrane transporter into presynaptic terminals, where it may be repackaged or degraded by monoamine oxidases. 5-HT transporters are one of the initial sites of action for antidepressants and psychomotor stimulants. Using an expression cloning strategy, we have isolated a cDNA for an antidepressant-sensitive, Na^+-dependent 5-HT reuptake protein from rat mast cells. The cDNA for the 5-HT transporter (5-HTT) predicts a 630-amino acid protein with 12 putative transmembrane domains and extensive homology to GABA, norepinephrine (NE), and dopamine (DA) transporters. 5-HTT mRNA is present in brain as well as peripheral tissues. While screening for the 5-HTT, we have also isolated a cDNA for a vesicular monoamine transporter (vMAT) with broad selectivity for 5-HT, DA, and NE. The vMAT cDNA predicts a protein of 515 amino acids with 12 transmembrane domains but no significant homology to any other known proteins. The structural features, mRNA localization, and pharmacology of these two 5-HT transport systems in relation to their contribution to 5-HT neurotransmission is discussed.

1. Background

In a serotonergic neuron, tryptophan hydroxylase is the rate limiting enzyme in the biosynthesis of serotonin (5-HT) from tryptophan. 5-HT synthesized in the cytoplasm is packaged into synaptic vesicles by a proton (H^+) antiporter which exchanges protons inside the vesicle for biogenic amine. Following appropriate stimulation, 5-HT is released into the synaptic cleft where it may act on postsynaptic receptors or on presynaptic receptors. Action of 5-HT is terminated by the "re-uptake" of neurotransmitter by a Na^+/Cl^--dependent symporter.

Tricyclic antidepressants revolutionized the treatment of depression. The study of the mechanism of action of antidepressants began with reserpine, one of the first effective drugs for hypertension [1]. Physicians observed that reserpine sometimes caused severe depression, even suicide. Reserpine causes the release and subsequent depletion of monoamines. Thus, these observations led to the proposal of the "amine hypothesis" of depression. Most antidepressants act to concentrate monoamines (NE, DA, 5-HT) in the synapse either by blocking their metabolism via monoamine oxidases or by inhibiting re-uptake into the presynaptic terminal by plasma membrane transporters. Many of the clinically effective antidepressants inhibit re-uptake of serotonin (for example, Prozac). Specific 5-HT uptake inhibitors may be useful in treating obsessive-compulsive disorder, schizophrenia, panic disorder, bulimia, and obesity [2].

2. Molecular Cloning of the Na^+/Cl^--Dependent 5-HT Transporter

Since the initial demonstration of norepinephrine uptake by Axelrod and co-workers [3,4], considerable effort has been directed at characterizing the proteins responsible for uptake of norepinephrine, dopamine, and 5-HT, both in normal and disease states. Biochemical purification has been hampered by the minute quantities of protein available and by the difficulty in working with membrane proteins. Thus, there has been little molecular information on the Na^+/Cl^+-dependent monoamine transporters.

In addition to neurons, other, non-neuronal cells have 5-HT re-uptake sites or transporters including platelets, mast cells, macrophages, uterine smooth muscle and enterochromaffin cells of the gut, and chromaffin cells of the adrenal medulla. Kanner and Bendahan [5] have described a rat basophilic leukemia cell line, RBL 2H3 which has characteristics of a mast cell such as IgE-stimulated release of histamine and 5-HT and re-uptake of 5-HT.

We used a novel cDNA expression cloning strategy to isolate a cDNA for the 5-HT transporter from the RBL 2H3 cell line [6]. This strategy relied on the accumulation of ^{3}H-5-HT in mammalian tissue culture cells into which pools of cDNA had been introduced by calcium phosphate-mediated transfection. Accumulated radioactivity was detected microscopically following fixation of ^{3}H-5-HT by acrolein (1:100 v/v) in 2.5% glutaraldehyde and exposure to photographic emulsion. Critical aspects of this screening procedure were the use of the cytomegalovirus (CMV) promoter and the method for fixing 5-HT in the cell for subsequent microscopic visualization. Using this screening strategy, we identified two positive pools of 13,000 recombinants. By successive rounds of subdividing one of these pools, we were able to isolate a single cDNA clone for the 5-HT transporter.

During the course of this work, transporters for GABA [7] and NE [8] were cloned. Comparison of the amino acid sequences revealed a 21 amino acid region (corresponding to amino acids 101-121 in figure 1) in which 20 amino acids were conserved. Assuming that this region might be conserved in the rest of this newly emerging gene family, we used a degenerate oligonucleotide to probe the pools of cDNA from the RBL 2H3 library. Interestingly, only those cDNA pools which conferred 5-HT uptake activity on tranfected cells hybridized to this consensus oligonucleotide. Using this probe to screen other cDNA libraries, we have isolated other members of this gene family including a bovine dopamine transporter [9] and a glycine transporter [10]. Numerous groups have used this same region for designing polymerase chain reaction (pcr) primers and have successfully isolated additional members of this gene family [for review, see 11]. Independently, Blakely and coworkers used a pcr-based strategy combined with *in situ* hybridization histochemistry to isolate a 5-HT transporter cDNA from a rat brain library [12].

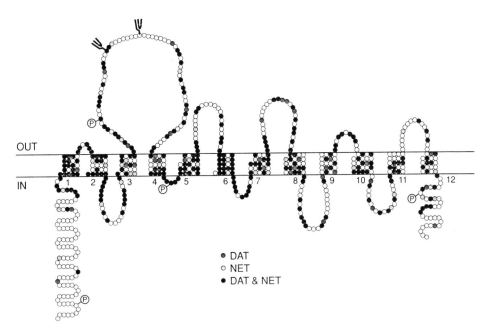

Figure 1. The Na+/Cl--dependent serotonin transporter. Putative sites of N-linked glycosylation face the outside of the cell. Potential protein kinase C phosphorylation sites are indicated. Shaded circles indicate conservation with other monoamine plasma membrane transporters.

The sequence of the cDNA predicts a protein of 630 amino acids with 12-13 putative transmembrane domains. Although there is significant homology between the amino acid transporters and the 5-HT transporter, it is clear that the monoamine transporters share greater homology with one another and constitute a subfamily within the larger Na+/Cl--dependent transporter family.

By Northern (RNA) blot analyses, there are high levels of a single 3.1 kb mRNA species in both brain (brainstem) and in peripheral tissues (gut, lung). Spleen and adrenal and intermediate levels of mRNA while stomach, uterus and kidney have low levels of mRNA. *In situ* hybridization histochemistry (ISHH) with a 5-HT transporter-specific oligonucleotide revealed mRNA in brain regions with serotonergic neurons and in the glia surrounding the pineal gland. In the periphery, mRNA was identified in the lamina propria of the stomach and duodenum, in adrenal chromaffin cells, in mast cells of the lung, and in macrophages in spleen.

^{3}H-5-HT uptake in tissue culture cells transfected with the 5-HT transporter cDNA was sensitive to inhibition by tricyclic antidepressants such as imipramine and amytrityline as well as the 5-HT transporter-selective drugs fluoxetine, paroxetine, and citalopram (table 1). Antidepressants more selective for norepinephrine and dopamine transporters, such as desipramine and mazindol had lower affinity for the 5-HT transporter. Interestingly, the most useful therapeutics like imipramine and amitriptyline have roughly equivalent potency at both 5-HT and NE transporters. Dopamine, norepinephrine, and reserpine, an inhibitor of vesicular uptake, do not block 5-HT uptake. The NE transporter uses DA and NE as substrates, and actually prefers DA. The 5-HT and DA transporters are more selective for their respective substrates. Cocaine exhibits high affinity for 5-HT, DA, and NE transporters, and so, is likely to effect neurotransmission by all three monoamine systems. Substituted amphetamines like MDMA (ecstasy) are more selective for the 5-HT transporter whereas amphetamine, itself, is more selective for NE and DA transporters. The properties of the cloned 5-HT transporter closely resembles 5-HT uptake in platelets [13] and in brain synaptosomes [14]. A single protein appears to be responsible for both antidepressant binding and 5-HT transport.

3. Molecular Cloning of a Vesicular Monoamine Transporter (vMAT)

As mentioned above, another type of transporter works by exchanging protons (H^+) for substrate in order to load synaptic vesicles or secretory vesicles with neurotransmitter [15]. The activity of this transport depends on a functional H^+-ATPase and Cl channel in the vesicle membrane. Energy stored in the proton gradient drives uptake of monoamines. Following re-uptake by the plasma membrane transporter, monoamines are either degraded by monoamine oxidases or re-vesicularized by transport into vesicles. Therefore, this transporter is crucial to the initial storage and recycling of monoamines.

While screening the RBL cDNA library for the plasma membrane 5-HT transporter by ^{3}H-5-HT accumulation in transfected tissue culture cells, a second pool of cDNA which directed 5-HT uptake was identified. The signal (number of silver grains) was weaker and not displaceable by antidepressants. However, the 5-HT uptake mediated by this cDNA pool was inhibited by reserpine. To isolate this cDNA, we used a recombinant vaccinia virus expressing T7 polymerase to infect cells which were then transfected with successively less complex pools of cDNA [16]. This method yields efficient and high levels of expression by transcribing transfected cDNA in the cytoplasm rather than relying on the transport of cDNA into the nucleus.

TABLE 1. Drug Affinities for Cloned Biogenic Amine Transporters

Inhibitor	5-HT[6]	NE[8]	DA[9]
Antidepressants			
paroxetine	3.1	312	-
citalopram	6.1	>1000	-
clomipramine	7.1	-	-
fluoxetine	33	-	2000
imipramine	209	65	1670
amitriptyline	262	100	-
zimelidine	382	-	-
mazindol	548	1.4	51
desipramine	1680	3.9	7800
doxepin	1850	-	-
nomifensine	-	7.7	536
bupropion	-	-	2740
nortryptiline	-	16.5	-
benztropine	-	822	-
Biogenic Amines			
serotonin	529	>10,000	>10,000
dopamine	>10,000	139	1,000
noradrenaline	>10,000	457	>10,000
Others			
S(+)fenfluramine	129	-	-
± MDMA	186	-	5800
d-amphetamine	3180	56	386
cocaine	1080	140	244, 2000
GBR12909	-	133	52, 10
MPP+	>10,000	-	>5,000
reserpine	>10,000	-	>10,000

Initial characterization showed that uptake in cells tranfected with this cDNA was slow and of very low affinity. We used digitonin to permeabilize plasma membranes while leaving membrane of organelles intact. This dissipated the Na^+ gradient which is the driving force for the plasma membrane, and consequently, abolished antidepressant-sensitive uptake. However, reserpine-sensitive uptake was rapid and of high affinity. Apparently, uptake in intact cells was a complex event requiring diffusion of the neutral substrate through the plasma membrane followed by uptake into an intracellular compartment. In cells permeabilized with digitonin, uptake was ATP-dependent, and abolished by FCCP, a proton-translocating ionophore, and by tri-n-butyltin, an inhibitor of vacuoloar H^+-ATPases. Taken together, these characteristics suggested that sequestration of 5-HT in an acidic compartment of a monkey kidney cell, CV-1, perhaps in endosomes.

The protein expressed from the vMAT1 cDNA has many of the features previously described for the bovine chromaffin granule and brain synaptic vesicle transporters. This transporter has a broad selectivity for monoamines (5-HT > DA > NE > epinephrine > histamine) with 5-HT (K_m = 1.3 μM) as the preferred substrate. Uptake is potently inhibited by reserpine (K_i = 12.5 nM) and tetrabenazine (K_i = 0.63 μM) in cells transfected with the vMAT1 cDNA. A glycoprotein with apparent molecular weight of 75 kD was photolabeled with 7-azido-8-(^{125}I)iodoketanserin [17] as compared to a 70 kD protein detected in bovine chromaffin granules. Treatment of membranes from cells expressing vMAT1 with glycosidases caused a reduction of apparent molecular weight by approximately 20 kD, suggesting that 25% of the mass is due to glycosylation.

Using ISHH, with specific oligonucleotide probes, vMAT1 mRNA was localized in the rat brain to monoamine-containing cells including the locus coeruleus (NE), substantia nigra (DA), and raphe nuclei (5-HT), each of which expresses a unique plasma membrane re-uptake transporter. Thus the specificity of monoaminergic neurons appears to be determined by the biosynthetic enzymes and the plasma membrane transporters while a vesicular transporter of broad selectivity is used. By Northern (RNA) blot analyses, three discrete sizes of mRNA were detected (4.0 kb, 2.9 kb, and 2.2 kb) with high levels of mRNA in RBL cells and intermediate levels in the stomach. Transcripts were detected in brainstem at low levels. In RBL cells and in the stomach, the 4.0 kb species predominated while in brainstem, the 2.9 kb species was the most abundant. It is possible that these three transcripts arise from the use of multiple polyadenylation sites present in the 3' untranslated region of the cDNA.

4. Two Vesicular Monoamine Transporters Are Expressed in a Tissue-Specific Manner

While our work was in press, Liu et al. [18] isolated a highly related transporter from PC12 cells by screening cDNA-transfected cells for resistance to the neurotoxin MPP^+. Interestingly, this cDNA was localized only to adrenal medulla but was not present in brain. A second cDNA clone isolated from rat brain was identical to the cDNA we isolated from RBL cells and was expressed in brain but not in adrenal medulla. These two proteins are highly conserved but clearly arise from two distinct genes (figure 2, table 2). There are 12 predicted transmembrane domains. It remains to be determined which of these transmembrane domains may be involved in binding and transport functions. Both vesicular transporters have a broad specificity for monoamines but display somewhat different rank order of substrate preference (table 2). Another distinguishing feature is the difference in affinity for tetrabenazine (vMAT1; K_i = nM, vMAT2; K_i = μM). Both proteins have putative phosphorylation sites for protein kinase A and protein kinase C which raises the possibility of regulating transport activity by altering phosphorylation states of the transporters.

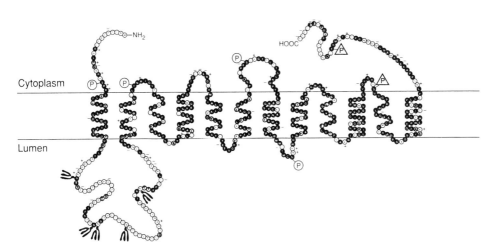

Figure 2. The vesicular monoamine transporter. Potential N-linked glycosylation sites face the vesicular lumen. Potential protein kinase sites are indicated (circles = protein kinase C sites, triangles = protein kinase A sites). Black circles indicate identity between vMAT1 and vMAT2; shaded circles indicate conserved amino acids.

Preliminary data from our laboratory indicates that vMAT1 is expressed in sympathetic ganglion cells and in chromaffin cells of the adrenal medulla while vMAT2 is found only in chromaffin cells. This further emphasizes that vMAT1 is localized in neurons. Since vMAT1 is also present in chromaffin cells, it seems redundant to have two proteins which are so similar in function. It is possible that differences in pharmacology, patterns of expression during development, association with specific monoamines, or ability to respond to specific stimuli may require separate genes.

TABLE 2. Summary of Vesicular Monoamine Transporters

	vMAT-1[16]	vMAT-2[18]
Amino Acids (Predicted)	515	520
Molecular Weight (Predicted)	≈57,000	≈57,200
^{125}I-AZIK Photolabelling	≈75,000	ND
Selectivity	5-HT>DA>NE>Epi>HIS	5-HT>Epi>DA>NE
Reserpine	13 nM	25 nM
Tetrabenazine	nM	μM
Glycosylation Sites	5	3
Phosphorylation Sites	2 PKA 4 PKC	1 PKA 2 PKC
Localization	CNS monoamine cells Mast Cells Other Immune Cells	Adrenal Chromaffin Granules
Homology	62% Identical (319/515) 72% Conserved (370/515)	

5. Regulation of Transporter Function

Changes in the function of either the plasma membrane transporter or the vesicular transporters may profoundly affect neurotransmission. 5-HT autoreceptors or receptors for other neurotransmitters on the presynaptic terminal may affect uptake by either of these transporters through second messengers. In addition, retrograde signalling by arachidonic acid metabolites as well as the gaseous second messengers NO and CO might regulate transporter activity. There are several potential phosphorylation sites on both the plasma membrane 5-HT transporter and the vesicular transporter. Several groups have shown that activation of protein kinases or treatment with arachidonic acid alter transport function. We have shown that phorbol esters which activate protein kinase C inhibit uptake by the 5-HT transporter both in RBL 2H3 cells and in cells transfected with the 5-HT transporter. Myers et al. [19] have shown a correlation between translocation of PKC and inhibition of 5-HT uptake in endothelial cells. Studies by Cool et al. [20] have indicated that cAMP enhances 5-HT uptake in human placental choriocarcinoma (JAR cells); however, this is not likely to be a direct effect by phosphorylation of the transporter because the effect is only detected after exposure to cAMP analogues for 16-18 hours.

There are several possible mechanisms by which protein kinases may regulate transporter function. Phosphorylation state may (1) change the K_m or V_{max} of the transporter, (2) alter the ion gradients in the cell, (3) effect other proteins (such as Na^+/K^+ ATPase, vesicular H^+ ATPase, or phosphatases), or (4) effect the distribution of transporters in active and inactive pools (intracellular trafficking). These issues will certainly be addressed in the future.

6. Future Directions

The cloning of the plasma membrane and vesicular transporters should allow numerous lines of investigation into how the individual components of the synapse interact to modulate communication between neurons. It will be important to understand how the transporter is regulated both in normal and disease states, and how antidepressants effect the normal regulation of transport. Comparison of rat and human sequences should point to conserved amino acids which may be important for ion and substrate binding and may suggest which transmembrane domains form the "pore" for transport [see 11]. Chromosomal localization and comparison of restriction fragment length polymorphisms (RFLP) or other polymorphisms will be necessary to determine whether alterations in these transporter genes are associated with psychiatric disorders. Availability of cell lines expressing each transporter should be useful for future drug development.

7. Acknowledgements

I am indebted to my collaborators Eva Mezey, Jeff Erickson, Lee Eiden, and Michael Brownstein.

8. References

1. Shore, P.A. and Giachetti, A. (1978) 'Reserpine: basic and clinical pharmacology', in L.L. Iversen, S.D. Iversen and S.H. Snyder (eds.), Handbook of Psychopharmacology, Plenum Press, New York, vol. 10, pp. 197-219.
2. Fuller, R.W. and Wong, D.T. (1990) 'Serotonin uptake and serotonin uptake inhibition', in P.M. Whitaker-Azmitia and S.J. Peroutka (eds.), The Neuropharmacology of Serotonin, New York Academy of Sciences, New York, vol. 600, pp. 68-80.
3. Axelrod, J., Whitby, L.G., and Hertting, G. (1961) 'Effect of psychotropic drugs on the uptake of ^{3}H-norepinephrine by tissues', Science 133, 383-384.
4. Glowinski, J., Kopin, I.J., and Axelrod, J. (1965) 'Metabolism of ^{3}H-norepinephrine in the rat brain', J. Neurochem. 12, 25-30.
5. Kanner, B.I. and Bendahan, A. (1985) 'Transport of 5-hydroxytryptamine in

membrane vesicles from rat basophilic leukemia cells', Biochim. Biophys. Acta 816, 403-410.
6. Hoffman, B.J., Mezey, E., and Brownstein, M.J. (1991) 'Cloning of a serotonin transporter affected by antidepressants', Science 254, 579-580.
7. Guastella, J., Nelson, N., Nelson, H., Czyzk, L., Keynan, S., Miedel, M.C., Davidson, N., Lester, H.A., and Kanner, B.I. (1990) 'Cloning and expression of a rat brain GABA transporter', Science 249, 1303-1306.
8. Pacholczyk, T., Blakely, R.D., and Amara, S.G. (1991) 'Expression cloning of a cocaine- and antidepressant-sensitive human noradrenaline transporter', Nature 350, 350-354.
9. Usdin, T.B., Mezey, E., Chen, C., Brownstein, M.J., and Hoffman, B.J. (1991) 'Cloning of the cocaine-sensitive bovine dopamine transporter', Proc. Natl. Acad. Sci. USA 88, 11168-11171.
10. Borowsky, B., Mezey, E., and Hoffman, B.J. (1993) 'Two glycine transporter variants with distinct localization in the CNS and peripheral tissues are encoded by a common gene', Neuron, in press.
11. Brownstein, M.J. and Hoffman, B.J. (1993) 'Neurotransmitter transporters', in Recent Progress in Hormone Research, in press.
12. Blakely, R.D., Berson, H.E., Fremeau Jr., R.T., Caron, M.G., Peek, M.M., Prince, H.K., and Bradley, C.C. (1991) 'Cloning and expression of a functional serotonin transporter from rat brain', Nature 354, 66-70.
13. Hyttel, J. (1982) 'Citalopram-pharmacological profile of a specific uptake inhibitor with antidepressant activity', Prog. Neuro-Psychopharmacol. Biol. Psychiat. 6, 277-295.
14. Wood, M.D., Broadhurst, A.M., and Wylie, M.G. (1986) 'Examination of the relationship between the uptake system for 5-hydroxytryptamine and the high affinity [^{3}H]imipramine binding site-1', Neuropharmacology 25, 519-525.
15. Johnson, R.G. (1988) 'Accumulation of biological amines into chromaffin granules: a model for hormone and neurotransmitter transport', Physiol Rev. 68, 232-307.
16. Erickson, J.D., Eiden, L.E., and Hoffman, B.J. (1992) 'Expression cloning of a reserpine-sensitive vesicular monoamine transporter', Proc. Natl. Acad. USA 89, 10993-10997.
17. Isambert, M.-F., Gasnier, B., Laduron, P.M., and Henry, J.-P. (1989) 'Photoaffinity labeling of the monoamine transporter of bovine chromaffin granules and other monoamine storage vesicles using 7-azido-8-[^{125}I]iodoketanserin', Biochemistry 28, 2265-2270.
18. Liu, Y., Peter, D., Roghani, A., Schuldiner, S., Prive, G.G., Eisenberg, D., Brecha, N., and Edwards, R.H. (1992) 'A cDNA that suppresses MPP$^+$ toxicity encodes a vesicular amine transporter', Cell 70, 539-551.
19. Myers, C.L., Lazo, J.S., and Pitt, B.R. (1989) 'Translocation of protein kinase C is associated with inhibition of 5-HT uptake by cultured endothelial cells', Am. J. Physiol. 257, L253-258.

20. Cool, D.R., Leibach, F.H., Bhalla, V.K., Mahesh, V.B., and Ganapathy, V. (1991) 'Expression and cyclic AMP-dependent regulation of a high affinity serotonin transporter in the human placental choriocarcinoma cell line (JAR)', J. Biol. Chem. 266, 15750-15757.

MOLECULAR BIOLOGY OF 5-HT$_1$ RECEPTORS

P. HARTIG, N. ADHAM, J. BARD, A. HOU-YU, H.-T. KAO, M. MACCHI, L. SCHECHTER, D. URQUHART, J. ZGOMBICK, M. DURKIN, R. WEINSHANK, and T. BRANCHEK
Synaptic Pharmaceutical Corporation
215 College Road
Paramus, New Jersey 07652
USA

ABSTRACT. Five human serotonin receptor subtypes belonging to the 5-HT$_1$ subfamily (5-HT$_{1A}$, 5-HT$_{1D\alpha}$, 5-HT$_{1D\beta}$ [and its species homologue: rat 5-HT$_{1B}$], 5-HT$_{1E}$, and 5-HT$_{1F}$) have now been cloned and characterized. The cloning of these subtypes has settled the controversy over the relationship between the 5-HT$_{1B}$ and 5-HT$_{1D}$ receptor sites (termed pharmacological subtypes) by providing the underlying genes (termed genetic subtypes) that encode these receptors. The characterization of two human genes (hu 5-HT$_{1D\alpha}$ and hu 5-HT$_{1D\beta}$) and one species homologue in the rat genome (rt 5-HT$_{1B}$) has shown that amino acid sequence homologies do not always accurately predict pharmacological properties. Two human subtypes, hu 5-HT$_{1D\alpha}$ and hu 5-HT$_{1D\beta}$ have been shown to be fairly divergent in deduced amino acid sequence yet they display nearly indistinguishable pharmacological properties that are characteristic of a 5-HT$_{1D}$ receptor. Conversely, the cloning from the rat genome of the species homologue of the hu 5-HT$_{1D\beta}$ receptor provided a gene with a very similar deduced amino acid sequence but quite distinct pharmacological properties characteristic of a 5-HT$_{1B}$ receptor.

In this review, the concepts of genetic and pharmacological subtypes are defined, the subfamily of five 5-HT$_1$ subtypes is reviewed, and a key question is examined: why are there five different serotonin receptors, all of which couple to inhibition of adenylate cyclase? Evidence is presented that coupling to different isoforms of G protein may be responsible for at least some of the diversity present in this receptor subfamily.

1. Subtypes and Species Homologue

In this rapidly evolving field, careful definition of three concepts will prove useful in reviewing receptor subtypes and clones. Two of these terms, "pharmacological subtype" and "genetic subtype," illustrate the classical and molecular biological approaches to receptor classification, while the term "species homologue" provides a means to identify "equivalent" genes in different species. The term "pharmacological subtype" will be used to define individual receptor subtypes identified by indirect criteria such as radioligand binding profiles (e.g. 5-HT$_{1A}$ binding site) or physiological responses (e.g. dog saphenous vein contractions). The intent of each of these classical approaches has been to identify individual molecular species, although the approaches have been indirect. Now that gene cloning is revealing the complete set of molecular subtypes, we have learned that most of our deductions about which receptor subtypes underlie these indirect assays have been surprisingly accurate.

The term "genetic subtype" will be used to define a single gene and its corresponding protein. The 7TM (7 transmembrane domains) superfamily of receptors is the receptor group responsible for most of the known actions of serotonin. In this superfamily, one gene almost always encodes for a single protein (a notable exception is the long and short forms of the dopamine D2 receptor produced by alternative splicing within a single D2 receptor gene). Since 7TM receptors are single polypeptides with one unique gene for each protein, we will define a "genetic subtype" as a single gene within a single animal species. The cloning of over 100 different genetic subtypes of 7TM receptors and 9 genetic subtypes of serotonin receptors (see below) has indicated that each genetic subtype of receptor is represented by an "equivalent" genetic subtype in all mammalian species. In other words, it appears that the entire set of genetic subtypes of 7TM receptors is represented in the genome of each mammalian species. This leads to the final concept of "species homologue" as the "equivalent" gene in a different mammalian species.

Once the entire set of genes is cloned from every experimental species and from the human genome, it should be a relatively simple matter to match each gene with its equivalent gene in a different species. For the next decade or two, before that task is completed, we must rely on certain generalizations to help us with these assignments. Within the 7TM superfamily it appears that a transmembrane domain homology of 95% or greater is diagnostic for species homologue of a single genetic subtype whereas different genetic subtypes typically exhibit less than 85% amino acid identity [1,2].

Based on these generalizations, we can now review the current status of serotonin receptor subtypes. Table 1 shows a list of pharmacological subtypes of serotonin receptors, along with their known genetic subtypes from molecular cloning studies. The properties of the 5-HT$_{1D}$ and 5-HT$_{1B}$ pharmacological subtypes, and two new subtypes, 5-HT$_{1E}$ and 5-HT$_{1F}$, will be covered in this review.

TABLE 1. List of serotonin receptor subtypes. Question marks indicate genetic subtypes whose corresponding pharmacological subtype is unknown or not extensively characterized. The 5-HT$_4$ receptor has not yet been reported as a genetic subtype. Subfamilies are grouped by the most commonly reported second messenger coupling as indicated by the corresponding G protein class.

Pharmacological Subtype	Genetic Subtype	Subfamily
5-HT$_{1A}$	5-HT$_{1A}$	
5-HT$_{1B}$	(rat 5-HT$_{1D\beta}$)	
5-HT$_{1D}$	5-HT$_{1D\alpha}$ and 5-	G$_i$
5-HT$_{1E}$	5-HT$_{1E}$	
?	5-HT$_{1F}$	
5-HT$_2$	5-HT$_2$	
?	5-HT$_{2F}$	G$_p$
5-HT$_{1C}$	5-HT$_{1C}$	
5-HT$_4$	?	G$_s$

2. The 5-HT$_{1D}$ Receptor Group

The group of serotonin 5-HT$_{1D}$ receptor clones provides an extreme example of how similar the pharmacological properties of two different genetic subtypes can be. It also illustrates the opposite extreme: two closely related species homologue can display widely divergent pharmacological properties. The cloning of this receptor group began in 1989 with the publication by Libert et al. [3] of a dog cDNA clone that they termed RDC4. Although this group did not identify any radioligand binding to this receptor, they speculated that it might be a serotonin receptor due to its sequence homology to the 5-HT$_{1A}$ receptor. Our group [4], and Maenhaut et al. [5] have isolated and characterized this dog clone, and identified it as a serotonin 5-HT$_{1D}$ receptor clone. We [6] and others [7-11] also isolated homologous clones from the human genome that we termed 5-HT$_{1D\alpha}$ and 5-HT$_{1D\beta}$. When we examined the affinities of 19 different compounds at each of these cloned, transfected human receptor subtypes, we found that the two clones were essentially indistinguishable in pharmacological properties for this large group of divergent serotonergic drugs subtypes in pure form would be likely to conclude that the same receptor subtype was present in each of the two tissues. Due to this very high degree of pharmacological similarity, we (and others) chose to name each of these two clones as a 5-HT$_{1D}$ subtype but with different Greek letters (α and β) to differentiate the separate genetic subtypes.

This example with 5-HT$_{1D}$ clones represents the exception rather than the rule. The degree of transmembrane homology (identity) between the two human 5-HT$_{1D}$ subtypes (77%) is similar to that of many related genetic subtypes of receptors (e.g. $α_{1A}$-, $α_{1B}$-, and $α_{1C}$-adrenergic or dopamine D1 and D5 receptors). Unlike the 5-HT$_{1D}$ clones, most related genetic subtypes show distinct pharmacological properties. Thus, the 5-HT$_{1D}$ receptor group represents an extreme example in which distinct receptor proteins are so similar in pharmacological properties that they would be exceedingly difficult to separate by classical approaches.

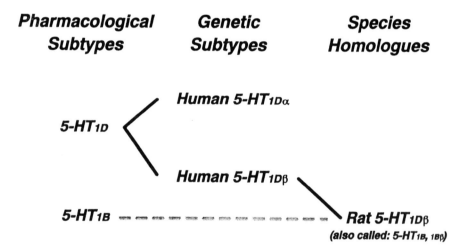

Figure 1. Relationships between pharmacological subtypes, genetic subtypes, and species homologue of the 5-HT$_{1D}$ receptor subfamily.

An example of two very closely related receptors with widely divergent pharmacological properties is provided by certain species homologue of this 5-HT$_{1D}$ receptor group. When we [12] and other [13,14] isolated the species homologue of the human 5-HT$_{1D\alpha}$ gene from the rat genome, we found that it displayed a markedly different pharmacological profile than its human homologue, even though it exhibited a typical degree of amino acid conservation for a species homologue: 94% overall and 96% in the transmembrane domains. The rat gene encoded a receptor with a 5-HT$_{1B}$ pharmacological profile, providing final proof to the speculation that the 5-HT$_{1B}$ and 5-HT$_{1D}$ receptors were species subtypes of the same receptor type [15]. Thus, one genetic subtype (hu 5-HT$_{1D\beta}$) is responsible for two different pharmacological subtypes of serotonin receptors: 5-HT$_{1D}$ in the human and 5-HT$_{1B}$ in the rat.

A summary of these relationships for the 5-HT$_{1D}$ receptor group is provided in figure 1. It should be emphasized again that these relationships represent the exception rather than the rule among 7TM receptors. In most cases, receptors with related amino acid sequences display related pharmacological profiles. The surprising relationships among 5-HT$_{1D}$ receptor clones clearly demonstrate that changes of a few key amino acids can dramatically alter the pharmacological properties of serotonin receptors, while fairly extensive changes can be tolerated at other sites in the receptor with very little effect on its pharmacology. Recently, site-directed mutagenesis studies have begun to elucidate the key amino acids of the 5-HT$_{1A}$ [16], 5-HT$_{1B}$ [17], and 5-HT$_2$ [18] receptors responsible for these effects.

3. The 5-HT$_{1E}$ and 5-HT$_{1F}$ Receptors

Three independent groups [19-21] have now identified and characterized a novel, intronless serotonin receptor clone whose pharmacological properties match those of a high affinity human cortical serotonin binding site first described by Leonhart et al. [22]. Like all other 5-HT$_1$ receptor subtypes, it displays high affinity for [^{3}H]5-HT (K_d = 10-15 nM) and couples to inhibition of adenylate cyclase activity. Unlike the other 5-HT$_1$ receptors, the 5-HT$_{1E}$ receptor exhibits weak affinity (pK$_i$ = 5.1) for 5-CT (5-carboxamidotryptamine) and most other serotonergic compounds. In contrast to the 5-HT$_{1D\alpha}$ and 5-HT$_{1D\beta}$ receptors, its affinity for sumatriptan is quite weak (pK$_i$ = 5.6).

The most recent addition to the family of 5-HT$_1$ receptor clones is the 5-HT$_{1F}$ receptor clone by Adham et al. [23] and as a mouse 5-HT$_{1E\beta}$ (a.k.a. 5-HT$_6$ [24]) receptor clone by Amlaiky et al. [25]. These clones represent species homologue of the same genetic subtype. This intronless clone is coupled to inhibition of adenylate cyclase activity, as are all other members of the 5-HT$_1$ subfamily. It is characterized by high affinity for serotonin (K_d = 9.2 nM [23]), intermediate affinity for sumatriptan (pK$_i$ = 7.6,7.1 [23,25]) and low affinity for 5-carboxamidotryptamine (pK$_i$ = 6.1,5.5 [23,25]). The 5-HT$_{1F}$ receptor is most homologous (amino acid sequence) to the 5-HT$_{1E}$ receptor and is found in the CNS (especially in the hippocampus [25]) and in the periphery, with especially high expression levels found in the uterus and in the mesentery [23]. Another serotonin receptor clone with high affinity for serotonin and 5-CT has been named 5-HT$_5$ receptor [24]. This new clone in the uterus and in the mesentery [23]. Another serotonin receptor clone with high affinity for serotonin and 5-CT has been named 5-HT_5 receptor [24]. This new clone is distinct from the 5-HT_{1F} subtype, exhibits multiple coding region introns, and has not yet been associated with a second messenger response.

A summary diagram (dendogram) indicating amino acid sequence homologies in the transmembrane regions between all 5-HT_2 and 5-HT_1 receptor clones is shown in figure 2. This figure clearly shows the natural division of the known serotonin clones into the 5-HT_1 subfamily containing the 5-HT_{1A}, 5-HT_{1B}, 5-$HT_{1D\alpha}$, 5-$HT_{1D\beta}$, 5-HT_{1E}, and 5-HT_{1F} receptors, and the 5-HT_2 subfamily containing the 5-HT_2, 5-HT_{1C} and the recently described 5-HT_{2F} receptors. It is pleasing to see that the pharmacological groupings of serotonin receptor subtypes into 5-HT_1 and 5-HT_2 subfamilies developed from years of pharmacological research are in agreement with the amino acid sequence relationships of the clones representing each of these receptors.

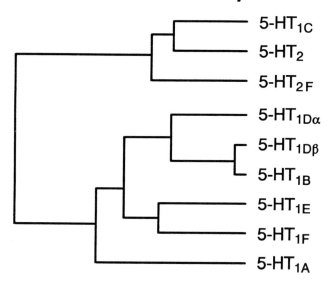

Figure 2. Dendogram of amino acid sequence homologies among serotonin receptor clones. Vertical lines are drawn to represent deduced amino acid sequence homologies between pairs of clones. The position of the vertical line along the x-axis represents decreasing degrees of sequence homology as you move from right to left.

4. Why Are There Five Serotonin 5-HT$_1$ Receptors Coupled to Adenylate Cyclase Inhibition?

A critical question for both our understanding of brain function and our search for new drugs is: why have five separate serotonin receptor genetic subtypes (5-HT$_{1A}$, 5-HT$_{1D\alpha}$, 5-HT$_{1D\beta}$ [and its species homologue 5-HT$_{1B}$], 5-HT$_{1E}$, and 5-HT$_{1F}$) been preserved in the human genome if all five are coupled to the same second messenger response? At least two hypotheses can be quickly rejected while three others remain to be explored. One possibility is that each subtype is designed to sense a different concentration of serotonin, thus providing an array of subtypes that trigger after different degrees of serotonergic synaptic firing (or exogenous release). The available data on the high affinity binding state of these subtypes does not support this idea. The high affinity component of serotonin binding to these human clones ranges from 1.9 nM for the 5-HT$_{1A}$ receptor to 9.7 nM for the 5-HT$_{1E}$ receptor, a narrow range which seems unlikely to signify different physiological roles. It seems more likely that the nervous system may use different degrees of receptor reserve to fine tune the midpoint for receptor activation (K$_A$), as has been noted for the 5-HT$_{1A}$ receptor [26].

Another obvious reason for subtypes is to differentiate receptors for presynaptic autoreceptor or heteroreceptor roles vs. postsynaptic activities. Recent *in situ* hybridization studies in our laboratory, however, indicate that all five serotonin 5-HT$_1$ receptor subtypes are found in raphe nuclei, suggesting that all five serve as autoreceptors [27].

The one area in which subtype differences are known to exist is in their tissue distributions. Both receptor binding autoradiography and *in situ* hybridization histochemistry have shown widespread CNS and selected peripheral organ distributions that differ among the subtypes. It can be anticipated that receptor genes will be shown to contain different tissue-specific regulatory sequences that control this tissue-specific activation of transcription. Various 5-HT$_1$ receptor genes may have arisen by gene duplication events, followed by evolution of different tissue and developmentally-specific regulatory elements. The differences we now see in the coding regions of these genes may simply represent subsequent random mutations of amino acids that are not important for receptor function.

Another hypothesis for the diversity of 5-HT$_1$ receptors is that subtle differences may exist in their coupling to second messenger systems. We and others have observed strong effects of receptor reserve on the apparent intrinsic activities of drugs in transfected cell systems including the finding that drugs with low intrinsic activity can behave as partial agonists with no apparent receptor reserve while strong or full agonists exhibit high degrees of receptor reserve in the same cell line [28]. It remains to be determined whether 5-HT$_1$ receptor subtypes may differ in these properties. Along this same direction, we may find that receptors that can couple to inhibition of adenylate cyclase activity in the laboratory are actually designed to directly couple to ion channel modulation in their native cell environment [29]. Finally, we have recently obtained direct indications that at least two of the 5-HT_1 receptors show selective interactions with different isoforms of G_i protein alpha subunits. Transfection of 5-HT_{1D} receptor clones into a series of cell lines led to preferential expression of functional coupling in subsets of the cell lines that expressed particular G_i-α subunits [30]. In particular, the 5-$HT_{1D\alpha}$ receptor appears to selectively recognize the G_i-$\alpha 3$ isotype while the 5-$HT_{1D\beta}$ and 5-HT_{1B} receptors do not preferentially couple to this isotype [30]. Thus, these two subtypes of 5-HT_{1D} receptor may have evolved to couple to different intracellular G protein pathways.

Further support for this idea is found in the amino acid sequences of the 5-HT_{1D} receptors. The G protein coupling domains of 7TM receptors have now been localized to the amino and carboxy terminal ends of intracellular loop 3 (between TM domain V and VI) and the amino end of the carboxy tail [31]. Examination of these domains in the 5-HT_{1D} receptor group shows that the first 20 amino acids of these three G coupling domains exhibit 95%, 100%, and 100% identity respectively between the human 5-$HT_{1D\beta}$ receptor and its species homologue, the 5-HT_{1B} receptor, yet only exhibit 50%, 65%, and 45% amino acid identity with the human 5-$HT_{1D\alpha}$ receptor. These data again suggest that these two genetic subtypes are differentiated by their coupling to different G_i-α subunit isotypes. Given the large array of over 30 different G_i-α subunits that have now been described, it is entirely possible that each 5-HT_1 receptor subtype will prove to couple to a different G protein. This and related information on different functional roles of the 5-HT_1 receptors may help explain why there appear to be more serotonin receptor subtypes than exist for any other neurotransmitter in the body.

5. References

1. Hartig, P.R., Branchek, T.A., and Weinshank, R.L. (1992) 'A subfamily of serotonin 5-HT_{1D} receptor genes', Trends Pharmacol. Sci. 13, 152-159.
2. Hartig, P.R., Adham, N., Zgombick, J., Weinshank, R., and Branchek, T. (1992) 'Molecular biology of the 5-HT_1 receptor subfamily', Drug Dev. Res. 26, 215-224.
3. Libert, F., Parmentier, M., Lefort, A., Dumont, J.E., and Vassart, G. (1989) 'Selective amplification and cloning of four new members of the G protein-coupled receptor family', Science 44: 569-572.
4. Zgombick, J., Weinshank, R., Macchi, M., Schechter, L., Branchek, T., and Hartig, P. (1991) 'Expression and pharmacological characterization of a canine 5-hydroxytryptamine$_{1D}$ receptor subtype', Mol. Pharm. 40, 1036-1042.
5. Maenhaut, C., Van Sande, J., Massart, C., Dinsart, F., Libert, F., Monferini, E., Giraldo, E., Ladinsky, H., Vassart, G., and Dumont, J.E. (1991) 'The orphan receptor cDNA RDC4 encodes a 5-HT_{1D} serotonin receptor', Biochem. Biophys. Res. Commun. 180, 1460-1468.
6. Weinshank, R.L., Zgombick, J.M., Macchi, M., Branchek, T.A., and Hartig,

P.R. (1992) 'The human serotonin 1D receptor is encoded by a subfamily of two distinct genes: $5\text{-HT}_{1D\alpha}$ and $5\text{-HT}_{1D\beta}$', Proc. Natl. Acad. Sci. USA, 89, 3630-3634.

7. Hamblin, M. and Metcalf, M. (1991) 'Primary structure and functional characterization of a human 5-HT_{1D}-type serotonin receptor', Molecular Pharmacology 40, 143-148.

8. Jin, H., Oksenberg, D., Askenazi, A., Peroutka, S.J., Rozmahel, R., Yang, Y., Palacios, J.M., and O'Dowd, B.F. (1992) 'Characterization of the human 5-Hydroxytryptamine$_{1B}$ receptor', J. Biol. Chem. 267, 5735-5738.

9. Demchyshyn, L., Sunahara, R.K., Miller, K., Teitler, M., Hoffman, B.J., Kennedy, J.L., Seeman, P., Van Tol, H.H.M., and Niznik, H.B. (1992) 'A human serotonin 1D receptor variant $5\text{-HT}_{1D\beta}$) encoded by an intronless gene on chromosome 6', Proc. Natl. Acad. Sci USA 89, 5522-5526.

10. Levy, F.O., Gudermann, T., Perez-Reyes, E., Birnbaumer, M., Kaumann, A.J., and Birnbaumer, L. (1992) 'Molecular cloning of a human serotonin receptor (S12) with a pharmacological profile resembling that of the 5-HT_{1D} subtype', J. Biol. Chem. 267, 7553-7562.

11. Veldman, S.A. and Bienkowski, M.J. (1992) 'Cloning and pharmacological characterization of a novel human 5-Hydroxytryptamine$_{1D}$ receptor subtype', Mol. Pharmacol. 42, 439-444.

12. Adham, N., Romanienko, P., Hartig, P., Weinshank, R.L., and Branchek, T. (1992) 'The rat 5-hydroxytryptamine$_{1B}$ receptor is the species homolog of the human 5-hydroxytryptamine$_{1D\beta}$ receptor', Mol. Pharmacol. 41, 1-7.

13. Voight, M.M., Laurie, D.J., Seeburg, P.H., and Bach, A. (1991) 'Molecular cloning and characterization of a rat brain cDNA encoding a 5-hydroxytryptamine$_{1B}$ receptor', EMBO J., 10: 4017-4023.

14. Maroteaux, L., Saudou, F., Amlaiky, N., Boschert, U., Plassat, J.L., and Hen, R. (1992) 'Mouse 5-HT_{1B} serotonin receptor: cloning, functional expression, and localization in motor control centers', Proc. Natl. Acad. Sci. USA 89, 3020-3024.

15. Hoyer, D. and Middlemiss, D.N. (1989) 'The pharmacology of the terminal 5-HT autoreceptors in mammalian brain: evidence for species differences', Trends in Pharmacological Sciences 10, 130-132.

16. Guan, X.-M., Peroutka, S.J., and Kobilka, B.K. (1992) 'Identification of a single amino acid residue responsible for the binding of a class of β-adrenergic receptor antagonists to 5-Hydroxytryptamine$_{1A}$ receptors', Mol. Pharmacol. 41, 695-698.

17. Kao, H.-T., Adham, N., Olsen, M.A., Weinshank, R.L., Branchek, T.A., and Hartig, P.R. (1992) 'Site-directed mutagenesis of a single residue changes the binding properties of the serotonin 5-HT_2 receptor from a human to a rat pharmacology', FEBS Lett. 307, 324-328.

18. Hamblin, M.W. (1992) 'Thr355Asn mutation of the human $5\text{-HT}_{1D\beta}$ receptor results in 5-HT_{1B}-like pharmacological specificity', Abst. 2nd International

19. Levy, F.O., Gudermann, T., Birnbaumer, M., Kaumann, A.J., and Birnbaumer L. (1992) 'Molecular cloning of a human gene (S31) encoding a novel serotonin receptor mediating inhibition of adenylyl cyclase' FEBS Lett. 296: 201-206.
20. Zgombick, J.M., Schechter, L.E., Macchi, M., Hartig, P.R., Branchek, T.A., and Weinshank, R.L. (1992) 'Human gene S31 encodes the pharmacologically defined serotonin 5-HT$_{1E}$ receptor', Mol. Pharmacol. 42, 180-185.
21. McAllister, G., Charlesworth, A., Snodin, C., Beer, M.S., Noble, A.J., Middlemiss, D.N., Iversen, L.L., and Whiting, P. (1992) 'Molecular cloning of a serotonin receptor from human brain (5-HT$_{1E}$): a fifth 5-HT$_1$-like subtype', Proc. Natl. Acad. Sci. USA 89, 5517-5521.
22. Leonhardt, S., Herrick-Davis, K., and Teitler, M. (1989) 'Detection of a novel serotonin receptor subtype (5-HT$_{1E}$) in human brain: interaction with a GTP-binding protein', J. Neurochem. 53, 465-471.
23. Adham, N., Kao, H.-T., Schechter, L.E., Bard, J., Olsen, M., Urquhart, D., Durkin, M., Hartig, P.R., Weinshank, R.L., and Branchek, T.A. (1992) 'Cloning of a novel human serotonin receptor (5-HT$_{1F}$): a fifth 5-HT$_1$ receptor subtype coupled to the inhibition of adenylate cyclase', Proc. Natl. Acad. Sci. USA, in press.
24. Amlaiky, N., Plassat, J.-L., Ramboz, S., Boschert, U., and Hen, R. (1992) 'The mouse 5-HT$_5$ and 5-HT$_6$ receptors: two new "5-HT$_{1D}$-like" serotonin receptors: cloning and expression', Soc. Neurosci Abst. 18: 212.
25. Amlaiky, N., Ramboz, S., Boschert, U., Plassat, J.-L., and Hen, R. (1992) 'Isolation of a mouse "5-HT$_{1E}$-like" serotonin receptor expressed predominantly in the hippocampus', J. Biol. Chem., in press.
26. Meller, E., Goldstein, M., and Bohmaker, K. (1990) 'Receptor reserve for 5-Hydroxytryptamine$_{1a}$-mediated inhibition of serotonin synthesis: possible relationship to anxiolytic properties of 5-Hydroxytryptamine$_{1A}$ agonists', Mol. Pharmacol. 37: 231-237.
27. Branchek, T, Hou-Yu, A., Durkin, M., Urquhart, D., and Weinshank, R. (1992) 'Both 5-HT$_{1D\alpha}$ and 5-HT$_{1D\beta}$ receptor subtypes appear to be autoreceptors: evidence form "in situ" hybridization studies', Abst. 2nd International Symposium on Serotonin: from cell biology to pharmacology and therapeutics, Giovanni Lorenzini Medical Foundation, Houston, 9.
28. Adham, N., Ellerbrock, B., Hartig, P., Weinshank, R.L., and Branchek, T. (1993) 'Receptor reserve masks partial agonist activity of drugs in a cloned rat 5-HT$_{1B}$ receptor expression system', Mol. Pharmacol., submitted.
29. Karschin, A., Ho, B.Y., Labarca, C., Elroy-Stein, O., Moss, B., Davidson, N., and Lester, H.A. (1991) 'Heterologously expressed serotonin 1A receptors couple to muscarinic K$^+$ channels in heart', Proc. Natl. Acad. Sci. USA 88:

5694-5698.
30. Branchek, T., Adham, N., Zgombick, J., Schechter, L., Hartig, P., and Weinshank, R. (1992) 'Differential host preference in functional coupling of cloned 5-HT_{1D} receptor subtypes in heterologous expression systems', Abst. 2nd International Symposium on Serotonin: from cell biology to pharmacology and therapeutics, Giovanni Lorenzini Medical Foundation, Houston, 34.
31. Dohlman, H.G., Thorner, J., Caron, M.G., and Lefkowitz, R.J. (1991) 'Model systems for the study of seven-transmembrane-segment receptors', Ann. Rev. Biochem. 60, 653-688.

5-HT$_{1B/D}$-RECEPTOR AGONISTS BLOCK C-FOS-LI WITHIN TRIGEMINAL NUCLEUS CAUDALIS IN RESPONSE TO NOXIOUS MENINGEAL STIMULATION

MICHAEL A. MOSKOWITZ and KAZUHIKO NOZAKI
Stroke Research Laboratory
Neurosurgery and Neurology
Massachusetts General Hospital
Harvard Medical School
Boston, Massachusetts 02114

ABSTRACT. Important neuronal actions of sumatriptan and ergot alkaloids have been identified mediated by prejunctional 5-HT$_{1B/D}$ heteroreceptors on trigeminovascular fibers. Evidence is now presented showing that 5-HT$_{1B/D}$ receptor agonists (CP-93, 129, sumatriptan and dihydroergotamine) dose-dependently block the expression of c-fos-like immunoreactivity, an indicator of neuronal activation, within lamina I, II$_o$ of trigeminal nucleus caudalis, a major nociceptive brain stem nucleus. C-fos-like immunoreactivity was induced by injecting autologous blood or carrageenin into the subarachnoid space. C-fos-like immunoreactivity was also blocked by administering the analgesic morphine. Rather than mediated via an action directly on brain stem neurons, 5-HT$_{1B/D}$'s effects were probably mediated by trigeminovascular heteroreceptors. Because blood is a potent spasmogen, we infer that the analgesic actions of agonists at 5-HT$_{1B/D}$ receptors are not dependent upon the pre-existence of vascular dilatation.

1. Introduction

C-fos-like immunoreactivity (LI) has been used successfully to map the distribution of neurons activated by noxious stimulation [1,2]. The number of cells expressing c-fos-like immunoreactivity provide a measure of neuronal activation, and this has been examined within laminae I, II$_o$ of trigeminal nucleus caudalis (TNC) [3,4], a region containing both the synapses of primary afferent fibers concerned with the transmission of nociceptive information and second and higher order neurons which transmit nociceptive information to rostral centers.

We recently adapted an animal model [5] to address the possible neuronal effects of 5-HT$_{1B/D}$ agonists based on the response which follows the instillation of autologous blood into the cisterna magna and subarachnoid space [3]. Blood in the subarachnoid space causes significant vasoconstriction, reductions in blood flow, and pain. We determined previously that blood infection promotes c-fos protein expression within cells of lamina I, II_o [3]. Their number was not related to the volume of injectate but rather correlated with the quantity of injected blood. Chronic sectioning of the trigeminal nerve innervating the meninges, or destroying small unmyelinated C-fibers by neonatal capsaicin treatment significantly reduced the number of expressing cells after blood injection, thereby underscoring the importance of inputs from the trigeminovascular system to this response.

2. Materials and Methods

Experimental details have been published [3,4].

2.1 BLOOD INSTILLATION INTO THE SUBARACHNOID SPACE

Male Sprague-Dawley rats (250-300 g) were anesthetized with pentobarbitone. A soft catheter was then introduced into the cisterna magna. Six hours after catheter placement, 0.3 ml of autologous nonheparinized arterial blood or mock cerebrospinal fluid (CSF) (pH 7.4) was injected into the cisterna magna. Sham animals were treated identically but received no intracisternal injection. Animals were then kept at 30° head-down for 30 minutes after which they remained in the horizontal prone position.

Some animals were treated as neonates with capsaicin (50 mg kg^{-1} s.c. within the first 48 hours of life) to destroy small unmyelinated sensory fibers.

2.2 DRUG PRETREATMENT

In subarachnoid hemorrhage experiments, sumatriptan (240 or 720 nmol kg^{-1}), CP-93,129 (46, 140, or 460 nmol kg^{-1}) [6], dihydroergotamine (86 nmol kg^{-1}) (1 ml kg^{-1}) were injected intravenously 60 and 10 minutes prior to blood instillation. Drug dosages were chosen based on amounts previously shown to activate prejunctional receptors and block neurogenic extravasation within the meninges [7]. Morphine sulphate (10 mg kg^{-1}) (0.67 ml kg^{-1}) was injected subcutaneously 20 minutes prior to blood instillation [2].

2.3 C-FOS IMMUNOHISTOCHEMISTRY

Animals survived for 2 hours after blood instillation. The brain stem with upper cervical spinal cords were sectioned coronally (50 um) on a freezing microtome after perfusion fixation and postfixation overnight in 4% paraformaldehyde. Free-floating sections were processed immunohistochemically with the avidin-biotin procedure using commercially available kits (Vectastain ABC; Vector Labs Burlingame, CA) as described previously [3]. The employed rabbit polyclonal antisera was directed against an *in vitro* translated product of c-fos gene (kindly provided by Dr. Dennis

Slamon, the Department of Hematology and Oncology at University of California, Los Angeles), and was preabsorbed against acetone-dried rat liver powder overnight to reduce nonspecific background staining. The staining pattern produced by this antisera is comparable to a commercially available monoclonal antibody raised against a synthetic peptide which consists of residue 4-17 of the c-fos protein. In some experiments, antisera were reused (maximum of 5 times) and collected and stored at 4°C after use.

Every second section was processed for immunohistochemistry and approximately 150 sections were examined in each animal. In order to quantitate c-fox protein expression, cells showing c-fos-LI within lamina I, II_o of TNC (from area postrema [AP] to cervical 1 segment; about 50 sections in each animal), nucleus of the solitary tract (NTS); about 10 sections in each animal, AP; about 5 sections in each animal were counted by a "blinded" observer.

2.4 STATISTICS

Data are expressed as the number of cells within each 50 um sections (mean ± sd). Statistical comparisons were performed using unpaired Student's t-test. Probability less than 0.01 was considered statistically significant.

2.5 DRUGS

Sumatriptan (Glaxo Ltd, Hertfordshire, England) was dissolved in saline; dihydroergotamine (DHE) (Sandoz; supplied as D.H.E.45 ampules [1 mg ml^{-1}]) were diluted in saline; CP-93,129 [6] (Pfizer, Inc., Groton, CT, USA) was dissolved in dimethylsulfoxide:saline 1:19. Morphine sulfate (Elkins-Sinn, Inc., Cherry Hill, NJ, USA [15 mg ml^{-1}] was used without dilution.

3. Results

Mortality was less than 5 percent. There were no significant differences in the monitored physiological responses (MABP, pulse, respiratory rate, intracranial pressure) between vehicle- and drug-treated groups before or after experimental subarachnoid hemorrhage. Arterial pCO2, pO_2, pH, Hct, blood glucose did not differ between groups. Cortical blood flow as determined by laser Doppler flowmetry began to decrease several seconds after blood instillation and was reduced by 50% 5 minutes after blood injection.

Infrequent positive cells were detected in the brain stem of sham animals. Mock CSF (0.3 ml) increased the numbers of positive cells per section in TNC (lamina I, II_o), in AP and NTS (16 ± 5, 21 ± 8, or 21 ± 4 [n=10] respectively).

C-fos-LI was found most intensely within cells of TNC (lamina I,II_o), AP, and NTS in response to 0.3 ml of blood. The mean number of c-fos containing cells were 37 ± 4, 41 ± 11, 34 ± 9 (n=9) per 50 um section in TNC (lamina I,II_o), AP,

NTS, respectively.

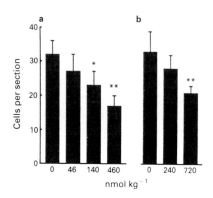

Figure 1. CP-93, 129 (a) and sumatriptan (b) dose-dependently decrease the numbers of immunoreactive cells within rat TNC [lamina I,II$_o$ after intracisternal blood injection. Drugs were administered intravenously 10 minutes prior to blood instillation. N=3-10 animals per group *p < 0.05; **p < 0.01 compared with vehicle treatment.

Pretreatment with sumatriptan (720 nmol kg^{-1}x2), dihydroergotamine (86 nmol kg^{-1} x2) or CP-93,129 (460 nmol kg^{-1}x2) significantly reduced the numbers of positive cells in TNC (lamina I,II$_o$) by 34%, 33%, or 39% respectively as compared to vehicle-treated animals (figure 1), but not in AP or NTS. Pretreatment with morphine (10 mg kg^{-1}) also markedly decreased the number of c-fos positive cells in TNC (lamina I,II$_o$) by 64% as compared to vehicle-treated animals (figure 2) and slightly in AP and NTS.

4. Discussion

5-HT$_{1B}$ receptors which mediate inhibition of the c-fos response to blood may be closely associated with, or directly on perivascular afferent fibers [8]. The spatial pattern of 5-HT$_{1B/D}$-induced inhibition was similar in capsaicin-treated animals in that neither capsaicin treatment nor the 5-HT$_{1B/D}$ agonists blocked c-fos expression in AP or NTS. Although we have not excluded the possibility that the inhibitory 5-HT$_{1B}$ receptor may be present within brain stem structures such as the raphe nucleus, periaqueductal grey area or TNC itself (5-HT$_{1B/D}$ subtypes have been localized within some of these areas [9]), we think this possibility is less likely for the following reasons. Firstly, inhibition of c-fos expression was dependent upon the integrity of trigeminovascular fibers. Hence, expressing cells in lamina I,II$_o$ after capsaicin-lesioning were not inhibited further by treatment with CP-93,129.

Secondly, a brain stem receptor-mediated mechanism would have been expected to block the c-fos response in other paradigms of nociception. However, we determined that the expression of c-fos-LI was not inhibited further by sumatriptan when the noxious chemical formalin was applied to the nasal mucosa [4]. Thirdly, we now have evidence that sumatriptan (and trigeminal nerve sectioning) inhibits the expression of c-fos-LI within TNC (but not in AP or NTS) induced by neocortical spreading depression [10]. In this instance, hydrophilic drugs such as sumatriptan do not readily penetrate the blood brain barrier as it might after subarachnoid hemorrhage.

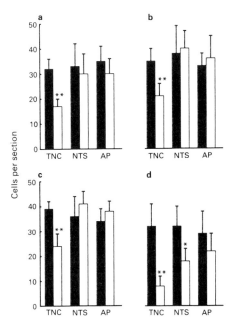

Figure 2. C-fos-LI within TNC, NTS, and AP after (a) CP-93, 129 (460 nmol kg^{-1}), (b) sumatriptan (720 nmol kg^{-1}), (c) dihydroergotamine (86 nmol kg^{-1}), or (d) morphine (15 umol kg^{-1}) administration. Vehicle treatments shown in dark columns. *p < 0.05; **p < 0.01 compared with vehicle treatment.

Based on the above, we suggest that the specificity of the 5-HT$_{1B/D}$ inhibitory response is conferred by the receptor population on innervating primary afferent fibers, and not by mechanisms intrinsic to brain stem. Indeed, recent gene amplification experiments using polymerase chain reaction technology (unpublished) and *in situ* hybridization confirm that the gene encoding the 5-HT$_{1B/D}$ receptor protein are expressed by trigeminal ganglion cells (unpublished). Based on the data presented, and previous results from our group [7,8,11], we suggest that prejunctional $5\text{-}HT_1$ receptors are coupled to at least two important functions which may diminish pain and sensitization in vascular headache, including blockade of neural transmission within trigeminovascular fibers (as reflected by inhibition of c-fos protein expression), and blockade of the neuroinflammatory responses. Nevertheless, evidence is consistent with the interpretation that both receptors reside on primary afferent fibers wherein both may be coupled, for example, to hyperpolarization mediated by opening of a ligand-gated potassium channel.

5. Acknowledgement

This work was supported by grant #NS 21558 by the National Institutes of Neurological Disease.

6. References

1. Hunt, S.P., Pini, A., and Evan, G. (1987) 'Induction of c-fos-like protein in spinal cord neurons following sensory stimulation', Nature 328, 632-634.
2. Presley, R.W., Menetrey, D., Levine, J.D., and Basbaum, A.I. (1990) 'Systemic morphine suppresses noxious stimulus-evoked fos protein-like immunoreactivity in rat spinal cord', J. Neurosci. 10, 323-335.
3. Nozake, K., Boccalini, P., and Moskowitz, M.A. (1992) 'Expression of c-fos-like immunoreactivity in brainstem after meningeal irritation by blood in the subarachnoid space', Neuroscience 49, 669-680.
4. Nozaki, K., Moskowitz, M.A., and Boccalini, P. (1992) 'CP-93,129, sumatriptan, dihydroergotamine block c-fos expression within rat trigeminal nucleus caudalis caused by chemical stimulation of the meninges', Br. J. Pharmacol. 106, 409-415.
5. Delgado, T.J., Brismar, J., and Svendgaard, N.A. (1985) 'Subarachnoid haemorrhage in the rat: Angiography and fluorescence microscopy of the major cerebral arteries, Stroke, 16 595-602.
6. Macor, J.E., Burkhart, C.A., Heym, J.H., Ives, J.L., Lebel, L.A., Newman, M.E., Nielsen, J.A., Ryan, K., Schulz, D.W., Torgersen, L.K., and Koe, B.K. (1990) '3-(1,2,5,6-Tetrahydropyrid-4-yi)pyrrolo[3,2-b]pyrid-5-one: a potent and selective serotonin ($5\text{-}HT_{1B}$ agonist and rotationally restricted phenolic analogue of 5-methoxy-3-(1,2,5,6-tetrahydropyrid-4-yi)indole', J. Med. Chem. 33, 2087-2093.
7. Matsubara, T., Moskowitz, M.A., and Byun, B. (1991) 'CP-93,129, a potent and selective $5\text{-}HT_{1B}$ receptor agonist blocks neurogenic plasma extravazation within rat but not guinea-pig dura mater', Br. J. Pharmacol. 104, 3-4.
8. Moskowitz, M.A. (1992) 'Neurogenic versus vascular mechanisms of sumatriptan and ergot alkaloids in migraine', Trends Pharmacol. Sci. 13, 307-311.
9. Hoyer, D. and Middlemiss, D.N. (1989) 'Species difference in the

pharmacology of terminal 5-HT autoreceptors in mammalian brain', Trends Pharmacol. Sci. 10, 130-132.
10. Moskowitz, M.A., Nozaki, K. and Kraig, R.P. (1992) 'Neocortical spreading depression provokes the expression of c-fos protein-like immunoreactivity within trigeminal nucleus caudalis via trigeminovascular mechanisms', J. Neurosci., in press.
11. Buzzi, M.G., Moskowitz, M.A., Peroutka, S.J., and Byun B. (1991) 'Further characterization of the putative 5-HT receptor which mediates blockade of neurogenic plasma extravazation in rat dura mater', Br. J. Pharmacol. 103, 1421-1428.

A REAPPRAISAL OF 5-HT RECEPTOR CLASSIFICATION

P.P.A. HUMPHREY, P. HARTIG[1], and D. HOYER[2]
Glaxo Institute of Applied Pharmacology
University of Cambridge
Department of Pharmacology
Tennis Court Road
Cambridge CB2 1QJ
UK,
[1]Synaptic Pharmaceutical Corporation
215 College Road
Paramus, New Jersey 07652-1410
USA, and
[2]Sandoz Pharma Ltd
Preclinical Research, 360/604
CH-4002 Basel
Switzerland

ABSTRACT. On the basis of a modern integrated approach to receptor characterization, the classification of 5-HT receptors has been reappraised by the Serotonin Club Receptor Nomenclature Committee. This has led to a modified nomenclature which is outlined. The new classification extends the classification of Bradley and colleagues, published in 1986, to include the 5-HT_4 receptor and several subtypes of the 5-HT_1 and 5-HT_2 receptor group. However, it is recognized that the scheme is still incomplete as several 5-HT receptors, including the newly cloned 5-HT_5 receptor, cannot yet be classified due to lack of adequate characterization data.

1. Current Classification of 5-HT Receptors

The accepted 5-HT receptor classification of Bradley and colleagues, recognizes three main groups of 5-HT receptor, 5-HT_1, 5-HT_2, 5-HT_3 [1,2]. This classification was based on the need to reconcile data from functional studies with that from radioligand binding studies and provide a common system of nomenclature for 5-HT receptors identified from all such studies. In this respect the scheme has been a success, obviating much potential confusion. Thus it amalgamated Peroutka and Snyder's 5-HT_1 and 5-HT_2 classification, based on binding data, with that of Gaddum's D (5-HT_2) and M (5-HT_3) classification based on data from studies of drug action [3,4], thereby recognizing that these receptors comprise three not four distinct receptor types. The scheme has proved to be robust and has provided a useful framework for five or so years, but it is now somewhat dated. Thus, with the major advances made in molecular biology, the protein structures of many 5-HT receptor molecules have been determined. It now only remains to be seen how well such data correlates with data generated from the study of drug actions. Undoubtedly, molecular biology has served to confirm the existence of the three groups of 5-HT receptor defined by Bradley and colleagues. It now remains to determine the significance of the 5-HT$_4$ receptor in molecular terms and to understand the functional importance of the recently cloned 5-HT receptors, named by the authors as 5-HT$_{1E}$, 5-HT$_{1F}$, 5-HT$_{2F}$, and 5-HT$_5$ [5-9].

2. A New Generic Approach to Receptor Characterization

Although molecular biology undoubtedly serves to enrich and extend our knowledge greatly, it nevertheless presents a short-term problem in trying to integrate our understanding of all the data at both the structural and functional level. Nevertheless, enzymologists have long recognized that structure correlates with function and it should therefore be expected that structural (molecular) data on receptors will largely correlate with data from operational studies, involving measurement of drug related parameters, and also with data measuring receptor transduction events, associated with receptor activation. It would therefore seem that each of three types of criteria are required to fully characterize a receptor: operational, structural and transductional.

Prior to the advent of molecular biology, receptors were necessarily characterized primarily on the basis of OPERATIONAL (drug related) data. This involves quantification of the actions of selective ligands (agonists and antagonists) to define the receptor mediating a particular function. Since many ligands have only a degree of selectivity, great care is needed in generating reliable quantitative data. Nevertheless, good estimates of affinity of antagonists or potency of agonists, will provide important and therapeutically useful information about the nature of the receptor protein with which they interact.

The cloning and sequencing of receptors provides a definitive "fingerprint" of the receptor. In general, comparison of amino acid (receptor protein) and nucleotide (gene) sequences, particularly in the transmembrane domains, seem to correlate well (but not invariably) with receptor types and subtypes defined operationally. Nevertheless, the STRUCTURAL definition of newly identified receptors, while definitive, is of restricted value in isolation since the operational significance needs to be defined in studies with the same ligands used to define operational characteristics in the physiological state.

The significance of receptor structure in the mediation of a cellular response is now becoming better understood. Thus, for example, it is now apparent from site-directed mutagenesis studies that the size and sequence of the third intracellular loop and carboxy tail of seven transmembrane domain receptors will dictate the nature of G-protein binding [10]. This in turn will indicate the preferred transduction system for native receptor in its natural tissue or in a wild-type cell. This suggests that a knowledge of TRANSDUCTIONAL (receptor-response coupling) mechanisms will provide important information relevant to the nature and structure of a given receptor type. This leads to the view that no one criteria alone should be used to classify receptors, each being equally important, and that a receptor can only be fully characterized and confidently named if comprehensive information on operational pharmacology, structure, and transduction is available.

3. A Reappraisal of the Classification of 5-HT Receptors

With this modern approach, the current state of knowledge on 5-HT receptors can be readdressed. Thus several new 5-HT receptor clones can be ascribed either to the 5-HT_1, or 5-HT_2 group, even in some cases (e.g. 5-HT_{1E} and 5-HT_{1F}) on the basis of very limited operational data. This exercise also provides the opportunity to finally recommend the change of name of the 5-HT_{1C} receptor to 5-HT_{2C}, knowing it to be will characterized on the basis of all three characterization criteria (i.e. operational, structural, and transductional) as a 5-HT_2 receptor subtype. It now also seems appropriate to acknowledge the 5-HT_4 receptor as being in a separate group though its cloning is eagerly awaited. Too little is currently known about the pharmacology of the recently cloned 5-HT_5 receptor to know whether its appellation is appropriate and is tacitly accepted only on a very provisional basis because of its low homology to other 5-HT receptor types [9]. The proposed new receptor classification and nomenclature agreed in Houston by the Serotonin Club Receptor Nomenclature Committee is outlined in Table 1.

The known 5-HT_1 receptor subtypes can all be grouped together within the new scheme with greater or lesser confidence in terms of sequence homology, absences of introns, a common transduction system (inhibition of adenylyl cyclase), and some degree of commonality in their pharmacological (operational) characteristics. However, on this basis the potent agonist activity of 5-carboxamidotryptamine (5-CT) can no longer be regarded as a definitive diagnostic for all 5-HT_1 receptors [1,2].

There is, however, one obvious nomenclature problem in the 5-HT_1 receptor group. Thus the rat and mouse 5-HT_{1B} receptors are species equivalents (97% homology) of the human $5\text{-HT}_{1D\beta}$ receptor as suspected for some time [11]. However, the 5-HT_{1B} subtype retains its name in the new scheme, since it has a typical "5-HT_{1D} profile" in terms of its operational pharmacology. In general, it would seem sensible that species homologues of receptors should almost invariably have the same name, even when marked species-related operational differences exist, as is the case for 5-HT_3 receptors, for example. However, we argue that a rare exception should be made in the case of the $5\text{-HT}_{1D\alpha}/5\text{-HT}_{1D\beta}$ receptor pair, since in man both receptors have similar operational characteristics, at least on the

Table 1 5-HT receptor family classification*

Superfamily	Group	Receptor subtypes	Transduction system	Gene structure	Comments/previous names
G-Protein	5-HT$_1$	5-HT$_{1A}$	Adenylyl Cyclase Inhibition	No introns in the coding regions	Originally identified as G21.
		5-HT$_{1B}$			Rodent homologue to human 5-HT$_{1D\beta}$.
		5-HT$_{1D\alpha}$			First cloned human 5-HT$_{1D}$ receptor.
		5-HT$_{1D\beta}$			Human 5-HT$_{1B}$ receptor but human homologue differs from rat homologue by having 5-HT$_{1D}$-like operational characteristics.
		5-HT$_{1E}$			More operational data required.
		5-HT$_{1F}$			More operational data required. Has been called 5-HT$_6$, 5-HT$_{1E\beta}$.
G-Protein	5-HT$_2$	5-HT$_{2A}$	Stimulation Phosphoinositide Metabolism	Introns present	"Classical" 5-HT$_2$ receptor.
		5-HT$_{2B}$			Rat fundus receptor, 5-HT$_{2F}$
		5-HT$_{2C}$			Previously called 5-HT$_{1C}$.
Ligand-gated	5-HT$_3$		Cation Channel	Only one sub-unit cloned to date.	Previously called M receptor. Species differences exist on basis of operational data (rat, mouse, guinea-pig).
G-Protein	5-HT$_4$		Adenylyl Cyclase Stimulation	Not yet cloned	Different to 5-HT$_1$ receptors on basis of operational and transductional criteria.

* New classification scheme agreed by Serotonin Club Receptor Nomenclature Committee in September at Houston (see Hartig, Hoyer and Humphrey, 1993, *Trends Pharmacol. Sci.* in press). In addition to the authors, D.E. Clarke, G.R. Martin, D.N. Middlemiss, E.J. Mylecharane and P.R. Saxena also actively participated in the meeting. See also references 1 and 2.

basis of the ligands tested to date, despite the fact that their sequence homology is only about 70% [12]. Admittedly, one could have designated the human 5-HT$_{1D\beta}$ gene product as a human 5-HT$_{1B}$ receptor (as suggested by some authors [13]), but this appellation does not account for its pharmacological features uniquely shared with the 5-HT$_{1D\alpha}$ receptor.

Since we now consider that 5-HT$_1$ receptors are defined as being negatively linked to adenylyl cyclase, the 5-HT$_1$-like receptor which mediates smooth muscle relaxation and is positively linked to adenylyl cyclase must not be considered as unclassifiable or as an "orphan" receptor for the time being, as its operational characteristics are very different from those of the 5-HT$_4$ receptor [14,15]. However, the 5-HT$_1$-like sumatriptan-sensitive receptor which mediates vascular smooth muscle contraction and is negatively coupled to adenylyl cyclase remains as a recognized 5-HT$_1$ receptor subtype [16]. However, until shown definitively whether it is or is not a 5-HT$_{1D}$ receptor, it is best described as a vascular 5-HT$_1$ subtype.

5-HT$_2$ receptors and subtypes satisfy all the new criteria for characterization: close sequence homology, the presence of similar intron/exon patterns, same transduction system (stimulation of phospholipase C), and similar operational profiles. At least three subtypes are known and suggested names agreed (see table 1) by the Serotonin Club Receptor Nomenclature Committee. The 5-HT$_{2A}$ receptor subtype refers to the classical 5-HT$_2$ receptor. The 5-HT$_{2B}$ subtype is the receptor mediating the contractile action of 5-HT described in the fundus many years ago [17] and which has recently been cloned (called 5-HT$_{2F}$ just before the scheme proposed here [8]). The 5-HT$_{2C}$ receptor corresponds to the previously known 5-HT$_{1C}$ receptor; it was felt that this was the least confusing nomenclature transition. It would seem sensible that the 5-HT$_{1C}$ appellation is no longer used to avoid potential confusion in the future.

5-HT$_3$ receptors form a group apart, since they belong to the ligand-gated ion channel receptor super-family: this has been inferred from electrophysiological studies and confirmed by cloning studies in which the cloned receptor subunit from murine NCB 20 cells has some homology with the nicotinic receptor α-subunit and when expressed, the receptor subunit has all the features of a ligand-gated channel [18]. Assuming that other subunits will be identified, it would seem that Greek subscripts could be used to account for the different subunits, as currently used for nicotinic receptors (e.g. 5-HT$_{3\beta}$ for the cloned β subunit). There is presently little information about intra-species subtypes; it will probably take several years to clone the numerous subunits and to determine which subunit combination confers a defined pharmacological profile.

4. Concluding Remarks

Molecular biology has confirmed that there are at least three groups of 5-HT receptor as proposed by Bradley and colleagues, on the basis of operational data.

In addition, all of the necessary evidence is now available for several subtypes of the first two of these groups. There is also evidence for further distinct groups of 5-HT receptor. However, more data is required before we can be confident of the precise identity of the 5-HT$_4$ and 5-HT$_5$ receptors. A number of orphan 5-HT receptors also require study before they can be properly classified.

5. References

1. Bradley, P.B., Engel, G., Feniuk, W., Fozard, J.R., Humphrey, P.P.A., Middlemiss, D.N., Mylecharane, E.J., Richardson, B.P., and Saxena, P.R. (1986) 'Proposals for the classification and nomenclature of functional receptors for 5-hydroxytryptamine', Neuropharmacol. 25, 563-576.
2. Humphrey, P.P.A. (1992) '5-Hydroxytryptamine receptors and drug discovery', Int. Acad. Biomed. Drug Res., Karger, Basel, vol. 1, pp. 129-139.
3. Gaddum, J.H. and Picarelli, Z.P. (1957) 'Two kinds of tryptamine receptor', J. Pharmacol. Chemother. 12, 323-328.
4. Peroutka, S.J. and Snyder, S.H. (1979) 'Multiple serotonin receptors: differential binding of [^{3}H]5-hydroxytryptamine, [^{3}H]lysergic acid diethylamide and [^{3}H]spiroperidol', Mol. Pharmacol. 16, 687-699.
5. McAllister, G., Charlesworth, A., Snodin C., Beer, M.S., Noble, A.J., Middlemiss, D.N., Iversen, L.L., and Whiting, P. (1992) 'Molecular cloning of a serotonin receptor from human brain (5-HT$_{1E}$); a fifth 5-HT$_1$-like subtype', Proc. Natl. Acad. Sci. USA 89, 5517-5521.
6. Amlaily, N., Ramboz, S., Boschert, U., Plassat, J.L., and Hen, R. (1992) 'Isolation of a mouse 5-HT$_{1E}$-like serotonin receptor expressed predominantly in hippocampus', J. Biol. Chem. 92, 267, 19761-19764.
7. Adham, N., Kao, H.-T., Schechter, L.E., Bard, J., Olsen, M. Urquhart, D., Durkin, M., Hartig, P.R., Weinshank, R.L., and Branchek, T.A. (1993) 'Cloning of another human serotonin receptor (5-HT$_{1F}$): a fifth 5-HT$_1$ receptor subtype coupled to the inhibition of adenylate cyclase', Proc. Natl. Acad. Sci. USA 89, in press.
8. Kursar, J.D., Nelson, D.L., Wainscott, D.B., Cohen, M.L., and Baez, M. (1992) 'Molecular cloning, functional expression and pharmacological characterization of a novel serotonin receptor (5-hydroxytryptamine$_{2F}$ from rat stomach fundus', Mol. Pharmacol. 92, 42, 549-557.
9. Plassat, J.L., Boschert, U., Amlaiky, N., and Hen, R. (1992) 'The mouse 5-HT$_5$ receptor reveals a remarkable heterogeneity within the 5-HT$_{1D}$ receptor family', EMBO Journal 11, 4779-4786.
10. Ostrowski, J., Kjelsberg, M.A., Caron, M.G., and Lefkowitz, R.J. (1992) 'Mutagenesis of the β_2-adrenergic receptor. How structure elucidates function', Annu. Rev. Pharmacol. Toxicol. 32, 167-183.
11. Hoyer, D. and Middlemiss, D.N. (1989) 'The pharmacology of the terminal 5-HT autoreceptors in mammalian brain: evidence for species differences',

Trends Pharmacol. Sci. 10, 130-132.
12. Weinshank, R.L., Zgombick, J.M., Macchi, M.J., Branchek, T.A., and Hartig, P.R. (1992) 'Human serotonin 1D receptor is encoded by a subfamily of two distinct genes: 5-HT$_{1D\alpha}$ and 5-HT$_{1D\beta}$, Proc. Natl. Acad. Sci., USA 89, 3630-3634.
13. Oksenberg, D., Marsters, S.A., Odowd, B.F., Jin, H., Havlik, S., Peroutka, S.J., and Ashkenazi, A. (1992) 'A single amino-acid difference confers major, pharmacological variation between human and rodent 5-HT(1B) receptors', Nature 92, 360, 161-163.
14. Sumner, M.J., Feniuk, W., and Humphrey, P.P.A. (1989) 'Further characterization of the 5-HT receptor mediating vascular relaxation and elevation of cyclic AMP in porcine isolated vena cava', Br. J. Pharmacol. 97, 292-300.
15. Bockaert, J., Fozard, J.R., Dumuis A., and Clarke, D.E. (1992) 'The 5-HT$_4$ receptor: a place in the sun', Trends Pharmacol. Sci. 13, 141-145.
16. Sumner, M.J., Feniuk, W., McCormick, J.D., and Humphrey, P.P.A. (1992) 'Studies on the mechanism of 5-HT$_1$ receptor-induced smooth muscle contraction in dog saphenous vein', Br. J. Pharmacol. 105, 603-608.
17. Vane, J.R. (1957) 'A sensitive method for the assay of 5-hydroxytryptamine', Br. J. Pharmacol. 12, 344-349.
18. Maricq, A.V., Peterson, A.S., Brake A.J., Myers, R.M., and Julius, D. (1991) 'Primary structure and functional expression of the 5-HT$_3$ receptor, a serotonin-gated ion channel', Science 254, 432-437.

CLASSIFICATION OF AGENTS USED IN SEROTONIN RESEARCH

R.A. GLENNON
Department of Medicinal Chemistry
Medical College of Virginia; Box 540
Virginia Commonwealth University
Richmond, Virginia 23298
USA

ABSTRACT. Due to a lack of site-selective serotonergic ligands, and a need to develop novel agents with greater selectivity, an attempt is made to catalog serotonergic agents from a structural perspective. Not surprisingly, it would appear that the same structure-types are associated with multiple populations of 5-HT receptors. Selectivity seems to be more a matter of what substituents are present in a molecule, rather than overall structure-type.

1. Introduction

Different populations of serotonin (5-HT) receptors have been identified and categorized. The functional effects associated with these receptors, or with specific serotonergic agents, have also been (or are currently being) examined and cataloged. However, relatively little has been done to classify serotonergic agents from a structural perspective. What structural classes of agents are important? Are any classes of agents selective for one population of 5-HT receptors versus another? Are particular structures specifically associated with agonist versus antagonist activity? Do any of the many different serotonergic agents belong to related, but not necessarily obviously similar, chemical families? Currently, there are few serotonergic agents that may be termed "site-selective" [1]; the above questions are pertinent to the drug design process and a complete understanding of the answers should aid the development of novel site-selective agents.

We have recently begun a process of structural categorization of serotonergic agents [2]. Many of the agents fall into several common categories regardless of their selectivity (or lack thereof) (table 1). That is, for the most part, few (if any) chemical classes of agents may be considered uniquely related to, or associated with, a particular population of 5-HT receptors, or are composed solely of agents with agonist or antagonist activity. Selectivity appears to be more a matter of substituent type rather than of general structural type.

TABLE 1. Some major chemical categories of serotonergic agents.†

• Indolealkylamines	• Aryloxyalkylamines
• Ergolines	• N-Alkylpiperidines
• Aminotetralins	• N-Alkylpiperazines
• Arylpiperazines	• Phenylalkylamines
o Simple arylpiperazines	• Keto compounds
o Long-chain arylpiperazines	• Tricyclics
	• Miscellaneous

† Bioisoteres, conformationally-restricted agents, and closely related structural analogs (e.g. partial structures) are not listed separately.

Table 1 lists some major chemical classes of serotonergic agents. Derivatives of these agents, occasionally depending upon the types of substituents present, typically bind at multiple populations of 5-HT receptors. The indolealkylamines, for example, are amongst the least selective ligands and many indolealkylamines, such as serotonin itself, bind at all populations of 5-HT receptors. In contrast, although ergolines are typically considered nonselective high-affinity serotonergic agents, they display little affinity for $5-HT_3$ and $5-HT_4$ receptors, whereas certain keto compounds, although generally regarded as being $5-HT_3$-selective, bind at $5-HT_4$, and other, sites. Because 5-HT is nonselective, it is logical that agents displaying the greatest structural similarity to 5-HT will be less selective than more distinctly structurally-unique agents. Simple arylpiperazines exemplify this point; 1-naphthylpiperazine (1-NP) is a high affinity nonselective ligand. This is not to say that structural modification can not result in enhanced selectivity; for example, N4-substituted 1-NP, and other related N4-substituted arylpiperazines, display significantly greater selectivity for $5-HT_{1A}$ or, depending upon chain length, $5-HT_2$ receptors than either 5-HT or 1-NP.

Space does not allow for a comprehensive discussion here; however, an attempt will be made to compare the structures of various serotonergic agents in at least a cursory manner.

5-HT   1-NP   5-HTQ

2. Indolealkylamines

Suffice it to say that 5-HT is an indolealkylamine and that such agents, like 5-HT itself, are generally nonselective. Attempts have been made to develop more selective 5-HT derivatives and some of these attempts have been successful. Perhaps one of the more striking example is 5-HT_Q or the N,N,N,-trimethyl quaternary amine analog of 5-HT. Not only does 5-HT_Q bind at 5-HT_3 sites with selectivity, it displays about ten times the affinity of 5-HT [3]. One advantage of selective indolealkylamine derivatives over other selective, but structurally unrelated agents, is germane to molecular modeling studies. Modeling studies commonly require identification of structural similarities between agents [4]; such similarities are not always obvious when structurally unique agents are compared. The use of selective 5-HT derivatives allows for a greater ease of structural comparison; although it does not necessarily mean that different modes of receptor binding are not involved, this likelihood would seem to be reduced if minimal structural changes are made to the 5-HT molecule.

3. Ergolines

Ergolines contain an indolealkylamine moiety within their tetracyclic framework and, consequently, are typically nonselective with regard to their binding. Because ergoline derivatives bind with high affinity at most populations of 5-HT receptors, and because they represent conformationally-restricted and stereochemically defined structures, they can serve as structural templates in drug design and molecular modeling studies. Interestingly, they bind only with low affinity at 5-HT_3 and 5-HT_4 receptors indicating that either (i) they possess the wrong aromatic-to-amine distance or conformation, or (ii) the added bulk or structural features present in ergolines are not tolerated by 5-HT_3 and 5-HT_4 receptors. Certain ergoline derivatives bind with some selectivity at 5-HT_{1A} or $5\text{-HT}_{2/1C}$ receptors [review:2,3].

4. Aminotetralins

Aminotetralins are generally associated with 5-HT_{1A} receptors (e.g. 8-OH-DPAT). However, aminotetralin derivatives, again depending upon the presence of various substituents, can bind at other populations of 5-HT receptors. For example, the aminotetralin derivative of 2,5-dimethoxyphenylisopropylamine (2,5-DMA), i.e. 2-amino-5,8-dimethoxytetralin ($5\text{-HT}_{1A} = 5\text{-HT}_{1C}$ Ki = ca 350 nM) binds at 5-HT_{1C} receptors with three times the affinity of 2,5-DMA and, unlike 2,5-DMA, also binds at 5-HT_{1A} receptors (unpublished results). Although lacking the high 5-HT_{1A} affinity of 8-OH-DPAT (5-HT_{1A} Ki = 2 nM), the dimethoxytetralin serves to illustrate the point that aminotetralins bind at multiple receptors and that the presence of substituents (such as the dipropyl group of 8-OH-DPAT) contribute to determination of affinity and selectivity. Aporphines, which also possess an aminotetralin moiety, bind at 5-HT_{1A} and 5-HT_{1D} receptors with low-nanomolar affinity [3].

5. Arylpiperazines

Simple arylpiperazines, i.e. those lacking an N4-substituent, are nonselective serotonergic agents of modest affinity. mCPP, for example, binds at 5-HT_{1A}, 5-HT_{1B}, 5-HT_{1C}, 5-HT_2, and 5-HT_3 receptors with Ki values of between 20-100 nM [2,3]. Another example, 1-NP, was mentioned above. Introduction of an N4 substituent (*viz* long-chain arylpiperazines) results in agents that display greater selectivity for 5-HT_{1A} receptors [see 2,3,5]. These agents also bind at 5-HT_2, D2 dopamine, and/or α_1-adrenergic receptors depending upon the nature of the aryl substituent, the length of the alkyl spacer group, and the identity of the terminus (figure 1).

Figure 1. General structure of a long-chain arylpiperazine.

6. Aryloxyalkylamines

The aryloxyalkylamine structure-type, $Ar\text{-}O\text{-}(CH_2)_n NRR'$, is generally associated with β-adrenergic antagonism. Indeed, the aryloxyalkylamines pindolol and propranolol are high affinity β-adrenergic antagonists that bind stereoselectivity at 5-HT_{1A} and 5-HT_{1B} receptors. Structural modification of these, and other, aryloxyalkylamines results in 5-HT_{1A}-selective agents that are virtually devoid of affinity for 5-HT_{1B} and β-adrenergic receptors [2,3]. Depending upon specific structural features, aryloxyalkylamines can also bind at 5-HT_2, D2 dopamine, and α-adrenergic receptors [2,3].

7. N-Alkylpiperidines

N-Alkylpiperidine derivatives can bind at multiple receptors including 5-HT_{1A}, 5-HT_{1C}, 5-HT_2, 5-HT_3, D2 dopamine, and adrenergic receptors. Unlike the N-alkylpiperidine ketanserin which binds with nearly equal affinity at 5-HT_{1C} and 5-

HT$_2$ receptors, spiperone and cisapride display 500- to 1000-fold selectivity for the latter population. Interestingly, spiperone (but not ketanserin) binds at 5-HT$_{1A}$ receptors, whereas cisapride (but not ketanserin or spiperone) binds at 5-HT$_3$ receptors. Again, selectivity is controlled by substituent-type, not by general structure-type.

Replacement of the piperidine 4-position carbon atom with a nitrogen results in N-alkylpiperazines that are isoteric with the N-alkylpiperidines. It would appear that the piperazine derivatives bind at 5-HT$_2$ sites with somewhat lower affinity than their corresponding piperidine analogs; however, incorporation of certain additional substituents compensates for the presence of the piperazine nitrogen (e.g. irindalone, tefludazine) [6].

8. Phenylalkylamines

The 5-HT$_2$ agonists DOB and DOI are examples of phenylalkylamines. These agents also bind with nearly equal affinity at 5-HT$_{1C}$ receptors [see 2,3]. Replacement of the bromo or iodo moiety of DOB/DOI with small alkyl groups results in retention of 5-HT$_2$ agonist or partial agonist activity; however, extension of the alkyl group results in phenylalkylamines with 5-HT$_2$ antagonist activity [2]. The phenylalkylamine moiety is also found in agents that bind at 5-HT$_{1A}$ receptors; however, structural requirements for 5-HT$_{1A}$ versus 5-HT$_2$ binding are, to a certain extent, mutually exclusive.

9. Keto Compounds

In order to classify the numerous structure-types that bind at 5-HT$_3$ receptors, it was found that most could be conveniently characterized under the broad category of keto compounds; these are agents having the general structural formula: R-C(=O)-X-(CH$_2$)$_n$-NRR' [see 2]. Subcategories within this classification (benzoic acid derivatives [benzoates, benzamides, reverse benzoates and benzamides], keto indoles [indole, carboxylic acid derivatives, reverse amides and esters, γ-carbolines, carbazoles, and bioisoteres] and related structural analogs [e.g. carbamates, ureas] and bioisoteres) help visualize the structural similarity amongst these agents. Although most of these agents are typically considered to be 5-HT$_3$-selective, some have now been shown to bind at other populations, particularly 5-HT$_4$ receptors. In fact, certain benzamides are selective for 5-HT$_4$ over 5-HT$_3$ receptors [7], again indicating that selectivity is substituent related.

10. Summary

Most serotonergic agents fall into a small number of chemical classes. Very few chemical classes of agents appear to be associated only with one population of receptors; indeed, for most classes, derivatives can be identified that bind at most populations of 5-HT receptors. Modification of these general structure-types by the incorporation of the appropriate substituents group effects should ultimately allow the rational design and synthesis of ligands with greater selectivity than those that are now available.

11. References

1. Glennon, R.A. and Dukat, M. (1991) 'Serotonin receptors and their ligands: A lack of selective agents', Pharmacol. Biochem. Behav. 40, 1009-1017.
2. Glennon, R.A. and Dukat, M. (1992) '5-HT receptor ligands - Update 1992', Current Drugs-Serotonin, 1, 1-45.
3. Glennon, R.A., Westkaemper, R.B., and Bartyzel, P. (1991) 'Medicinal chemistry of serotonergic agents', in S.J. Peroutka (ed.) Serotonin Receptor Subtypes, Wiley-Liss, New York, pp. 19-64.
4. Westkaemper, R.B. and Glennon, R.A. (1991) 'Approaches to molecular modeling studies and specific application to serotonin ligands and receptors', Pharmacol. Biochem. Behav. 40, 1019-1031.
5. Glennon, R.A. (1992) 'Concepts for the design of $5\text{-}HT_{1A}$ serotonin agonists and antagonists', Drug. Dev. Res. 26, 251-274.
6. Bøgeso, K.P., Arnt, J., Boeck, V., Vibeke Christensen, A., Hytel, J., and Jensen, K.G. (1988) 'Antihypertensive activity of a series of 1-piperazino-3-phenylindans with potent $5\text{-}HT_2$ antagonist activity', J. Med. Chem. 31, 2247-2256.
7. Flynn, D.I., Zabrowksi, D.I., Becker, D.P., Nosal, R., Villamil, C.I., Gulikson, G.W., Moummi, C., and Yang, D.C. (1992) 'SC-53116: The first selective agonist at the newly identified serotonin $5\text{-}HT_4$ receptor subtype', J. Med. Chem. 35, 1486-1489.

ELECTROPHYSIOLOGY OF CENTRAL SEROTONIN NEUROTRANSMISSION

P. BLIER, G. PIÑEYRO, T. DENNIS, and C. DE MONTIGNY
Neurobiological Psychiatry Unit
McGill University
1033 Pine Avenue West
Montreal, Quebec
Canada H3A 1A1

ABSTRACT. Several classes and subtypes of serotonin (5-HT) binding sites have been identified using radioligand binding techniques. The existence of these multiple 5-HT receptors has recently been confirmed by cloning techniques [1,2]. There now also exists *in vivo* and *in vitro* electrophysiological evidence for distinct functional roles of these different 5-HT receptors. The present review will principally bear on 5-HT$_{1A}$ receptors but will also provide a brief review of the single-cell recording experiments on the other types of 5-HT receptors, and on the 5-HT uptake carrier, emphasizing whenever possible the clinical relevance of the data.

1.   5-HT$_{1A}$ Receptors

5-HT$_{1A}$ binding sites are present in high concentrations on the cell body of 5-HT neurons in the raphe nuclei, where they exert a negative feedback control on their firing activity [3,4]. They are also located on postsynaptic neurons in various regions of the brain including the hippocampus, the lateral septum, and the hypothalamus. Their activation by selective agonists, such as 8-OH-DPAT, produces a hyperpolarization of the neuronal membrane, which *in vivo* underlies the slowing of firing activity recorded extracellularly [5,6,7]. This hyperpolarization results from the opening of potassium channels which are coupled directly to G$_{i/o}$ proteins [5]. Although these two populations of 5-HT$_{1A}$ receptors exhibit a similar radioligand binding profile [8], electrophysiological studies have produced unequivocal evidence that they are endowed with different properties. First, the potency of 5-HT is much greater on postsynaptic hippocampus neurons than in the dorsal raphe (figure 1). In sharp contrast, 5-HT$_{1A}$ agonists such as 8-OH-DPAT and gepirone are much more potent on dorsal raphe neurons (figure 1). This differential responsiveness of these pre- and postsynaptic neurons could be attributed, at least in part, to the fact these 5-HT$_{1A}$ agonists are full agonists in the dorsal raphe while being partial agonists in the hippocampus, as suggested by the lesser degree of membranal hyperpolarization in the latter than in the former neurons [4,5]. This conclusion can also be drawn from *in vivo* extracellular recording experiments. The microiontophoretic ejection of gepirone sufficient to produce at least an 80% decrease in firing activity (which was restored with quisqualic acid) prevented the suppressant effect of a concomitant application of 5-HT in the hippocampus, but not in the raphe (figure 1). The lack of antagonistic effect of these 5-HT$_{1A}$ agonists in the dorsal raphe could theoretically be attributable to 5-HT neurons having a large receptor reserve, whereas hippocampus pyramidal neurons would lack such a reserve. Therefore, if a marked suppression of firing of a hippocampus neuron is produced by 8-OH-DPAT, all its 5-HT$_{1A}$ receptors would be occupied leaving no spare receptors to then be activated by 5-HT to produce a further suppression of firing. However, we have shown that the continuous application of 5-HT in the same paradigm will not prevent the suppressant effect of a further application of 5-HT [9]. In support of this contention, we have also observed that the 5-HT$_{1A}$ agonist BMY 42568 exhibited full agonistic activity in this extracellular recording studies on hippocampus pyramidal neurons [10].

Evidence for the distinct properties of pre- and postsynaptic 5-HT$_{1A}$ agonists has also been obtained using antagonists. On the one hand, the buspirone analogue BMY 7378 readily blocks the suppressant effect of microiontophoretically applied 5-HT onto CA$_3$ pyramidal neurons while being ineffective in the same paradigm on 5-HT neurons [11]. On the other hand, acute systemic administration of spiperone readily blocks the effect of microiontophoretically applied 5-HT and 8-OH-DPAT onto 5-HT neurons, while producing no immediate effect on CA$_3$ pyramidal neurons [12]. Another line of evidence for the distinct properties of 5-HT$_{1A}$ receptors was provided by studies assessing the effect of sustained administration of 5-HT$_{1A}$ agonists on the responsiveness of somatodendritic 5-HT$_{1A}$ autoreceptors and of postsynaptic 5-HT$_{1A}$ receptors [3,9]. Following 14-day treatments with gepirone, tandospirone, or flesinoxan, presynaptic 5-HT$_{1A}$ autoreceptors in the dorsal raphe desensitize whereas their postsynaptic congeners in the hippocampus do not. A last line of evidence concerns the G-protein coupling of the two populations of receptors. While both the receptors located on the cell body of 5-HT neurons and of hippocampus pyramidal neurons are inactivated by cholera toxin, an agent which causes a persistent activation of G$_s$ proteins [13]. That different 5-HT$_{1A}$ receptors may in fact exist was rendered even more likely by the reports of three mRNA bands on Northern blots with a cDNA probe of the rat 5-HT$_{1A}$ receptor and the capacity of a human 5-HT$_{1A}$ receptor to couple negatively but not positively to adenylyl cyclase [14,15].

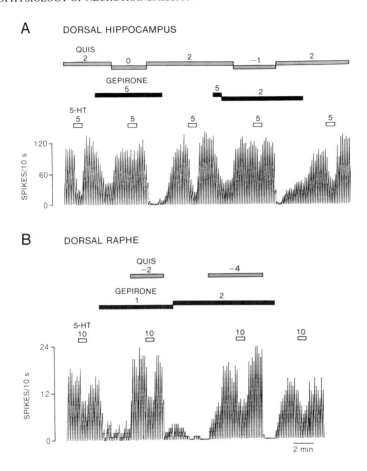

Figure 1. Integrated firing rate histograms of a CA_3 dorsal hippocampus pyramidal neuron (A) and of dorsal raphe 5-HT neuron (B) showing the effect of the concurrent microiontopherotic application of gepirone on the responsiveness to the application of 5-HT. Quisqualate was used to maintain and/or restore the firing rate of the neurons. Time base applies to both traces.

The 5-HT_{1A} agonists presently available activate both somatodendritic 5-HT_{1A} autoreceptors and the postsynaptic 5-HT_{1A} receptors. Since the activation of the former leads to a decrease of 5-HT release [16] and that of the latter increase 5-HT signal transfer at postsynaptic sites, it thus becomes evident that the concurrent activation of both populations of 5-HT_{1A} receptors by the acute or short-term systemic administration of a 5-HT_{1A} agonist leads to opposing effects on the efficacy of 5-HT transmission. Since these drugs have a delayed onset of action in both generalized anxiety and depressive disorders [17], it appears that adaptative changes in the 5-HT system, as a result of the sustained activation of 5-HT$_{1A}$ receptors, could better explain the clinical efficacy of these drugs. Indeed, as mentioned above, the somatodendritic but not the postsynaptic 5-HT$_{1A}$ receptors desensitize following a 14-day treatment with a 5-HT$_{1A}$ agonist. We have thus postulated that the tonic activation of postsynaptic 5-HT$_{1A}$ receptors is enhanced at that time. This would result from a normal release of 5-HT, due to the recovery of the firing activity of 5-HT neurons, superimposed on the constant activation of these normosensitive postsynaptic receptors by the exogenous 5-HT$_{1A}$ agonist. It is important to mention that the same net effect takes place following long-term treatment with tricyclic antidepressant drugs or electroconvulsive shocks. Indeed, these treatments leave unaltered the function of 5-HT neurons but progressively enhance the reponsiveness of 5-HT$_{1A}$ receptors in the hippocampus and several other brain structures [18].

2. 5-HT$_{1B}$ Receptors

As for 5-HT$_{1A}$ receptors, 5-HT$_{1B}$ receptors have been identified both on 5-HT neurons and on postsynaptic neurons of the nigro-striatal system, at least in the rat and mouse brains [19]. On 5-HT neurons, they have been shown to modulate the release of 5-HT. Numerous *in vitro* superfusion studies have documented their capacity to attenuate the evoked [^{3}H]5-HT release from preloaded slices when they are activated by 5-HT itself or various agonists, such as TFMPP, mCPP, and RU 24969 [20]. Electrophysiological experiments from our laboratory have confirmed that they exert a negative feedback role of 5-HT release *in vivo* [11]. The intravenous administration of methiothepin, a 5-HT$_{1/2}$ antagonist, enhances the inhibitory effect of the electrical stimulation of the 5-HT pathway on the firing activity of postsynaptic hippocampus pyramidal neurons, whereas the 5-HT$_{1C/2}$ antagonist ketanserin is inactive. That this enhancing effect of methiothepin is mediated by an antagonism of 5-HT$_{1B}$ receptors is indicated by the lack of effect of the 5-HT$_{1A}$ agonist 8-OH-DPAT and the reversal of the suppressant effect of the 5-HT$_{1A/B}$ agonist RU 24969 by methiothepin. It is noteworthy that, despite the fact that methiothepin is effective in displacing 5-HT$_{1A}$ ligand in binding experiments, its acute administration does not modify the responsiveness of the pyramidal neurons to microiontophoretic application of 5-HT and 8-OH-DPAT [12]. Although the binding of agonists to 5-HT$_{1B}$ sites is dependent upon G$_{i/o}$ proteins, the terminal 5-HT$_{1B}$ autoreceptor in the rat brain is not inactivated by pertussis toxin [21]. These apparently discrepant results are most likely attributable to 5-HT$_{1B}$ receptor heterogeneity; when carrying out radioligand binding studies on a given brain structure both pre- and postsynaptic receptors are labelled, the latter type accounting for the vast majority of binding sites. The properties of such presynaptic receptors can thus be put into evidence only by using physiological techniques which can assess their function.

3. 5-HT$_{1C}$ Receptors

In the rat piriform cortex, 5-HT produces an excitatory action on pyramidal neurons. Indeed, 5-HT increases the number of spikes evoked by a constant depolarizing current pulse, an effect that is blocked by the 5-HT$_{1C/2}$ antagonist ritanserin but not by spiperone, an antagonist with nanomolar affinity for 5-HT$_2$ sites but with a thousand-fold lower affinity for 5-HT$_{1C}$ sites [22]. These *in vitro* results are fully consistent with the identification of mRNA for the 5-HT$_{1C}$ receptor in cortical pyramidal cells [23].

4. 5-HT$_{1D}$ Receptors

To our knowledge, electrophysiological experiments have not yet been carried out on neuronal responses mediated by this receptor subtype. However, it appears that 5-HT$_{1D}$ receptors subserve in most species the role that 5-HT$_{1B}$ receptors exert in rats and mice. Furthermore, recent *in vitro* superfusion experiments carried out on preloaded guinea pig brain slices indicate that the terminal 5-HT$_{1D}$ autoreceptor desensitized following long-term treatment with a selective 5-HT reuptake inhibitor, as is the case with the terminal 5-HT$_{1B}$ autoreceptor in the rat brain [18, 24]. Since the terminal 5-HT autoreceptor is of the 5-HT$_{1D}$ subtype in the human brain [25], it appears that the desensitization of this terminal autoreceptor, which allows more 5-HT to be released per impulse, could well underlie the therapeutic effect of 5-HT reuptake blockers in major depression.

5. 5-HT$_2$ Receptors

Several single-cell studies have documented 5-HT$_2$ responses in various regions of the CNS, including cerebral cortex, nucleus accumbens, locus coeruleus, facial motor nucleus, and spinal cord. In general, intracellular recording experiments have indicated that 5-HT produces a slow depolarization, but it is not always clear whether this effect is mediated by 5-HT$_2$ or 5-HT$_{1C}$ receptors in the above-mentioned structures. In the piriform cortex, however, the activating effect of 5-HT on interneurons is blocked by low concentrations of spiperone [22]. In the facial motor nucleus, microiontophoretic application of 5-HT does not by itself induce firing activity in anesthetized rats but increased the excitatory effect of co-application of glutamate. It is noteworthy that long-term treatments with tricyclic antidepressant drugs increased this response to 5-HT, just as is the case for the 5-HT$_{1A}$ receptor in the rat dorsal hippocampus [18, 26].

6. 5-HT$_3$ Receptors

A rapid excitatory response to a 5-HT and to the 5-HT$_3$ agonist 2-methyl-5-HT has been identified in cultures of mouse hippocampal and striatal neurons [27]. In the medial prefrontal cortex, an inhibitory effect of 5-HT and 2-methyl-5-HT on the firing activity has recently been described in anesthetized rats. Although the latter effect has a slow onset of action and does not appear to attenuate progressively over time, as for most 5-HT$_3$ responses, it is blocked by several 5-HT$_3$ antagonists [28]. Nevertheless, given the recent observation that 5-HT$_3$ receptor activation increases 5-HT release in the frontal cortex [29], it will be interesting to determine whether this suppressant effect of 2-methyl-5-HT results from a 5-HT$_3$-mediated release of endogenous 5-HT.

7. "5-HT$_4$" Receptors

In rat hippocampus slices, 5-HT, aside from producing a rapid 5-HT$_{1A}$-mediated hyperpolarization, also elicits a slow excitatory response [30]. In the presence of the 5-HT$_{1A}$ antagonist BMY 7378, this depolarizing effect of 5-HT is mimicked by 5-methoxytryptamine and 5-carboxyamidotryptamine, two drugs with a documented agonistic effect at the so-called 5-HT$_4$ receptor (which is positively coupled to adneylyl cyclase), but not by the 5-HT$_3$ agonists 2-methyl-5-HT and phenylbiguanide. It is blocked by BRL 24924 (renzapride), zacopride, and cisapride, three benzamide derivatives with antagonistic activity at the "5-HT$_4$" receptor. It therefore appears that 5-HT$_4$-like receptors can exert slow excitatory responses in the central nervous system.

8. The 5-HT Uptake Carrier

Although the 5-HT uptake carrier is not a 5-HT receptor *per se*, it presents several characteristics of 5-HT receptors. As for the 5-HT autoreceptors, namely the 5-HT$_{1A}$ and 5-HT$_{1B/D}$ presynaptic receptors, we have recently provided *in vivo* electrophysiological evidence that long-term administration of a 5-HT reuptake blocker leads to an attenuation of the 5-HT reuptake system in the rat brain [18,31]. Forty-eight hours after a 21-day treatment with paroxetine administered subcutaneously with osmotic minipumps, the capacity of acute paroxetine to prolong the recovery of the firing activity of dorsal hippocampus neurons to microiontophoretic application of 5-HT is attenuated. The uptake of [^{3}H]5-HT into rat hippocampus slices, following a 48-hour washout period, was also markedly attenuated following the same treatment with paroxetine. Finally, the density of 5-HT uptake sites was also significantly decreased following long-term paroxetine administration, as indicated by a reduced B$_{max}$ and an unaltered K$_d$ of [^{3}H]paroxetine binding.

In conclusion, electrophysiological techniques have identified 5-HT responses attributable to the activation of nearly all subtypes of 5-HT receptors. In general, the activation of 5-HT$_1$ receptors produces inhibitory responses whereas that of 5-HT$_2$, 5-HT$_3$, and "5-HT$_4$" receptors produces excitatory responses. Furthermore, this experimental approach had raised the possibility that within the same animal species, different subtypes of 5-HT$_{1A}$ receptors could exist even before this receptor was first cloned [32,33]. It is nevertheless the molecular biology approach which will confirm or refute the possibility that the different properties of pre- or postsynaptic 5-HT$_{1A}$ receptors are in fact attributable to receptors having different amino acid sequences. It is important to note that in the case of the 5-HT$_{1D}$ subclass, two distinct genes in humans have already been identified [34]. Whatever conclusions such studies will provide, it is nevertheless remarkable that the mechanism of action of drugs used in human therapeutics has been traced back to subtypes of 5-HT receptors and that this information is presently used to devise more effective treatments.

9. References

1. Hartig, P.R., Adham, N., Zgombick, J., Weinshank, R., and Branchek, T. (1992) 'Molecular biology of the 5-HT$_1$ receptor subfamily', Drug Dev. Res., 26, 215-224.
2. Maricq, A.V., Peterson, A.S., Brake, A.J., Myers, R.M., and Julius, D. (1991) 'Primary structure and functional expression of the 5-HT$_3$ receptor, a serotonin-gated ion channel', Science, 254, 432-437.
3. Blier, P. and de Montigny, C. (1990) 'Differential effect of gepirone on presynaptic and postsynaptic serotonin receptors: Single-cell recording studies', J. Clin. Psychopharmacol., 10 (Suppl. 3), 13S-20S.
4. Sprouse, J.S. and Aghajanian, G.K. (1987) 'Electrophysiological responses of serotoninergic dorsal raphe neurons to 5-HT$_{1A}$ and 5-HT$_{1B}$ agonists', Synapse, 1, 3-9.
5. Andrade, R., Malenka, R.S., and Nicoll, R.A. (1986) 'A G protein couples serotonin and GABA$_B$ receptors to the same channels in hippocampus', Science, 234, 1261-1265.
6. Joels, M., Shinnick, G., Gallagher, P., and Gallagher, J.P. (1988) 'Actions of serotonin recorded intracellularly in rat dorsal lateral septal neurons', Synapse, 2, 45-53.
7. Newberry, N.R. (1992) '5-HT$_{1A}$ receptors activate a potassium inductance in rat ventromedial hypothalamic neurons', Eur. J. Pharmacol., 210, 209-212.
8. Beer, M., Kennett, G.A., and Curzon, G. (1990) 'A single dose of 8-OH-DPAT reduces raphe binding of [^{3}H]8-OH-DPAT and increases the effect of raphe stimulation on 5-HT metabolism', Eur. J. Pharmacol., 178, 179-187.
9. Godbout, R., Chaput, Y., Blier, P., and de Montigny, C. (1991) 'Tandospirone and its metabolite, 1-(2-pyrimidinyl)-piperazine (1-PP): I-Effects of acute and long-term tandospirone on serotonin neurotransmission', Neuropharmacology, 30, 697-698.
10. de Montigny, C., Lista, A., and Blier, P. (1992) 'Electrophysiological studies on the effects of the 5-HT$_{1A}$ agonist/5-HT reuptake blocker BMY 42568 on the 5-HT system', Soc. Neurosci. Abst., 18, 90.

11. Chaput, Y. and de Montigny, C. (1988) 'Effects of the 5-HT$_1$ receptor antagonist, BMY 7378, on 5-HT neurotransmission: Electrophysiological studies in the rat CNS', J. Pharmacol. Exp. Ther., 246, 359-370.
12. de Montigny, C., Lista, A., and Blier, P. (1990) 'Pre- and postsynaptic 5-HT$_{1A}$ receptors exhibit different electrophysiological properties: I-Effects of spiperone', Soc. Neurosci. Abst., 16, 462.4.
13. Blier, P., Lista, A., and de Montigny, C. (1990) 'Pre- and postsynaptic 5-HT$_{1A}$ receptors exhibit different electrophysiological properties: II-Effects of pertussis and cholera toxins', Soc. Neurosci. Abst., 16, 462.5.
14. Albert, P.R., Zhou, Q.Y., Van Tol, H.H.M., Bunzow, J.R., and Civelli, O. (1990) ' Cloning functional expression and mRNA tissue distribution of the rat 5-hydroxytryptamine$_{1A}$ receptor gene', J. Biol. Chem., 265, 5825-5832.
15. Bertin, B., Freissmuth, M., Breyer, R.M., Schütz, W., Strosberg, A.D., and Mariello, S. (1992) 'Functional expression of the human serotonin 5-HT$_{1A}$ receptor in *Escherichia coli*', J. Biol. Chem., 267, 8200-8206.
16. Huston, P.M., Sarna, G.S., O'Connell, M.T., and Curzon, G. (1989) 'Hippocampal 5-HT synthesis and release *in vivo* is decreased by infusion of 8-OH-DPAT into the nucleus raphe dorsalis', Neurosci. Lett., 100, 276-280.
17. Robinson, D.S., Rickels, K., Feighner, J., Fabre, L.F., Gammans, R.E., Shrotriya, R.C., Alms, D.R., Andary, J.J., and Messina, M.E. (1990) 'Clinical effects of the 5-HT$_{1A}$ partial agonists in depression', J. Clin. Psychopharmacol., 10 (Suppl. 3) 67S-76S.
18. Blier, P., de Montigny, C., and Chaput, Y. (1990) 'A role for the serotonin system in the mechanism of action of antidepressant treatments: Preclinical evidence', J. Clin. Psychiat., 51 (Suppl. 4), 14-20.
19. Vergé, D., Doval, G., Marcinkiewicz, M., Patry, A., El Mestikawy, D., Gozlan, H., and Hamon, M. (1986) 'Quantitative autoradiography of multiple 5-HT$_1$ receptor subtypes in the brain of control or 5,7-dihydroxytryptamine-treated rats', J. Neurosci., 6, 3474-3482.
20. Starke, K., Göthert, M., and Kilbinger, H. (1989) 'Modulation of neurotransmitter release by presynaptic autoreceptors', Physiol. Rev., 69, 864-969.
21. Blier, P. (1991) 'Terminal serotonin autoreceptor function in the rat hippocampus is not modified by pertussis and cholera toxins', Naunyn-Schmiedeberg's Arch. Pharmacol., 344, 160-166.
22. Sheldon, P.W. and Aghajanian, G.K. (1990) 'Serotonin (5-HT) excites interneurons via 5-HT$_2$ receptors and pyramidal cells via 5-HT$_{1C}$ receptors in rat piriform cortex', Soc. Neurosci. Abst., 16, 1036.
23. Mengod, G., Nguyen, H., Lee, H., Waeber, C., Lubbert, H., and Palacios, J.M. (1990) 'The distribution and cellular localization of 5-HT$_{1C}$ receptor mRNA in the rodent brain examined by *in situ* hybridization histochemistry', Neuroscience, 35, 577-592.
24. Blier, P. and Bouchard, C. (1992) 'Enhancement of serotonin release in the

guinea pig hypothalamus following long-term treatment with antidepressant drugs', Clin. Neuropharmacol., 15 (Suppl. B), 66B.
25. Galzin, A.M., Poirier, M.F., Lista, A., Chodkiewiecz, J.P., Blier, P., Ramdine, R., Loo, H., Roux, F.X., Redondo, A., and Langer, S.Z. (1992) 'Characterization of the 5-HT autoreceptor modulating the release of [^{3}H]-5-HT in slices of the human frontal cortex', J. Neurochem., in press.
26. Menkes, D.B., Aghajanian, G.K., and McCall, R.B. (1980) 'Chronic antidepressant treatment enhances α-adrenergic and serotonergic responses in the facial nucleus', Life Sci., 27, 45-55.
27. Yakel, J.L. and Jackson, M.B. (1988) '5-HT$_3$-like receptors in cultured hippocampus and a clonal cell line', Neuroscience, 1, 615-621.
28. Ashby, C.R., Minabe, Y., Edwards, E., and Wang, R.Y. (1992) '5-HT$_3$-like receptors in the rat medial prefrontal cortex: An electrophysiological study', Brain Res., 550, 181-191.
29. Blier, P. and Bouchard, C. (1992) 'Functional characterization of a 5-HT$_3$ receptor which modulates the release of 5-HT in the guinea pig brain', Br. J. Pharmacol., in press.
30. Andrade, R. and Chaput, Y. (1991) '5-Hydroxytryptamine$_4$-like receptors mediate the slow excitatory response to serotonin in the rat hippocampus', J. Pharmacol. Exp. Ther., 257, 930-937.
31. Piñeyro, G., de Montigny, C., and Blier, P. (1992) 'Attenuated effect of paroxetine following its long-term administration: An electrophysiological study in the rat hippocampus', Soc. Neurosci. Abst. 18, 95.2.
32. Blier, P. and de Montigny, C. (1987) 'Modification of 5-HT neuron properties by sustained administration of the 5-HT$_{1A}$ agonist gepirone: Electrophysiological studies in the rat brain', Synapse, 1, 470-480.
33. Fargin, A., Raymond, J.R., Lohse, M.J., Kolbilka, B.K., Caron, M.G., and Lefkowitz, R.J. (1988) 'The genomic clone G-21 which resembles a β-adrenergic receptor sequence evokes the 5-HT$_{1A}$ receptor', Nature, 335, 358-360.
34. Weinshank, R.L., Zgombick, J.M., Macchi, M.J., Branchek, T.A., and Hartig, P.R. (1992) 'Human serotonin 1D receptor is encoded by a subfamily of two distinct genes: 5-HT$_{1D\alpha}$ and 5-HT$_{1\beta}$', Proc. Natl. Acad. Sci., 89, 3630-3634.

IMMUNOHISTOCHEMICAL VISUALIZATION OF 5-HT$_{1A}$ RECEPTORS IN THE RAT BRAIN

H. GOZLAN[1], K. KIA[2], J. GINGRICH[1], M. RIAD[1], E. DOUCET[1], C. GERARD[1], D. VERGE[2], and M. HAMON[1]
[1]INSERM U288
Faculty of Medicine Pitié-Salpêtrière
91, Boulevard de l'Hôpital
75634 Paris Cedex 13 and
[2]CNRS UA 1199
Department of Cytology
Université Pierre et Marie Curie
7, Quai Saint Bernard
75005 Paris
France

ABSTRACT. Anti-5-HT receptor antibodies were raised by injecting rabbits with either a synthetic peptide corresponding to the aa 243-268 portion of the third intracellular loop of the rat 5-HT$_{1A}$ receptor, or a fusion protein made of the complete loop (aa 216-350) coupled to glutathione-S-transferase. Immunohistochemical staining with these antibodies allowed the visualization of 5-HT$_{1A}$ receptors mainly in the limbic areas and anterior raphe nuclei, as expected from previous autoradiographic data with selective 5-HT$_{1A}$ receptor radioligands. Double immunostaining with anti-5-HT$_{1A}$ receptor antibodies and anti-5-HT or anti-choline acetyl-transferase antibodies demonstrated that these receptors are located in the plasma membrane of practically all serotoninergic neurones in anterior raphe nuclei and a small fraction of cholinergic neurones in the median septum.

1. Introduction

A least 8 different serotonin (5-HT) receptors have been identified in the central nervous system (CNS), and among them the 5-HT$_{1A}$ subtype has been the focus of intense interest. This receptor has been specifically studied for almost 10 years thanks to the development of [^{3}H]8-OH-DPAT as a selective radioligand [1]. In addition, it soon appeared that 5-HT$_{1A}$ receptor (partial) agonists had anxiolytic and/or antidepressant-like properties in several species including man [2]. The cloning of the 5-HT$_{1A}$ receptor gene allowed the demonstration that this receptor belongs to the G protein-coupled superfamily with 7 transmembrane domains [3].

Comparison of its amino acid sequence with that of the other members of the same superfamily indicated some homology, especially within the transmembrane domains. However, the aminoacid sequence of the third intracellular loop is highly selective of the 5-HT_{1A} receptor. Accordingly, this portion of the protein was selected for raising polyclonal antibodies [4], in an attempt to visualize the 5-HT_{1A} receptor at the cellular and ultrastructural levels, and identify the cell types which express it in the rat CNS. For this purpose, we used two strategies. The first one consisted of injecting rabbits with a synthetic peptide corresponding to a limited portion of the third intracellular loop of the 5-HT_{1A} receptor protein. For the second strategy, the cDNA corresponding to the complete loop was inserted into a plasmid for the expression in bacteria of a fusion protein made of the third intracellular loop coupled to glutathione-S-transferase. This fusion protein was then used as an antigen for raising anti-5-HT_{1A} receptor antibodies in rabbits. Data obtained with these antibodies are described herein.

2. Materials and Methods

2.1 PRODUCTION OF ANTI-RAT 5-HT_{1A} RECEPTOR ANTIBODIES

The synthetic peptide GTSSAPPPKKSLNGQPGSGDWRRCAE corresponding to amino acid residues 243-268 of the third intracellular loop of the rat 5-HT receptor was coupled to bovine serum albumin with 1-ethyl-3-(3-dimethylaminopropyl)carbodiimide and emulsified with Freund's adjuvant to be injected into male white New Zealand rabbits. The antiserum collected after the fifth booster injection was purified by affinity chromatography on a column made of the synthetic peptide coupled to agarose. The purified antibodies were kept at - 30°C with 50% (v/v) glycerol, and used at a final dilution of 1/1000 (as compared to the crude antiserum).

The portion of the rat gene corresponding to the third intracellular loop of the 5-HT receptor (aa residues 216-350) was amplified by polymerase chain reaction (PCR) using the following primers: AGGATCCGAATTCCTCTA-CGGGCGCATCTTCAGA (5') and AGGATCCCTCGAGTCAC-CCAGAGTCTTCACCGTCTTC (3'). The amplified material was purified using Sephaglass beads (Pharmacia), restricted with Bam HI and subcloned into pSK+ bluescript plasmid (Stratagene). A clone of the proper orientation was selected and the sequence verified by the double stranded dideoxy chain termination method of Sanger (Sequenase, USB Corp.). The gene fragment was then inserted into a pGEX-KG plasmid for the coupled expression of the encoded protein (corresponding to the complete sequence of the third intracellular loop of the 5-HT receptor) and gluthation-S-transferase in XL-1 blue bacteria (Stratagene). After 2 hours at 37°C in the presence of 0.1 mM isopropyl-thiogalactoside to induce transcription (through the promoter pTAC), the bacterial pellet was harvested by centrifugation and lysed by sonication in phosphate buffered saline (PBS) containing

10 mM EDTA, 1% triton X-100, and 1 mM phenylmethylsulfonyl fluoride. The particulate material was spun down at 30,000 x g for 15 minutes at 4°C, and the supernatant was incubated with a 50% slurry of glutathione coupled to agarose (Sigma) for 10 minutes at room temperature. The mixture was poured into a column, which was then washed with PBS containing 10 mM EDTA, and then eluted with 20 mM reduced glutathione in 50 mM Tris-HCl, pH 8.0. The identity of the fusion protein eluted from the gel was verified by SDS-PAGE on a discontinuous 12% slab gel. Aliquots of the gel eluate (corresponding to 200 µg of fusion protein) were then emulsified with Freund's adjuvant to be injected into rabbits. Antiserum collected after the fourth booster injection was partially purified by a combination of protein A agarose chromatography and affinity chromatography on a column made of glutathione-S- transferase crosslinked to glutathione agarose by fomaldehyde. The fraction passing through this gel was used for immunohistochemical staining at a final dilution of 1/1000 (relative to the crude antiserum).

2.2 IMMUNOSTAINING PROCEDURES

Adult male rats (Sprague-Dawley) were anesthetized with chloral hydrate (400 mg/kg IP) and intracardially perfused through the ascending aorta with 4% paraformaldehyde in 0.12 M sodium phosphate buffer, pH 7.4, for 10 minutes. The brains were dissected out and postfixed by immersion in the same fixative for 90-120 minutes. Coronal brain sections were then prepared and washed in 0.12 M phosphate buffer containing 30% sucrose and 1% sodium azide for 48 hours at 4°C. A standard avidin-biotin-peroxidase complex method was used for the immunocytochemical staining. Free floating sections were first incubated in PBS containing gelatin (0.1%), lysine (0.1 M), sodium azide (0.1%), and triton X-100 (0.25%) for 1 hour to reduce nonspecific binding of the antibodies. After washing, the sections were incubated overnight at room temperature with anti-5-HT receptor antibodies (1/1000). This step was followed by incubation with biotinylated goat-anti-rabbit IgG (1/250) for 1 hour. The sections were then washed and incubated for 1 hour with the avidin-biotin-horseradish peroxidase complex (1/100). The peroxidase reaction was finally performed for 20 minutes in 0.1 M Tris-HCl, pH 7.6, containing 0.03% diaminobenzidine tetrahydrochloride and 0.005% hydrogen peroxide.

3. Results and Discussion

The injection of rabbits with either the synthetic peptide corresponding to the aa 243-268 portion of the third intracellular loop of the 5-HT$_{1A}$ receptor or the fusion protein gave rise to specific anti-5-HT$_{1A}$ receptor antibodies. Both antisera were able to immunoprecipitate 5-HT$_{1A}$ receptor binding sites solubilized by CHAPS from rat hippocampal membranes [5]. In contrast, no immunoprecipitation of solubilized 5-HT_{1B}, 5-HT_{1C}, 5-HT_2, and 5-HT_3 receptors could be found with these antisera. Immunoblot experiments indicated that a major band of 63 kDa, as expected for the native 5-HT_{1A} receptor protein, was specifically recognized by the antipeptide antiserum [6]. Furthermore, immuno-autoradiographic labelling obtained by revealing antipeptide and antifusion protein antibodies bound to fresh brain sections with donkey [^{35}S]IgG-anti-rabbit IgG corresponded exactly to that found with selective 5-HT_{1A} receptor radioligands [4,6].

Specific immunostaining of rat brain sections was achieved with both antisera. Indeed, no immunostaining could be found with preimmune serum or antisera which were first saturated by either the synthetic peptide or the fusion protein, respectively. Immunostaining was intense in the dorsal raphe nucleus, dentate gyrus, and CA1 area of the hippocampus, frontal and entorhinal cortex, and septum, particularly its medio-lateral and dorsal-lateral parts (figure 1). Specific staining

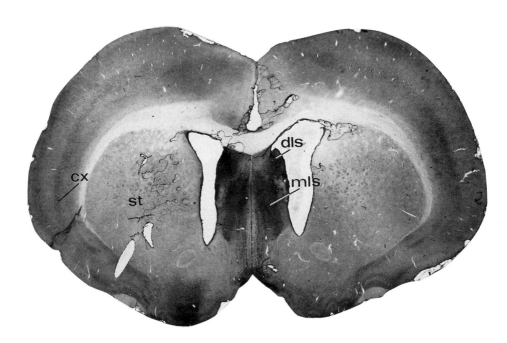

Figure 1. Immunostaining of 5-HT receptors in the rat septum by antipeptide antibodies. An intense immunoreaction can be observed in the medio-lateral (mls) and dorso-lateral (dls) parts of the septum. A specific immunostaining is also found in the cerebral cortex (layers IV - V, cx). In contrast, no immunoreactive material is detected in the striatum (st).

was also observed in layers IV-V throughout the cerebral cortex, the inferior and superior colliculi, central gray, interpeduncular nucleus, reticularis nucleus of the thalamus, median raphe nucleus and within two distinct superficial layers of the dorsal horn in the spinal cord. In contrast, the choroid plexus, dorsal subiculum, striatum, substantia nigra, and cerebellum were devoid of any specific immunolabelling. Previous autoradiographic studies with selective radioligands (notably [^{3}H]8-OH-DPAT) gave exactly the same regional distribution for the 5-HT$_{1A}$ receptor binding sites [7], further confirming that the antibodies raised against either the synthetic peptide or the fusion protein specifically bound to the 5-HT$_{1A}$ receptor protein. In particular, the absence of immunostaining within the substantia nigra and choroid plexus clearly indicated that these antibodies did not recognize 5-HT$_{1B}$ and 5-HT$_{1C}$ receptors [8].

Thanks to these antibodies, the visualization of 5-HT$_{1A}$ receptors could be achieved at the cellular level. Most of the immunoreactive material appeared to be localized at the periphery of the soma and processes of cells (figure 2), that according to their sizes and shapes, as well as the morphology of their processes, were neuronal in nature. No obvious astrocytic labelling was encountered, in agreement with data obtained in cultures from the brain stem of rat fetuses where neuron-like cells identified by the presence of specific markers (neuron-specific enolase and the microtubule-associated protein MAP-2) could be labelled by the antipeptide antibodies, whereas astrocytes expressing the glial fibrillary acidic protein did not bind these antibodies (unpublished observations). In most cases, the labeling was continuous and outlined the whole neurones. In some cases, discontinuities were apparent, as if the labelling was distributed in closely related patches. A typical example of the cellular distribution of the immunoreactive material is illustrated in figure 2 for the dorsal raphe nucleus, and similar observations were made in the median septum, amygdala and entorhinal cortex. Within the CA1 area of Ammon's horn of the hippocampus, immunostaining was particularly dense in the layers bordering the pyramidal cell layer, indicating that the density of 5-HT$_{1A}$ receptors is markedly higher in the dendritic field than on the soma of these cells.

Finally, double immunostaining experiments were performed by combining the visualization of 5-HT$_{1A}$ receptors with that of either 5-HT or choline acetyltransferase. Within the dorsal raphe nucleus, it thus appeared that practically all 5-HT-containing neurones are endowed with 5-HT$_{1A}$ receptors [9], further confirming that the latter act as somato-dendritic 5-HT autoreceptors for serotoninergic neurones [10]. Indeed, the selective lesion of these neurones by the microinjection of 5,7-dihydroxytryptamine into the anterior raphe nuclei results in the local disappearance of both the immunostaining by antipeptide antibodies and the *in situ* hybridization labelling by a specific probe of the 5-HT$_{1A}$ receptor mRNA [11], clearly demonstrating that 5-HT$_{1A}$ receptors are synthesized by and expressed on serotoninergic raphe neurones. At the level of the median septum, especially its

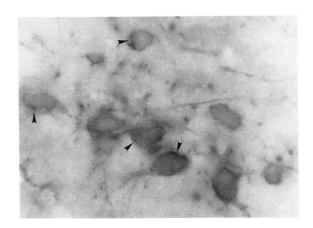

Figure 2. Cellular distribution of 5-HT1A receptors labelled by anti-fusion protein antibodies within the dorsal raphe nucleus. The strongest immunoreactivity appears around the perikarya. Arrow heads point to cells where the immunostaining has a patchy distribution.

anterior part, some cholinergic neurones identified by anticholine acetyltransferase antibodies (Boerhinger-Mannheim) could be specifically labelled by anti-5-HT_{1A} receptor antibodies, giving anatomical support to the modulation of the septo-hippocampal cholinergic system [12] by 5-HT_{1A} receptor ligands. As only a limited number of cholinergic cells could be labelled by both the anticholine acetyltransferase antibodies and the anti-5-HT_{1A} receptor antibodies, it can be concluded that cholinergic neurones do not constitute a homogeneous cell population within the rat septum.

In conclusion, the availability of the sequence of the 5-HT_{1A} receptor protein has permitted the production of specific antibodies for its visualization at the cellular level, and the identification of cell types expressing this receptor. Current studies are aimed at using these antibodies for the visualization of the 5-HT_{1A} receptor protein at the ultrastructural level. Furthermore, the same strategies could be applied for the production of antibodies against the other 5-HT receptor subtypes.

4. Acknowledgements

This research has been supported by grants from INSERM and Bayer-Pharma.

5. References

1. Gozlan, H., El Mestikawy, S., Pichat, L., Glowinski, J., and Hamon, M. (1983) 'Identification of presynaptic serotonin autoreceptors using a new ligand: 3H-PAT', Nature 305, 140-142.
2. Hamon, M., Gozlan, H., El Mestikawy, S., Emerit, M.B., Bolanos, F., and Schechter, L. (1990) 'The central 5-HT receptors: pharmacological, biochemical, functional and regulatory properties', Ann. N.Y. Acad. Sci. 600, 114-131.
3. Albert, P.R., Zhou, Q.Y., Van Tol, H.H.M., Bunzow, J.R., and Civelli, O. (1990) 'Cloning, functional expression and mRNA tissue distribution of the rat 5-hydroxytryptamine$_{1A}$ receptor gene', J. Biol. Chem. 265, 5825-5832.
4. El Mestikawy, S., Riad, M., Laporte, A.M., Vergé, D., Daval, G., Gozlan, H., and Hamon, M. (1990) 'Production of specific anti-rat 5-HT receptor antibodies in rabbits injected with a synthetic peptide', Neurosci. Lett. 118, 189-192.
5. El Mestikawy, S., Taussig, D., Gozlan, H., Emerit, M.B., Ponchant, M., and Hamon, M. (1989) 'Chromatographic analyses of the serotonin 5-HT receptor solubilized from the rat hippocampus', J. Neurochem. 53, 1555-1566.
6. Riad, M., El Mestikawy, S., Vergé, D., Gozlan, H., and Hamon, M. (1991) 'Visualization and quantification of central 5-HT receptors with specific antibodies', Neurochem. Int. 19, 413-423.
7. Vergé, D., Daval, G., Marcinkiewicz, M., Patey, A., El Mestikawy, S., Gozlan, H., and Hamon, M. (1986) 'Quantitative autoradiography of multiple 5-HT1 receptor subtypes in the brain of control or 5,7-dihydroxytryptamine-treated rats', J. Neurosci. 6, 3474-3482.
8. Hoyer, D. (1991) 'The 5-HT receptor family: Ligands, distribution and receptor-effector coupling', in R.J. Rodgers and S.J. Cooper (eds.), 5-HT Agonists, 5-HT3 Antagonists and Benzodiazepines: Their Comparative Behavioural Pharmacology, John Wiley & Sons Ltd, Chichester, pp. 31-57.
9. Sotelo, C., Cholley, B., El Mestikawy, S., Gozlan, H., and Hamon, M. (1990) 'Direct immunohistochemical evidence of the existence of 5-HT autoreceptors on serotoninergic neurons in the midbrain raphe nuclei', Eur. J. Neurosci. 2, 1144-1154.
10. Sprouse, J.S. and Aghajanian, G.K. (1987) 'Electrophysiological responses of serotonergic dorsal raphe neurons to 5-HT and 5-HT1B agonists', Synapse 1, 3-9.
11. Miquel, M.C., Doucet, E., Riad, M., Adrien, J., Vergé, D., and Hamon, M. (1992) 'Effect of the selective lesion of serotoninergic neurons on the regional distribution of 5-HT receptor mRNA in the rat brain', Mol. Brain. Res. 14, 357-362.
12. Fibiger, H.C. and Vincent, S.T. (1987) 'Anatomy of central cholinergic neurons', in H.Y. Meltzer (ed.), Psychopharmacology, the third generation of progress, Raven Press, New York, pp. 211-218.

IN VITRO FUNCTIONAL CORRELATES OF 5-HT$_1$-LIKE RECEPTORS

D.N. MIDDLEMISS, M.S. BEER, and L.O. WILKINSON
Merck Sharp and Dohme Research Laboratories
Neuroscience Research Centre, Terlings Park,
Eastwick Road
Harlow   CM20 2QR
UK

ABSTRACT. 5-HT$_1$-like receptors have been subdivided into 5 main subtypes denoted 5-HT$_{1A}$, 5-HT$_{1B}$, 5-HT$_{1C}$, 5-HT$_{1D}$, and 5-HT$_{1E}$. Each of these receptor subtypes has been cloned and the 2nd messenger-effector system to which they are linked identified. In addition *in vitro* functional correlates of the activation of these receptors in brain or peripheral tissues have been proposed and the evidence for these propositions is critically discussed in this review.

1. Introduction

The most widely used classification of mammalian 5-HT receptors is that proposed in 1986 by Bradley et al. [1] which defined three main types of receptor, 5-HT$_1$-like, 5-HT$_2$, and 5-HT$_3$. The purpose of the present review is to examine the progress in identifying the *in vitro* functional correlates of the 5-HT$_1$-like subclass of mammalian 5-HT receptors. In this respect we have choosen to use the commonly accepted nomenclature of their being five subtypes of 5-HT$_1$-like receptor (5-HT$_{1A}$-5-HT$_{1E}$), as described by radioligand binding studies.

2.   Functional Correlates of 5-HT$_{1A}$ Receptors

2.1   BIOCHEMICAL CORRELATES

Activation of the 5-HT$_{1A}$ receptor has been linked to several effector systems. Thus, 5-HT$_{1A}$ receptor activation has been linked to inhibition of adenylate cyclase in guinea pig, rat, and calf hippocampus, and in cultured mouse cortical and hippocampal neurones [2-4]. 5-HT$_{1A}$ receptor activation has also been reported to stimulate adenylate cyclase in rat and guinea pig hippocampus, [5,6] and produce an increase in K$^+$ conductance in rat hippocampal pyramidal cells, lateral septum, and raphe nuclei via a G protein which interacts directly with the ion channel [see 7 for review]. Serotonin 5-HT$_{1A}$ receptors also appear to inhibit muscarinic receptor-stimulated phosphoinositide turnover in rat hippocampus by stimulating a phospholipase A2 [8].

There is considerable variation in the absolute potency of 5-HT_{1A} receptor agonists in these models, and receptor agonists in one model can act as receptor antagonists in another. These findings led to suggestions of heterogeneity in the 5-HT_{1A} receptor profile [4,9,10]. However, the human and rat 5-HT_{1A} receptors have been cloned [11-13], and the subsequent expression of these receptors in several different cell types has indicated clearly that 1) the cloned rat and human 5-HT_{1A} receptors can both couple to multiple effector systems [14-17], and 2) differences in receptor density can convert partial agonists to silent antagonists and alter the EC_{50} of agonist response [16]. In addition, when the 5-HT_{1A} receptor is expressed in HeLa cells, the cloned receptor appears to be more efficiently coupled to adenylate cyclase inhibition than to calcium mobilization via phospholipase C activation, suggesting that agonist efficacy in these functional models may also be influenced by the efficiency of the receptor coupling to the second messenger which mediates the response [16]. Differences in agonist potencies and efficacy in different preparations can thus be explained without the suggestion of receptor heterogeneity. However, the observation that the brain 5-HT_{1A} receptor may be coupled to Gs [5,6], but the cloned human 5-HT_{1A} receptor expressed in escherichia coli is incapable of coupling to Gs alpha [18] indicates that the question of 5-HT_{1A} receptor heterogeneity is as of yet unresolved.

2.2 PHARMACOLOGICAL CORRELATES

5-HT_{1A} receptors have been identified as the 5-HT somatodendritic autoreceptor mediating inhibition of 5-HT neuronal firing, and a decrease in 5-HT synthesis in raphe slice preparations [19,20]. Many compounds which act as partial agonists or antagonists in the inhibition of adenylate cyclase (described above) act as full agonists in the raphe, leading to the suggestion that this response has a large receptor reserve [20]. In contrast, 5-HT_{1A} receptors mediating a reduction in action potential amplitude in frog dorsal root ganglion [21], and those responsible for hyperpolarization of hippocampal pyramidal cells [22] may have a low receptor reserve, since 8-OH-DPAT, which is a full agonist in all other models of 5-HT_{1A} receptor activation, is an antagonist in both of these preparations. It has also been shown that 5-HT_{1A} receptor activation inhibits acetylcholine release and contractile responses produced by field stimulation in the guinea pig ilium myenteric plexus preparation [23,24].

3. Functional Correlates of 5-HT_{1B} Receptors

3.1 BIOCHEMICAL CORRELATES

The 5-HT_{1B} receptor, originally defined on the basis of radioligand binding studies, is negatively coupled to adenylate cyclase activity in the rat brain [25], a chinese hamster lung fibroblast cell line [26] and in the cloned mouse [27] and rat [28] receptors. Each of these receptors is characterized by a high affinity for the 5-HT_1-like receptor agonist, 5-carboxamidotryptamine and for beta adrenoceptor antagonists such as propranolol and cyanopindolol. Indeed the latter drug is the most potent, albeit nonselective, 5-HT_{1B} receptor antagonist [25]. Cyanopindolol is also the most potent receptor antagonist identified to date at the terminal 5-HT autoreceptor in the rat cortex whereas the prototypic 5-HT_{1A} receptor ligand, 8-OH-DPAT, is essentially inactive as an agonist at this receptor [29]. On this basis, together with studies of a range of 5-HT receptor agonists and antagonists, the terminal 5-HT autoreceptor in the rat has also been convincingly demonstrated to be of the 5-HT_{1B} subclass of 5-HT receptors [29]. Another 5-HT receptor which controls the release of acetylcholine in the rat hippocampal slice, albeit via a heteroreceptor, has also been shown to belong to the 5-HT_{1B} subclass [30], indicating that these receptors are often found prejunctionally on central nerve terminals.

3.2 PHARMACOLOGICAL CORRELATES

5-HT_{1B} receptors have been shown to be functionally coupled in the periphery as well as the brain and appear to be located presynaptically in at least one case, that is on sympathetic nerve terminals on the rat vena cava [31]. They have also been identified as subserving the 5-HT mediated potentiation of electrically-stimulated contractions of the mouse urinary bladder strip, a response which is also blocked by the 5-HT_{1B} receptor antagonists propranolol and cyanopindolol [32,33].

4. Functional Correlates of 5-HT_{1C} Receptors

4.1 BIOCHEMICAL CORRELATES

The most well-established functional response mediated by 5-HT_{1C} receptors is the stimulation of phospholipase C resulting in phosphotylinositol turnover [34]. This response is insensitive to tetrodotoxin as well as cyclooxygenase and lipoxygenase inhibitors suggesting a direct action, i.e. not mediated via arachidonic acid metabolism.

The gene encoding the 5-HT_{1C} receptor has been cloned from mouse choroid plexus and translates a typical G-protein linked receptor structure. 5-HT sensitive Ca^{2+}-activated Cl^- channels have been demonstrated in Xenopus oocytes transfected with the cloned 5-HT_{1C} receptor. Mouse fibroblast transfected cells also display a marked increase in intracellular Ca^{2+} in response to 5-HT. Xenopus oocytes coinjected with cloned 5-HT_{1C} receptor mRNA and rat or mouse-brain-derived K+ channel mRNA manifest an inward current associated with a decrease in membrane conductance in response to 5-HT. This response however is Ca^{2+} independent with the rat K^+ channel and Ca^{2+} dependent with the mouse K^+ channel [35-41].

5-HT_{1C} receptors are therefore thought to mediate their response by an initial activation of phospholipase C followed by an increase in inositol triphosphate levels which results in an inward current either via a rise in intracellular free calcium and consequent opening of a calcium-dependent chloride channel/closing of a Ca^{2+}-dependent K^+ channel or via the closing of Ca^{2+}-independent K^+ channels.

4.2 PHARMACOLOGICAL CORRELATES

The 5-HT sensitive contractile response seen in rat stomach fundus was initially thought to be mediated by 5-HT_{1C} receptors due to the excellent correlation between agonist potencies and their 5-HT_{1C} binding affinities in brain tissue [36]. Baez et al. [35], however, have been unable to demonstrate the presence of 5-HT_{1C} receptor mRNA in this tissue despite amplification by PCR techniques and this coupled with discrepant findings for the rank order of potencies seen with the antagonists ritanserin, LY 53857, and SCH23390 indicates that this response may not be mediated via 5-HT_{1C} receptors but via a novel 5-HT receptor subtype [42].

Despite the studies using cloned receptors described above the electrophysiological effects of the 5-HT_{1C} receptor in brain has only recently been demonstrated by Sheldon and Aghajanian [43]. These voltage clamp studies investigated the 5-HT blocking actions of the antagonists spiperone, ritanserin, and LY 53857 and the excitatory effects of the partial agonist mCPP in pyramidal cells of the rat piriform cortex and conclude that 5-HT_{1C} receptors mediate excitatory effects in this localized population of neurons.

5. Functional Correlates of 5-HT_{1D} Receptors

5.1 BIOCHEMICAL CORRELATES

The 5-HT_{1D} receptor subserves many of the functional effects in higher species that are seen in the rat and mouse for the 5-HT_{1B} receptor [44]. Thus the 5-HT_{1D} receptor is found to be negatively linked to adenylate cyclase activity in the calf [45] and guinea pig [46] brain, the dog saphenous vein [47], and indeed in the cloned dog [48] and human [49] receptors. There is, however, one report of a cloned canine 5-HT_{1D} receptor (RCD4) which has been reported to be positively linked to cyclase [50]. In the case of the cloned human 5-HT_{1D} receptor, two distinct gene products, 5-$HT_{1D\alpha}$ and 5-$HT_{1D\beta}$ have been isolated although both are reported to be negatively linked to adenylate cyclase activation [49]. In general the pharmacological specificity of the cloned human 5-$HT_{1D\alpha}$ and 5-$HT_{1D\beta}$ receptors is similar although they can be distinguished by the drug ketanserin which displays a 71-fold selectivity for the 5-$HT_{1D\alpha}$ subtype [51].

The similarities of the 5-HT_{1D} to the 5-HT_{1B} receptor is also apparent in that the inhibitory terminal 5-HT autoreceptor in several higher species including guinea pig

[52, 53], pig [54], rabbit [53, 55], and human [56] appears to conform to the 5-HT_{1D} receptor subtype. This inhibitory terminal 5-HT autoreceptor may not, however, be of a single receptor subtype since metitepine, a 5-HT_{1D} receptor antagonist, can distinguish two receptors mediating inhibition of 5-HT release in the guinea pig hippocampus [57], a phenomenon which has previously been noted at the rat terminal 5-HT_{1B} autoreceptor [58].

5.2 PHARMACOLOGICAL CORRELATES

Functional 5-HT_{1D} receptors have been reported to exist on the endothelium-intact pig coronary artery [59] and guinea pig jugular vein [60], the rabbit [61, 62] and human [63, 64] saphenous vein, and the human pial arteriole [65]. In each of these preparations metitepine was more potent a receptor antagonist than metergoline which is at variance with the reported affinity of these drugs at 5-HT_{1D} receptors measured by radioligand binding studies [66]. The origin of this discrepancy is not at present clear but may reside in the heterogeneity of 5-HT_{1D} receptor subtypes or indeed to the use of racemic metitepine in these studies.

There have also been a number of reports of the actions of the moderately selective 5-HT_{1D} receptor agonist, sumatriptan [67], at so-called 5-HT_1-like receptors which bear a resemblance to 5-HT_{1D} receptors. These include the dog saphenous vein [68], the dog [69], rabbit [70], sheep [71], primate [69], and human [72] basilar artery, the isolated perfused rat kidney [73], the guinea pig iliac artery [74], and the human coronary artery [75]. The exact relationship of these responses to the 5-HT_{1D} receptor is not fully resolved since in many cases an extensive range of receptor agonists was not used and only limited antagonist studies were carried out. In addition where the receptor antagonists metitipine and metergoline were utilized, it is often the former drug which is the most potent blocker which, as discussed above, is in direct conflict with the data from radioligand binding studies at 5-HT_{1D} receptors. Clearly these difficulties of classification of the receptor(s) mediating these responses will only be clarified by the discovery of potent and selective 5-HT_{1D} receptor antagonists for the two subtypes of receptor which have been cloned.

6. Functional Correlates of 5-HT_{1E} Receptors

6.1 BIOCHEMICAL CORRELATES

The most recently identified member of the 5-HT_1-like family of receptors, the 5-HT_{1E} receptor, was first descibed by Leonhardt et al. following radioligand binding studies in human cerebral cortex [76]. The gene encoding the 5-HT_{1E} receptor has been cloned, stably expressed in HEK 293 cells and the receptor shown to be functionally coupled to an inhibitory adenylate cyclase [77]. 5-HT and 5-CT are both full agonists in inhibiting forskolin-stimulated adenylate cyclase, 5-HT displaying a high potency ($pEC_{50}=7$) and 5-CT a characteristically low potency ($pEC_{50}=4.3$).

Other functional correlates of this receptor subtype remain to be defined.

6. References

1. Bradley, P.B., Engel, G., Fenuik, W., Fozard, J.R., Humphrey, P.P.A., Middlemiss, D.N., Mylecharane, E.J., Richardson, B.P., and Saxena, P.R. (1986) 'Proposals for the classification and nomenclature of functional receptors for 5-hydroxytryptamine', Neuropharmacol. 25, 563-576.
2. De Vivo, M. and Maayani, S. (1986) 'Characterization of the 5-hydroxytryptamine$_{1A}$ receptor-mediated inhibition of forskolin-stimulated adenylate cyclase activity in guinea pig and rat hippocampal membranes', J. Pharm. Exp. Ther. 238, 248-253.
3. Schoeffter, P. and Hoyer, D. (1988) 'Centrally acting hypotensive agents with affinity for 5-HT$_{1A}$ binding sites inhibit forskolin-stimulated adenylate cyclase activity in calf hippocampus', Br. J. Pharmacol. 85, 975-985.
4. Dumuis, A., Sebben, M., and Bockaert, J. (1988) 'Pharmacology of 5-hydroxytryptamine-1A receptors which inhibit cAMP production in hippocampal and cortical neurons in primary culture', Mol. Pharmacol. 33, 178-186.
5. Markstein, R., Hoyer, D., and Engel, G. (1986) '5-HT$_{1A}$-receptors mediate stimulation of adenylate cyclase in rat hippocampus', Naunyn-Schmiedeberg's Arch. Pharmacol. 333, 335-341.
6. Schenker, A., Maayani, S., Weinstein, H., and Green, J.P. (1987) 'Pharmacological characterization of two 5-hydroxytryptamine receptors coupled to adenylate cyclase in guinea pig hippocampal membranes', Mol. Pharmacol. 31, 357-367.
7. Andrade, R. and Chaput, Y. (1991) 'The electrophysiology of serotonin receptor subtypes' in S.J. Peroutka (ed.), Serotonin Receptor Subtypes: Basic and Clinical Aspects, Wiley-Liss, Inc., New York, pp. 103-124.
8. Claustre, Y., Benavides, J., and Scatton, B. (1991) 'Potential mechanisms involved in the negative coupling between serotonin 5-H$_{1A}$ receptors and carbachol-stimulated phosphoinositide turnover in the rat hippocampus' J. Neurochem. 56, 1276-1285.
9. Radja, F., Daval, G., Hamon, M., and Verge, D. (1992) 'Pharmacological and physiochemical properties of pre- versus postsynaptic 5-hydroxytryptamine$_{1A}$ receptor binding sites in the rat brain: a quantitative study' J. Neurochem. 58, 1338-1346.
10. Marszaled, W., Scroggs, R.S., and Anderson, E.G. (1988) 'Serotonin-induced reduction of the calcium-dependent plateau in frog dorsal root ganglion cells is blocked by serotonergic agents acting at 5-hydroxytryptamine$_{1A}$ sites', J. Pharm. Exp. Ther. 247, 399-404.
11. Albert, P.R., Zhou, Q.Y., Van Tol, H.H.M, Bunzow, J.R., and Civelli, O. (1990) 'Cloning, functional expression, and mRNA tissue distribution of the

rat 5-hydroxytryptamine$_{1A}$ receptor gene', J. Biol. Chem. 265, 5825-5832.
12. Kobilka, B.K., Frielle, T., Collins, S., Yang-Feng, T., Kobilka, T.S., Francke, U., Lefkowitz, R.J., and Caron, M.G. (1987) 'An intronless gene encoding a potential member of the family of receptors coupled to guanine nucleotide regulatory proteins', Nature 329, 75-79.
13. Fargin, A., Raymond, J.R., Lohse, M.J., Kobilka, B.K., Caron, M.G., and Lefkowitz, R.J. (1988) 'The genomic clone G-21 which resembles a beta-adrenergic receptor sequence encodes the 5-HT$_{1A}$ receptor', Nature 335, 358-360.
14. Fargin, A., Yamamoto, K., Cotecchia, S., Goldsmith, P.K., Spiegel, A.M., LaPetina, E.G., Caron, M.G., and Lefkowitz, R.J. (1991) 'Dual coupling of the cloned 5-HT$_{1A}$ receptor to both adenylyl cyclase and phospholipase C is mediated via the same Gi protein', Cell Signalling 3, 547-557.
15. Fargin, A., Raymond, J.R., Regan, J.W., Cotecchia, S., Lefkowitz, R.J., and Caron, M.G. (1989) 'Effector coupling mechanisms of the cloned 5-HT$_{1A}$ receptor', J. Biol. Chem. 264, 14848-14852.
16. Boddeke, H.W.G.M., Fargin, A., Raymond, J.R., Schoeffter, P., and Hoyer, D. (1992) 'Agonist/antagonist interactions with cloned human 5-HT$_{1A}$ receptors: variations in intrinsic activity studies in transfected HeLa cells', Naunyn-Schmiedeberg's Arch. Pharmacol. 345, 257-263.
17. Liu, Y.F. and Albert, P.R. (1991) 'Cell-specific signalling of the 5-HT$_{1A}$ receptor', J. Biol. Chem. 266, 23689-23697.
18. Bertin, B., Freissmuth, M., Breyer, R.M., Schultz, W., Strosberg, A.D., and Marullo, S. (1992) 'Functional expression of the human serotonin 5-HT$_{1A}$ receptor in escherichia coli', J. Biol. Chem. 267, 8200-8206.
19. Sprouse, J. and Aghajanian, G.K. (1987) 'Electrophysiological responses of serotoninergic dorsal raphe neurons to 5-HT1A and 5-HT1B agonists', Synapse 1, 3-9.
20. Sawada, M. and Nagatsu, T. (1986) 'Stimulation of the serotonin autoreceptor prevents the calcium-calmodulin-dependent increase of serotonin biosynthesis in rat raphe slices', J. Neurochem. 46, 963-967.
21. Marszalec, W., Scroggs, R.S., and Anderson, E.G. (1988) 'Serotonin-induced reduction of the calcium-dependent plateau in frog dorsal root ganglion cells is blocked by serotonergic agents acting at 5-hydroxytryptamine1A sites', J. Pharm. Exp. Ther. 247, 399-404.
22. Colino, A. and Halliwell, J.V. (1986) '8-OH-DPAT is a strong antagonist of 5-HT action in rat hippocampus', Eur. J. Pharmacol. 130, 151-152.
23. Fozard, J.R. and Kilbinger, H. (1985) '8-OH-DPAT inhibits transmitter release from guinea-pig cholinergic neurones by activating 5-HT$_{1A}$ receptors', Br. J. Pharmacol. 86, 601P.
24. Galligan, J.J. (1992) 'Differential inhibition of cholinergic and noncholinergic neurogenic contractions by 5-hydroxytryptamine1A receptor agonists in guinea pig ileum', J. Pharm. Exp. Ther. 260, 306-312.

25. Bouhelal, R., Smounya, L., and Bockaert, J. (1988) '5-HT$_{1B}$ receptors are negatively coupled with adenylate cyclase in rat substantia nigra', Eur. J. Pharmacol. 151, 189-196.
26. Senwen, K., Magnaldo, I., and Ponyssegur, J. (1988) 'Serotonin stimulates DNA synthesis in fibroblasts acting through 5-HT$_{1B}$ receptors coupled to a G-protein', Nature 335, 254-256.
27. Maroteaux, L., Saudou, F., Amlaiky, N., Boschert, U., Plassat, J.L., and Hen, R. (1992) 'Mouse 5HT1B serotonin receptor: Cloning, functional expression, and localization in motor control centers', Neurobiol. 89, 3020-3024.
28. Adham, N., Romanienko, P., Hartig, P., Weinshank, R.L., and Branchek, T. (1991) 'The rat 5-hydroxytryptamine$_{1B}$ Receptor is the species homologue of the human 5-hydroxytryptamine1Dalpha receptor', Mol. Pharmacol. 41, 1-7.
29. Middlemiss, D. N. and Hutson, P. H. (1990) 'The 5HT$_{1B}$ receptors', Ann. NY. Acad. Sci. 600, 132-148.
30. Maura, G. and Raiteri, M. (1986) 'Cholinergic terminals in rat hippocampus possess 5-HT$_{1B}$ receptors mediating inhibition of acetylcholine release', Eur. J. Pharmacol., 129, 333-337.
31. Molderings, G.J., Fink, K., Schlicker, E., and Gothert, M. (1987) 'Inhibition of noradrenaline release via presynaptic 5-HT$_{1B}$ receptors of the rat vena cava', Naunyn-Schmiedeberg's Arch. Pharmacol. 336, 245-250.
32. Holt, S.E., Cooper, M., and Wyllie, J.H. (1986) 'On the nature of the receptor mediating the action of 5-hydroxytryptamine in potentiating responses of the mouse urinary bladder strip to electrical stimulation', Naunyn-Schmiedeberg's Arch. Pharmacol. 334, 333-340.
33. Wyllie, J.H. (1989) 'Blockade by cyanopindolol of 5-HT receptors in mouse urinary bladder' in Serotonin from Cell Biology to Pharmacology and Therapeutics, Serotonin Club, Florence, Italy, 56.
34. Conn, P.J., Sanders-Bush, E., Hoffman, B.J., and Hartig, P.R. (1986) 'A unique serotonin receptor in choroid plexus is linked to phosphatidylinositol turnover', Neurobiol. 83, 4086-4088.
35. Baez, M., Yu, L., and Cohen, M.L. (1990) 'Pharmacological and molecular evidence that the contractile response to serotonin in rat stomach fundus is not mediated by activation of the 5-hydroxytryptamine$_{1C}$ receptor', Mol. Pharmacol. 38, 31-37.
36. Cohen, M.L. (1989) '5-hydroxytryptamine and non-vascular smooth muscle contraction and relaxation', in J.R. Fozard (ed.), The Peripheral Actions of 5-Hydroxytryptamine, Oxford University Press, Oxford, England, pp. 201-218.
37. Lubbert, H., Hoffman, B.J., Snutch, T.P., Van Dyke, T., Levine, A.J., Hartig, P.R., Lester, H.A., and Davidson, N. (1987) 'cDNA cloning of a serotonin 5-HT$_{1C}$ receptor by electrophysiological assays of mRNA-injected xenopus oocytes', PNAS 84, 4332-4336.
38. Julius, D., MacDermott, A.B., Axel, R., and Jessell, T.M. (1988) 'Molecular characterization of a functional cDNA encoding the serotonin 1c receptor',

39. Parker, I., Panicker, M.M., and Miledi, R., (1990) 'Serotonin receptors expressed in xeonopus oocytes by mRNA from brain mediate a closing of K⁺ membrane channels', Mol. Brain Res. 7, 31-38.
40. Panicker, M.M., Parker, I., and Miledi, R. (1991) 'Receptors of the serotonin 1C subtype expressed from cloned DNA mediate the closing of K⁺ membrane channels encoded by brain mRNA', Neurobiol. 88, 2560-2562.
41. Hoger, J.H., Walter, A.E., Vance, D., Yu, L., Labarca, C., Lester, H.A., and Davidson, N. (1990) 'Cloning, expression and modulation by serotonin of a mouse brain potassium channel', Soc. Neurosci. Abst. 16, 671.
42. Foguet, M., Hoyer, D., Pardo, L.A., Parekh, A., Kluxen, F.W., Kalkman, H. O., Stuhmer, W., and Lubbert, H. (1992) 'Cloning and functional characterisation of the rat stomach fundus serotonin receptor', EMBO 11, 3481-3487.
43. Sheldon, P.W. and Aghajanian, G.K. (1991) 'Excitatory responses to serotonin (5-HT) in neurons of the rat piriform cortex: Evidence for mediation 5-HT$_{1C}$ receptors in pyramidal cells and 5-HT$_2$ receptors in interneurons', Synapse 9, 208-218.
44. Hoyer, D. and Middlemiss, D.N. (1989) 'Species differences in the pharmacology of terminal 5-HT autoreceptors in mammalian brain', TIPS, 10, 130-132.
45. Schoeffter, P., Waeber, C., Palacios, J.M., and Hoyer, D. (1988) 'The 5-hydroxytryptamine 5-HT$_{1D}$ receptor subtype is negatively coupled to adenylate cyclase in calf substantia nigra', Naunyn-Schmiedeberg's Arch. Pharmacol. 337, 602-608.
46. Waeber, C., Schoeffter, P., Palacios, J.M., and Hoyer, D. (1989) '5-HT$_{1D}$ receptors in guinea-pig and pigeon brain. Radioligand binding and biochemical studies', Naunyn-Schmiedeberg's Arch. Pharmacol. 340, 479-485.
47. Sumner, M.J. and Humphrey, P.P.A. (1990) 'Sumatriptan (GR43175) inhibits cyclic-AMP accumulation in dog isolated saphenous vein', Br. J. Pharmacol. 99, 219-220.
48. Zgombick, J.M., Weinshank, R.L., Macchi, M., Schechter, L.E., Branchek, T.A., and Hartig, P.R. (1991) 'Expression and pharmacological characterization of a canine 5-hydroxytryptamine$_{1D}$ receptor subtype', Mol. Pharmacol. 40, 1036-1042.
49. Hartig, P.R., Branchek, T.A., and Weinshank, R.L., (1992) ' A subfamily of 5-HT$_{1D}$ receptor genes', TIPS 13, 152-159.
50. Maenhaut, C., Van Sande, J., Massart, C., Dinsart, C., Libert, F., Monferini, E., Giraldo, E., Ladinsky, H., Vassart, G., and Dumont, J.E. (1991) ' The orphan receptor cDNA RDC4 encodes a 5-HT$_{1D}$ serotonin receptor', Biochem. Biophys. Res. Comm. 180, 1460-1468.
51. Weinshank, R.L., Branchek, T., and Hartig P.R. (1991) 'Nucleic acid encoding 5-HT$_{1D}$ receptors and their antibodies', Patent Application

WO9117174-A1.
52. Middlemiss, D.N., Bremer, M.E., and Smith, S.M. (1988) 'A pharmacological analysis of the 5-HT receptor mediating inhibition of 5-HT release in the guinea-pig frontal cortex', Eur. J. Pharmacol. 157, 101-107.
53. Limberger, N., Deicher, R., and Starke, K. (1991) 'Species differences in presynaptic serotonin autoreceptors: mainly 5-HT_{1B} but possibly in addition 5-HT_{1D} in the rat, 5-HT_{1D} in the rabbit and guinea-pig brain cortex', Naunyn-Schmiedeberg's Arch Pharmacol. 343, 353-364.
54. Schlicker, E., Fink, K., Gothert, M., Hoyer, D., Molderings, G., Roschke, I., and Schoeffter, P. (1989) 'The pharmacological properties of the presynaptic serotonin autoreceptor in the pig brain cortex conform to the 5-HT_{1D} receptor subtype', Naunyn-Schmiedeberg's Arch. Pharmacol. 340, 45-51.
55. Feuerstein, T.J., Lupp, A., and Hertting, G. (1992) 'Quantitative evaluation of the autoinhibitory feedback of release of 5-HT in the caudate nucleus of the rabbit where an endogenous tone on $alpha_2$-adrenoceptors does not exist', Neuropharmacol. 31, 15-23.
56. Galzin, A.M. and Langer, S.Z. (1991) 'Modulation of 5-HT release by presynaptic inhibitory and facilitatory 5-HT receptors in brain slices', Adv. Biosci. 82, 59-62.
57. Wilkinson, L.O. and Middlemiss, D.N. (1992) 'Metitepine distinguishes two receptors mediating inhibition of ^{3}H-5-hydroxytryptamine release in guinea pig hippocampus', Naunyn-Schmiedeberg's Arch. Pharmacol. 345, 696-699.
58. Moret, C. and Briley, M. (1986) 'Dihydroergocristine-induced stimulation of the 5-HT autoreceptor in the hypothalamus of the rat', Neuropharmacol. 25, 169-174.
59. Schoeffter, P. and Hoyer, D. (1990) '5-hydroxytryptamine (5-HT)-induced endothelium-dependent relaxation of pig coronary arteries is mediated by 5-HT receptors similar to the 5-HT_{1D} receptor subtype', J. Pharm. Exp Ther. 252, 387-395.
60. Gupta, P. (1992) 'The endothelial 5-HT receptor that mediates relaxation in guinea-pig isolated jugular vein resembles the 5-HT_{1D} subtype' Br. J. Pharmacol. 106, 703-709.
61. Martin, G.R. and MacLennan, S.J. (1990) 'Analysis of the 5-HT receptor in rabbit saphenous vein exemplifies the problems of using exclusion criteria for receptor classification', Naunyn-Schmiedeberg's Arch Pharmacol. 342, 111-119.
62. Ormandy, G.C., Wilson, D.J., Wren, P., Barrett, V.J., and Prentice, D.J. (1992) 'Comparison of the 5-HT_{1D}-like receptors in the saphenous veins and CNS of the dog and rabbit', Br. J. Pharmacol., in press.
63. Bax, A.W., Van Heuven-Nolsen, D., Bos, E., Simoons, M.L., and Saxena, P.R. (1992) '5-hydroxytryptamine-induced contractions of the human isolated saphenous vein: Involvement of 5-HT_2 and 5-HT_{1D}-like receptors, and a comparison with grafted veins', Naunyn-Schmiedeberg's Arch. Pharmacol.

345, 500-508.
64. Molderings, G.J., Werner, K., Likungu, J., and Gothert, M. (1990) 'Inhibition of noradrenaline release from sympathetic nerves of the human saphenous vein via presynaptic 5-HT receptors similar to the 5-HT$_{1D}$ subtype', Naunyn-Schmiedeberg's Arch. Pharmacol. 342, 371-377.
65. Hamel, E. and Bouchard, D. (1991) 'Contractile 5-HT$_1$ receptors in human isolated pial arterioles: correlation with 5-HT$_{1D}$ binding sites', Br. J. Pharmacol. 102, 227-233.
66. Heuring, R.E. and Peroutka, S.J. (1987) 'Characterization of a novel ^{3}H-5-hydroxytryptamine binding site subtype in bovine brain membranes', J. Neurosci. 7, 894-903.
67. Schoeffter, P. and Hoyer, D. (1989) 'How selective is GR 43175? Interactions with functional 5-HT$_{1A}$, 5-HT$_{1B}$, 5-HT$_{1C}$ and 5-HT$_{1D}$ receptors', Naunyn-Schmiedeberg's Arch. Pharmacol. 340, 135-138.
68. Humphrey, P.P.A., Feniuk, W., Perren, M.J., Connor, H.E., Oxford, A.W., Coates, I.H., and Butina, D. (1988) 'GR43175, a selective agonist for the 5-HT$_1$-like receptor in dog isolated saphenous vein', Br. J. Pharmacol. 94, 1123-1132.
69. Connor, H.E., Feniuk, W., and Humphrey, P.P.A. (1989) 'Characterization of 5-HT receptors mediating contraction of canine and primate basilar artery by use of GR43175, a selective 5-HT$_1$-like receptor agonist', Br. J. Pharmacol. 96, 379-387.
70. Parsons, A.A. and Whalley, E.T. (1989) 'Evidence for the presence of 5-HT$_1$-like receptors in rabbit isolated basilar arteries', Eur. J. Pharmacol. 174, 189-196.
71. Gaw, A.J., Wadsworth, R.M., and Humphrey, P.P.A. (1990) 'Pharmacological characterisation of postjunctional 5-HT receptors in cerebral arteries from the sheep', Eur. J. Pharmacol. 179, 35-44.
72. Parsons, A.A., Whalley, E.T., Feniuk, W., Connor, H.E., and Humphrey, P.P.A., (1989) '5-HT$_1$-like receptors mediate 5-hydroxytryptamine-induced contraction of human isolated basilar artery', Br. J. Pharmacol. 96, 434-449.
73. Bond, R.A., Craig, D.A., Charlton, K.G., Ornstein, A.G., and Clarke, D.E. (1989) 'Partial agonistic activity of GR43175 at the inhibitory prejunctional 5-HT$_1$-like receptor in rat kidney', J. Auton. Pharmac. 9, 201-210.
74. Sahin-Erdemli, I., Hoyer, D., Stoll, A., Seiler, M.P., and Schoeffter, P. (1991) '5-HT$_1$-like receptors mediate 5-hydroxytryptamine-induced contraction of guinea-pig isolated iliac artery', Br. J. Pharmacol. 102, 386-390.
75. Chester, A.H., Martin, G.R., Bodelsson, M., Arneko-Noboin, B., Tadjkarimi, S., Tornebrandt, K., and Yacoub, M.H. (1990) '5-hydroxytryptamine receptor profile in healthy and diseased human epicardial coronary arteries', Cardiovascular Res. 24, 932-937.
76. Leonhardt, S., Herrick-Davis, K., and Titeler, M. (1989) 'Detection of a novel serotonin receptor subtype (5-HT$_{1E}$) in human brain: Interaction with

a GTP-binding protein', J. Neurochem. 53, 465-471.
77. McAllister, G., Charlesworth, A., Snodin, C., Beer, M.S., Noble, A.J., Middlemiss, D.N., Iversen, L.L., and Whiting, P. (1992) 'Molecular cloning of a serotonin receptor from human brain ($5HT_{1E}$): A fifth $5HT_{1\text{-like}}$ subtype', Neurobiol. 89, 5517-5521.

5-HT$_{1D}$ RECEPTOR SUBTYPES

D. HOYER, A. T. BRUINVELS, and J.M. PALACIOS
Preclinical Research
Bldg. 360, Room 604,
SANDOZ Pharma Ltd
CH 4002 Basel
Switzerland

ABSTRACT. In spite of the title, this paper will address issues about the possible similarities and differences between what has initially been designated as 5-HT$_{1B}$, 5-HT$_{1D}$, and 5-HT$_1$-like receptors, since in view of recent cloning work on 5-HT$_{1B}$, 5-HT$_{1D\alpha}$, 5-HT$_{1D\beta}$, 5-HT$_{1F}$, and 5-HT$_5$ receptors, these entities cannot be easily separated in a variety of species.

1. Historical Aspects

The story begins with the pharmacological characterization of 5-HT$_{1B}$ sites in rat brain, using [^{125}I]CYP, a potent ß-adrenoceptor antagonist [1]. Pazos and Palacios [2] described the very distinct pattern of distribution of 5-HT$_{1A}$ and 5-HT$_{1B}$ sites in rat brain: high densities of 5-HT$_{1B}$ sites were found in nigro-striatal regions and basal ganglia (substantia nigra, subiculum, globus pallidus, caudate-putamen). It was then demonstrated that the terminal 5-HT autoreceptor in rat brain displayed a 5-HT$_{1B}$ profile [3].

However, it was noticed that in contrast to 5-HT$_{1A}$, 5-HT$_{1C}$, and 5-HT$_2$ sites, 5-HT$_{1B}$ sites were not to be found in pig [1,4] or human brain [5]. The apparent absence of 5-HT$_{1B}$ sites was also reported for other species. Then, Heuring and Peroutka [6] described a site called 5-HT$_{1D}$ in bovine brain (where no 5-HT$_{1B}$ sites were to be found). Subsequently, 5-HT$_{1D}$ sites were described in human brain using radioligand binding [7] or autoradiography [8]. Interestingly, the distribution of 5-HT$_{1D}$ sites in human brain was very similar to that of 5-HT$_{1B}$ sites in rat brain. These studies were extended to other species (pigeon, pig, guinea pig, calf, and man) and invariably, a typical 5-HT$_{1D}$ profile and distribution were observed [9-11]. Thus, 5-HT$_{1B}$ sites were only found in rat, mouse, hamster, and opossum (see [12], for a review).

Another similarity between 5-HT$_{1B}$ and 5-HT$_{1D}$ sites was the second messenger system: thus, 5-HT$_{1B}$ receptor activation leads to inhibition of adenylate cyclase activity in rat substantia nigra [13], whereas in substantia nigra from calf and guinea pig this activity had a typical 5-HT$_{1D}$ profile [11,14,15]. Similarly, the terminal autoreceptor in porcine brain was shown to be of the 5-HT_{1D} type [16]. This was also the case in guinea pig, rabbit, and human brain [12, 17, 18].

Considering the similarities between 5-HT_{1B} and 5-HT_{1D} receptors with respect to distribution and function, Hoyer and Middlemiss [19] suggested that the two receptors represented species equivalents, that is similar gene products which express different pharmacological profile depending on the species. We will see below, that this suggestion was partly correct but that the situation is still more complex.

2. Molecular Biology of $5\text{-HT}_{1B}/5\text{-HT}_{1D}$ Receptors

Most 5-HT receptors which have been described in binding and/or functional studies have been cloned, sequenced, and functionally expressed (see Humphrey, Hartig and Hoyer, this volume, [20]). Primers derived from putative canine RDC4 receptor which has limited homology (55%) with the human 5-HT_{1A} receptor [21] were used in PCR to find the human equivalent (93% homology in TSR); this gene displayed a classical 5-HT_{1D} receptor pharmacology [22,23] and has 377 amino acids. Eventually, when transfected into mammalian cells, RDC4 displayed also a 5-HT_{1D} type pharmacology [24,25]. A rat equivalent (95% homology in TSR) has been cloned [26]; this receptor has a 5-HT_{1D}-like pharmacology. It seems that the RDC4-related gene products appear to be expressed at very low levels [see 21,23,27] and indeed mRNA levels are very low. These clones (derived from RDC4) were named $5\text{-HT}_{1D\alpha}$ by Hartig et al. [28], and it is important to note that so far, none of these cloned receptors does express a 5-HT_{1B} pharmacology.

Somewhat surprisingly, in the same species, a second receptor relatively similar in terms of sequence (77% in TSR) was cloned. The human gene which has 390 predicted amino acids (named $5\text{-HT}_{1D\beta}$) expresses a 5-HT_{1D} pharmacology which is indistinguishable (see table 1) from that of the $5\text{-HT}_{1D\alpha}$ clone [23, 27-30]. Voigt et al. [31] and Adham et al. [32] identified the rat homologue to the human $5\text{-HT}_{1D\beta}$ clone. The rat clone has 96% homology in TSR with the human clone, but expresses the typical 5-HT_{1B} profile. Similarly, a mouse 5-HT_{1B} receptor has been cloned [33]; these receptors (human $5\text{-HT}_{1D\beta}$ and rat/mouse 5-HT_{1B}) represent species homologues, as suggested earlier based on their distribution in the brain of a variety of species [19]. Thus it appears that 5-HT_{1B} receptors which have only been described in a few species (rat, mouse, hamster, and opossum) are the equivalent of 5-HT_{1D} receptors which have been found in most of other mammals studied and birds. The obvious question is, which of the two receptors ($5\text{-HT}_{1D\alpha}$ or $5\text{-HT}_{1D\beta}$) is relevant for the so-called 5-HT_{1D} effects which have been observed in higher species? The complexity may even be greater since the cloning of 5-HT_{1F} [34,35] and 5-HT_5 receptors [36], both of which have a pharmacological profile which is reminiscent of the 5-HT_{1D} profile.

TABLE 1. Affinity values of various drugs for the sites labelled with [^{125}I]CYP in rat brain or [^{125}I]GTI in a brain of other species. Comparison with affinity values obtained with rat 5-HT$_{1D\beta}$ and 5-HT$_{1D\alpha}$ and human 5-HT$_{1D\alpha}$ and 5-HT$_{1D\beta}$ clones.

DRUG	Rat Cx	Rat 1Dβ	Rat 1Dα	GPC x	Monk Caud	Rab Brain	Hum Caud	Hum 1Dα	Hum 1Dβ
5-CT	8.3	8.14-9.45	9.43	9.28	9.28	9.08	9.04	9.13	8.79
5-HT	7.6	7.80-9.24	8.60	8.78	8.72	8.31	8.38	8.39	8.39
sumatriptan	6.4	6.33-7.35	8.02	7.71	8.56	8.19	8.39	8.44	8.17
DHE	8.4	8.38-10.56	10.19	8.86	8.96	8.55	8.42	9.87	7.74
ergotamine	8.7	10.52	10.11	8.58	8.74	8.6	8.42		
methysergide	5.8	5.74		7.32	7.79	7.72	7.90	8.42	7.63
yohimbine	5.5	6.82	7.51	7.32	8.05	7.29	8.00	7.63	7.61
metergoline	7.4	6.89-9.05	9.12	8.10	8.65	8.12	8.23	9.22	7.44
rauwolscine	5.2	5.24						7.7	7.45
CGS 12066	6.9	6.96-8.13	8.31	7.86	8.34	8.07	8.23	8.17	8.71
ICYP	9.5	9.57	7.44	7.45	7.59	7.35	7.70		
8-OH-DPAT	4.2	<6-<5	7.00	6.99	6.87	6.96	7.14	6.88	6.64
methiothepin	7.3	7.89-7.91	9.23	7.81	7.73	6.86	7.36	8.24	7.77
SDZ 21-009	9.4			7.12	6.55	7.11	7.14		
mianserin	5.2	<6	7.52	6.90	7.18	6.81	6.72		
SCH 23390	4.9			6.40	6.11	6.42	6.49		
CP 93129	7.8			6.41	6.76	6.52	6.57		
ketanserin	5.7			5.77	5.91	5.89	5.71		5.09
corynanthine	5.5			5.57	5.71	6.55	5.00		
isamoltane	7.0			5.71	5.34	5.63	5.70		
ipsapirone	3.9			5.39	5.36	5.59	5.65		
(-)pindolol	7.8	6.82		5.95	5.90	4.95	5.78	5.36	5.35
spiperone	5.3	<6	7.21	5.83	5.14	6.05	4.87	5.98	5.08
buspirone	3.9			4.55	5.07	5.03	4.98		
mesulergine	4.9	<6	7.03	5.12	5.30	6.06	5.45	5.77	5.89

Abbreviations: Rat Cx, rat cortex; Rat$_{1D\beta}$, rat 5-HT$_{1D\beta}$ clone; Rat$_{1D\alpha}$, rat 5-HT$_{1D\alpha}$ clone; GPC, guinea pig cortex; Hum Caud, human caudate; Monk Caud, monkey caudate; Rab Brain, rabbit brain; Hum $_{1D\alpha}$ and Hum $_{1D\beta}$, human 5-HT$_{1D\alpha}$ and 5-HT$_{1D\beta}$ clones. Radioligand binding data are from the author's laboratory. Binding data obtained in cloned cells were obtained from refs 22, 23, 26, 27, 31, 32.

3. Pharmacological Characteristics of $5\text{-}HT_{1B}$ and $5\text{-}HT_{1D}$ Receptors

In the early days of $5\text{-}HT_{1D}$ research, one of the main problems was the absence of selective ligands. This is still largely true for antagonists, since none with adequate potency and selectivity has been described for either $5\text{-}HT_{1B}$ or $5\text{-}HT_{1D}$ receptors. Few if any ligands show selectivity for either of these receptors. However, the situation may be even more complicated than expected, since the presence of two similar receptors has been described by cloning work in species where the presence of these receptors was assumed to be mutually exclusive. Nevertheless, receptors with the typical $5\text{-}HT_{1B}$ pharmacology have only been reported in a few rodents (rat, mouse, hamster) and opossum. Claims that RU 24969, TFMPP, and mCPP are $5\text{-}HT_{1B}$-selective agonists have not been confirmed [37]. However, the agonist CGS 12066 appears to carry some $5\text{-}HT_{1B}$ selectivity, and the recently described CP 93129 is indeed $5\text{-}HT_{1B}$ selective [38]. Sumatriptan posesses limited $5\text{-}HT_{1D}$ selectivity [39,40], while 5-benzyloxy-tryptamine is equieffective at $5\text{-}HT_{1B}$ and $5\text{-}HT_{1D}$ receptors. With regard to antagonists, the situation is less favorable. There is no documented selective antagonist for $5\text{-}HT_{1D}$ receptors. However, a few compounds enable a better definition of these receptors. Thus, PAPP was shown to distinguish between the two receptors [37]. On the other hand, it was established that some ß-blockers were more active at $5\text{-}HT_{1B}$ than at $5\text{-}HT_{1D}$ receptors, whereas the opposite was true for yohimbine and rauwolscine [37]. These compounds may act as agonists or antagonists depending on the model selected (see tables 1 and 2).

4. Localization of $5\text{-}HT_{1B}$ and $5\text{-}HT_{1D}$ Receptors

The basal ganglia [2], especially the globus pallidus and the pars reticulata of the substantia nigra also show high densities of $5\text{-}HT_1$ sites. In the rat brain these sites are of the $5\text{-}HT_{1B}$ type, as assesssed by their pharmacological profile. In contrast to the situation in rats and mice, [^{3}H]5-HT binding in the basal ganglia of the other mammals displays a pharmacological profile characteristic of $5\text{-}HT_{1D}$ sites. Further, there is no evidence for the presence of $5\text{-}HT_{1B}$ sites in guinea pig, pig, calf, rabbit, dog, monkey, human, and even pigeon brain. In autoradiographic studies, sumatriptan displaced [^{3}H]5-HT binding from $5\text{-}HT_{1B}$ sites in rat brain, and $5\text{-}HT_{1D}$ sites in monkey and human brain [41]. Until recently, $5\text{-}HT_{1D}$ binding had been performed using [^{3}H]5-HT in the presence of high concentration of ligands (e.g. 8-OH-DPAT and mesulergine) which block $5\text{-}HT_{1A}$ and $5\text{-}HT_{1C}$ binding. However, under these conditions, the binding has been reported to be heterogeneous [see 10,42,43] and to comprise the putative $5\text{-}HT_{1E}$ binding site. However, using the $5\text{-}HT_{1B}/5\text{-}HT_{1D}$ ligand [^{125}I]GTI, we and others have shown that the distribution of $5\text{-}HT_{1B}$ sites in rat brain was comparable to that of $5\text{-}HT_{1D}$ sites in guinea pig and human brain [44, 45, 46, 47].

TABLE 2. Potency of selected ligands at 5-HT_{1B} and 5-HT_{1D} receptors.

Receptor	Localization within CNS	Rank order of potency[1]			
		Agonists		Antagonists	
5-HT_{1B}	Substantia nigra	RU 24969	(8.4)	cyanopindolol	(8.2)
	Basal ganglia	5-CT	(7.9)	methiothepin	(8.1)
	Subiculum (rodent specific)	CGS 12066	(7.6)	SDZ 21009	(8.0)
		metergoline	(7.2)	isamoltane	(7.3)
		TFMPP	(6.9)	propranolol	(6.9)
		methysergide	(6.7)	pindolol	(6.8)
		mCPP	(6.5)	quipazine	(6.2)
		sumatripan	(6.0)	yohimbine	(6.1)
		LY 165163	(5.8)	rauwolscine	(6.0)
		DP-5-CT	(5.8)	mianserin	(6.0)
		8-OH-DPAT	(4.9)	spiperone	(4.4)
5-HT_{1D}	Substantia nigra	5-CT	(8.1)	methiothepin	(7.7)
	Basal Ganglia	LY 165163	(7.6)	mianserin	(6.5)
	Subiculum (guinea pig, pig,	metergoline	(7.5)	quipazine	(5.7)
	calf, monkey, human)	methysergide	(7.0)	mCPP	(5.1)
		sumatripan	(7.0)	spiperone	(4.8)
		CGS 12066	(7.1)	isamoltane	(4.4)
		rauwolscine	(6.9)	buspirone	(4.2)
		yohimbine	(6.8)		
		CYP	(6.8)		
		RU 24969	(6.8)		
		DP-5-CT	(6.6)		
		SDZ 21009	(5.9)		
		8-OH-DPAT	(5.8)		
		TFMPP	(5.8)		

[1] Agonists: pEC_{50} values; antagonists: pK_B or pA_2 values. Models: 5-HT_{1B} and 5-HT_{1D}, inhibition of forskolin stimulated adenylate cyclase activity in rat and calf substantia nigra respectively.

In the rat brain, lesions experiments have shown that these receptors could be presynaptically localized on the terminals of the striatal intrinsic neurons which innervate the substantia nigra pars reticulata, as destruction of caudate neurons or transection of the medial forebrain bundle results in a dramatic decrease of binding in the substantia nigra pars reticulata. On the other hand the lesion of dopaminergic neurons in the substantia nigra pars compacta does not induce a decrease of $5\text{-HT}_{1D}/5\text{-HT}_{1B}$ binding sites. Thus, 5-HT_{1D} and 5-HT_{1B} receptors, although different pharmacologically, appear to serve the same type of functions and are localized in the same areas in the mammalian brain.

Using [^{3}H]5-HT and [^{125}I]GTI, we were recently able to localize in autoradiographic studies performed in rat brain what appears to represent 5-HT_{1D}-

like receptors [48]. However, the 5-HT_{1D}-like sites appear to codistribute with 5-HT_{1B} sites. In a large variety of rat brain regions, 5-HT_{1D}-like binding was either absent or represented only a very minor component when compared to 5-HT_{1B} binding. Thus, there was no single brain area where 5-HT_{1D}-like sites were prominent. This is in keeping with results from *in situ* hybridization, in which 5-$HT_{1D\alpha}$ mRNA levels appear to be very low when compared to the levels of 5-$HT_{1D\beta}$ mRNA [27] and the apparent lack of success of Weinshank et al. [23] to clone the rat 5-$HT_{1D\alpha}$ receptor. Xiong and Nelson [49] have reported the presence of 5-HT_{1R} sites in rabbit brain; however, in an extensive study (see table 1), we have concluded that this site is very similar in its pharmacological profile to 5-HT_{1D} sites identified in other species [50]. This is compatible with the results of Limberger et al. [18] who identified the terminal autoreceptor of the rabbit brain to be a member of the 5-HT_{1D} family.

5. 5-HT_{1B} and 5-HT_{1D} Receptors and Function

5-HT_{1B} receptors are negatively coupled to adenylate cyclase in homogenates of rat substantia nigra [13], which possess a high density of almost exclusively 5-HT_{1B} sites. In this preparation (table 2), the rank order of potency of both agonists and antagonists correlates best with affinity values for 5-HT_{1B} binding sites [13,51]. Similar findings have been reported in a hamster lung cell line in which the mitogenic effects of 5-HT could be related to inhibition of adenylate cyclase activity and displays a 5-HT_{1B} profile. Cells transfected with rat or mouse 5-HT_{1B} receptors have been shown to carry inhibition of adenylate cyclase activity [32,33].

Activation of 5-HT_{1D} receptors leads to inhibition of forskolin-stimulated adenylate cyclase activity in calf (table 2) and guinea pig substantia nigra [11,14,15], which contain a high proportion of 5-HT_{1D} sites [8,9]. Most studies performed with cells transfected with 5-HT_{1D} receptors (both 5-$HT_{1D\alpha}$ and 5-$HT_{1D\beta}$ types) show that these receptors are indeed negatively coupled to adenylate cyclase activity. However, the canine RDC4 clone, depending on the type of cell system used can act in a stimulatory [25] or inhibitory fashion [25] on adenylate cyclase.

A similar functional role has been proposed to 5-HT_{1B} and 5-HT_{1D} receptors. This is based on distribution and second messenger studies performed in a variety of species, but also on functional models. Terminal 5-HT autoreceptors mediating inhibition of 5-HT release which have been identified in a variety of species represent probably the best example for the convergence between 5-HT_{1B} and 5-HT_{1D} receptors. It is generally accepted that they belong to the 5-HT_1 receptor class. Several studies have established the terminal 5-HT autoreceptors of the rat cortex to be of the 5-HT_{1B} type and there was indeed a highly significant correlation between the potencies of drugs for the rat autoreceptors and their affinities at 5-HT_{1B} binding sites [3,18]. On the other hand, it has become clear that 5-HT autoreceptors from other species (rabbit, guinea pig, pig, man) are not of the 5-HT_{1B} subtype [12, 16-18]. These autoreceptors may instead be of the 5-HT_{1D} type, since the potencies of a variety of agonists and antagonists at the 5-HT_{1D} receptor mediating inhibition of adenylate cyclase correlated very significantly with their effects on [^{3}H]5-HT release in pig cortex slices [16]. Similar findings have been reported in guinea pig and rabbit brain [17,18]. These studies strongly suggest that the terminal 5-HT autoreceptor is of the 5-HT_{1D} type in pig, guinea pig, human, and rabbit brain, although receptor heterogeneity cannot be excluded [52, 53]. 5-HT_{1B} heteroreceptors modulating acetylcholine release in rat brain have been described. In the periphery, inhibition of noradrenaline release was assigned to 5-HT_{1B} receptors in rat, but 5-HT_{1D} receptors in human saphenous vein [54,55].

Other functional correlates for these receptors have been described in vascular tissues: there is strong evidence that plasma extravasation produced by stimulation of the trigeminal ganglion and inhibited by 5-HT receptor agonists, is mediated by 5-HT_{1B} receptors in rats and probably by 5-HT_{1D} receptors in guinea pig [56-60]: indeed, these effects are produced by nonselective $5\text{-HT}_{1B}/5\text{-HT}_{1D}$ receptor agonists (5-CT, sumatriptan, ergotamine, DHE, 5-benzyloxytryptamine) in both species; by contrast, the selective 5-HT_{1B} receptor agonist CP 93129 is active in rat and inactive in guinea pig.

Endothelium-dependent relaxation in the pig coronary artery is mediated by 5-HT_{1D} receptors, based on the rank order of potency of a variety of agonists and antagonists [61]. Similarly, Gupta [62] reported on a relaxing 5-HT_{1D} receptor in guinea pig jugular vein. In the guinea pig iliac artery, a 5-HT_{1D}-like receptor which mediates contraction has also been described [63,64]. Similarly, 5-HT_{1D} or 5-HT_{1D}-like receptors mediating contraction have been reported in the porcine basilar artery, in human pial arterioles, in canine coronary artery, canine, rabbit, and human saphenous vein [65-68].

Few central effects have been convincingly demonstrated to be mediated by these receptors; the major difficulty comes from the quasi-absence of selective antagonists and the poor brain penetration of the few selective ligands, which have only recently become available. The 5-HT_{1B} receptor agonist RU 24969 has clear effects on locomotion; the hyperlocomotor activity produced by RU 24969 can be antagonized by propranolol. Penile erection in rat appears to involve, at least in part, 5-HT_{1B} receptors; similarly, hypophagia may be due to activation of 5-HT_{1B} receptors, although in both cases a 5-HT_{1C} receptor-mediated component may be possible. Higgins et al. [69] recently reported on a 5-HT_{1D} receptor-mediated turning behavior in guinea pig, produced by centrally administered 5-HT_{1D} receptor agonists, e.g. 5-CT and sumatriptan. These effects were susceptible to blockade by metergoline and methiothepin.

6. Outlook

It seems clear that 5-HT_{1B} receptor-mediated effects are due to activation of the $5\text{-HT}_{1D\beta}$ gene product in rodents. On the other hand, although the rat $5\text{-HT}_{1D\alpha}$ receptor has been cloned and 5-HT_{1D}-like binding was found in rat brain, there are at present no convincing reports about 5-HT_{1D} receptor-mediated functional correlates in rat. Similarly, it is difficult to decide whether functional effects reported earlier to be mediated by 5-HT_{1D} receptors in higher species are due to activation of the $5\text{-HT}_{1D\alpha}$ or $5\text{-HT}_{1D\beta}$ gene products. However, the combination of *in situ* hybridization, cloning work, and radioligand binding suggest that the expression levels of $5\text{-HT}_{1D\alpha}$ gene products seem to be significantly lower than those of $5\text{-HT}_{1D\beta}$ gene products. This is clear in rat brain, but it remains to be seen whether or not the situation in nonrodent and human brain is similar. However, now that even more receptors have been cloned, e.g. 5-HT_{1F} and 5-HT_5 [34-36] which have relatively high affinity for compounds such as sumatriptan, it has to be realized that effects which have been suggested to be mediated by 5-HT_{1D} and 5-HT_1-like receptors [see 20], may in fact be the result of the activation of these new receptors. These questions may be ultimately answered once ligands selective for either of these receptors will become available.

7. References

1. Hoyer, D., Engel, G., and Kalkman, H.O. (1985) 'Characterization of the 5-HT_{1B} recognition site in rat brain: binding studies with [^{125}I]iodocyanopindolol', Eur. J. Pharmacol. 118, 1-12.
2. Pazos, A. and Palacios, J.M. (1985) 'Quantitative autoradiographic mapping of serotonin receptors in the rat brain. I. Serotonin-1 receptors', Brain Res. 346, 205- 230.
3. Engel, G., Göthert, M., Hoyer, D., Schlicker, E., and Hillenbrand, K. (1986) 'Identity of inhibitory presynaptic 5-hydroxytryptamine (5-HT) autoreceptors in the rat brain cortex with 5-HT_{1B} binding sites', Naunyn Schmiedeberg's Arch. Pharmacol. 332, 1-7.
4. Hoyer, D., Engel, G., and Kalkman, H.O. (1985) 'Molecular pharmacology of 5-HT_1 and 5-HT_2 recognition sites in rat and pig brain membranes: radioligand binding studies with [^{3}H]5-HT, [^{3}H]8-OH-DPAT, (-)[^{125}I]iodocyanopindolol, [^{3}H]mesulergine and [^{3}H]ketanserin', Eur. J. Pharmacol. 118, 13-23.
5. Hoyer, D., Pazos, A., Probst, A., and Palacios, J.M. (1986) 'Serotonin receptors in the human brain. I: Characterization and autoradiographic localization of 5-HT_{1A} recognition sites. Apparent absence of 5-HT_{1B} recognition sites', Brain Res. 376, 85-96.
6. Heuring, R.E. and Peroutka, S.J. (1987) 'Characterization of a novel ^{3}H-5-hydroxytryptamine binding site subtype in bovine brain membranes', J. Neurosci. 7, 894-903.
7. Hoyer, D., Waeber, C., Pazos, A., Probst, A., and Palacios, J.M. (1988) 'Identification of a 5-HT_1 recognition site in human brain membranes different from 5-HT_{1A}, 5-HT_{1B} and 5-HT_{1C} sites', Neurosci. Letts, 85, 357-362.

8. Waeber, C., Dietl, M.M., Hoyer, D., Probst, A., and Palacios, J.M. (1988) 'Visualization of a novel serotonin recognition site (5-HT$_{1D}$) in the human brain by autoradiography', Neurosci. Lett. 88, 11-16.
9. Waeber, C., Dietl, M.M., Hoyer, D., and Palacios, J.M. (1989) '5-HT$_1$ receptors in the vertebrate brain: regional distribution examined by autoradiography', Naunyn Schmiedeberg's Arch. Pharmacol. 340, 486-494.
10. Waeber, C., Schoeffter, P., Palacios, J.M., and Hoyer, D. (1988) 'Molecular pharmacology of 5-HT$_{1D}$ recognition sites: radioligand binding studies in human, pig and calf brain membranes', Naunyn Schmiedeberg's Arch. Pharmacol. 337, 595-601.
11. Waeber, C., Schoeffter, P., Palacios, J.M., and Hoyer, D. (1989) '5-HT$_{1D}$ receptors in the guinea-pig and pigeon brain: radioligand binding and biochemical studies', Naunyn Schmiedeberg's Arch. Pharmacol. 340, 479-485.
12. Waeber, C., Schoeffter, P., Hoyer, D., and Palacios, J.M. (1990) 'The serotonin 5-HT1D receptor: A progress Review', Neurochem. Res., 15, 567-582.
13. Bouhelal, R., Smounya, L., and Bockaert, J. (1988) '5-HT$_{1B}$ receptors are negatively coupled with adenylate cyclase in rat substantia nigra', Eur. J. Pharmacol. 151, 189-196.
14. Hoyer, D. and Schoeffter, P. (1988) '5-HT$_{1D}$ receptor-mediated inhibition of forskolin-stimulated adenylate cyclase activity in calf substantia nigra', Eur. J. Pharmacol. 147, 145-147.
15. Schoeffter, P., Waeber, C., Palacios, J.M., and Hoyer, D. (1988) 'The serotonin 5-HT$_{1D}$ receptor subtype is negatively coupled to adenylate cyclase in calf substantia nigra', Naunyn Schmiedeberg's Arch. Pharmacol. 337, 602-608.
16. Schlicker, E., Fink. K,, Göthert, M., Hoyer, D., Molderings, G., Roschke, I., and Schoeffter, P. 'The pharmacological properties of the (1989). Presynaptic 5-HT autoreceptor in the pig brain cortex conform to the 5-HT$_{1D}$ receptor subtype', Naunyn Schmiedeberg's Arch. Pharmacol. 340, 45-51.
17. Middlemiss, D.N., Bremer, M.E., and Smith, S.M. (1988) 'A pharmacological analysis of the 5-HT receptors mediating inhibition of 5-HT release in the guinea-pig frontal cortex', Eur. J. Pharmacol. 157, 101-107.
18. Limberger, N., Deicher, R., and Starke, K. (1991) 'Species differences in presynaptic serotonin autoreceptors: mainly 5-HT$_{1B}$ but possibly in addition 5-HT$_{1D}$ in the rat, 5-HT$_{1D}$ in the rabbit and guinea-pig brain cortex', Naunyn Schmiedeberg's Arch. Pharmacol. 343, 353-364.
19. Hoyer, D. and Middlemiss, D.N. (1989) 'The pharmacology of the terminal 5-HT autoreceptors in mammalian brain: evidence for species differences', Trends Pharmacol. Sci. 10, 130-132.
20. Humphrey, P.P.A., Hartig, P.R., and Hoyer, D. (1993) 'A re-appraisal of 5-HT receptor classification', in Serotonin from cell biology to pharmacology and therapeutics, P.M. Vanhoutte, R. Paoletti, and P.R. Saxena, (eds.)

Kluwer Academic Publishers, Dordrecht, in press.
21. Libert, F., Parmentier, M., Lefort, A., Dinsart, C., Van Sande, J., Maenhaut, C., Simons, M.J., Dumont, J.E., and Vassart, G. (1989) 'Selective amplification and cloning of four new members of the G-protein-coupled receptor family', Science 244, 569-572.
22. Hamblin, M.W. and Metcalf, M.A. (1991) 'Primary structure and functional characterization of a human 5-HT_{1D} type serotonin receptor', Mol. Pharmacol. 40, 143-148.
23. Weinshank, R.L., Zgombick, J.M., Macchi, M.J., Branchek, T.A., and Hartig, P.R. (1992) 'Human serotonin 1D receptor is encoded by a subfamily of two distincts genes: 5-$HT_{1D\alpha}$ and 5-$HT_{1D\beta}$', Proc Natl Acad Sci USA, 89, 3630-3634.
24. Maenhaut, C., Van Sande, J., Massart, C., Dinsart, C., Libert, F., Monferini, E., Giraldo, E., Ladinsky, H., Vassart, G., and Dumont, J.E. (1991) 'The orphan receptor cDNA RDC4 encodes A 5-HT 1D serotonin receptor', Biochem. Biophys. Res. Comm. 180, 1460-1468.
25. Zgombick, J.M., Weinshank, R.L., Macchi, M., Schechter, L.E., Branchek, T.A., and Hartig, P.L. (1991) 'Expression and pharmacological characterization of a canine 5-hydroxytryptamine$_{1D}$ receptor subtype', Mol. Pharmacol. 40, 1036-1042.
26. Hamblin, M.W., McGuffin, R.W., Metcalf, M.A., Dorsa, D.M., and Merchan, K.M. (1992). 'Distinct 5-HT_{1B} and 5-HT_{1D} serotonin receptors in rat: structural and pharmacological comparison of the two cloned receptors', Moll. Cell. Neurosci. 3, 578-587.
27. Jin, H., Oksenberg, D., Askenazi, A., Peroutka, S.J., Duncan, A.M.V., Rozmahel, R., Yang, Y., Mengod, G., Palacios, J.M., O'Dowd, B.F. (1992) 'Characterization of the human 5-hydroxytryptamine$_{1B}$ receptor', J. Biol. Chem. 267, 5735-5738.
28. Hartig, P.R., Branchek, T.A., and Weinshank, R.L. (1992) 'A subfamily of serotonin 5-HT_{1D} receptor genes', TIPS 13, 152-159.
29. Levy, F.O., Gudermann, T., Peres-Reyes, E., Birbaumer, M., Kaumann, A.J., and Birnbaumer, L. (1992) 'Molecular cloning of a human serotonin receptor (S12) with a pharmacological profile resembling that of the 5-HT_{1D} subtype', J. Biol. Chem. 267, 7553-7562.
31. Voigt, M.M., Laurie, D.J., Seeburg, H., and Bach, A. (1991) 'Molecular cloning and characterization of a rat brain cDNA encoding a 5-hydroxytryptamine 1B receptor', EMBO J. 10, 4017-4023.
32. Adham, N., Romanienko, P., Hartig, P., Weinshank, R.L., and Branchek, T. (1992) 'The rat 5-hydroxytryptamine$_{1B}$ receptor is the species homologue of the human 5-hydroxytryptamine$_{1D\beta}$ receptor', Mol. Pharmacol. 41, 1-7.
33. Maroteaux, L., Saudou, F., Amlaiky, N., Boschert, U., Plassat, J.L., and Hen, R. (1992) 'The mouse 5HT_{1B} serotonin receptor : cloning, functional expression and localisation in motor control centers', Proc. Natl. Acad. Sci.

USA 89, 3020-3024.
34. Amlaiky, N., Ramboz, S., Boschert, U., Plassat, J.L., and Hen, R. (1992) 'Isolation of a mouse "5HT1E-like" serotonin receptor expressed predominantly in hippocampus', J. Biol. Chem. 267, 19761-19764.
35. Adham, N., Kao, H.T., Schechter, L., Bard, J., Olsen, M., Urquhart, D., Durkin, M., Hartig, P., Weinshank, R., and Branchek, T. (1993) 'Cloning of another human serotonin receptor (5-HT$_{1F}$): a fifth 5-HT$_1$ receptor subtype coupled to the inhibition of adenylate cyclase', Proc. Natl. Acad. Sci. USA 90, 408-412.
36. Plassat, J.L., Boschert, U., Amaliky, N., and Hen, R. (1992) 'The mouse 5-HT$_5$ receptor reveals a remarkable heterogeneity within the 5-HT$_{1D}$ receptor family', EMBO J. 11, 4779-4786.
37. Schoeffter, P. and Hoyer, D. (1989) 'Interactions of arylpiperazines with 5-HT$_{1A}$, 5-HT$_{1B}$, 5-HT$_{1C}$ and 5-HT$_{1D}$ receptors: do discriminatory 5-HT$_{1B}$ ligands exist', Naunyn Schmiedeberg's Arch. Pharmacol. 339, 675-683.
38. Macor, J.E., Burkhart, C.A., Heym, J.H., Ives, J.L., Lebel, L.A., Newman, M.E., Nielsen, J.A., Ryan, K., Schulz, D.W. et al., (1990) '3-1 2 5 6 tetrahydropyrid-4-ylpyrrolo-3 2-B-pyrid-5-one a potent and selective serotonin 5-HT-1B agonist and rotationally restricted phenolic analogue of 5 methoxy-3-1 2 5 6-tetrahydropyrid-4-ylindole', J. Med. Chem. 33, 2087-2093.
39. Schoeffter, P. and Hoyer, D. (1989) 'How selective is GR 43175? interactions with functional 5-HT$_{1A}$, 5-HT$_{1B}$, 5-HT$_{1C}$ and 5-HT$_{1D}$ receptors', Naunyn Schmiedeberg's Arch. Pharmacol. 340, 135-138.
40. Peroutka, S.J. and McCarthy, B.G. (1989) 'Sumatriptan (GR 43175) interacts selectively with 5-HT$_{1B}$ and 5-HT$_{1D}$ sites', Eur. J. Pharmacol. 163, 133-136.
41. Waeber, C., Hoyer, D., and Palacios, J.M. (1989) 'GR 43175: A preferential 5-HT(1D) agent in monkey and human brains as shown by autoradiography', SYNAPSE (USA) 4, 168-170.
42. Leonhardt, S., Herrick-Davis, K., and Titeler, M. (1989) 'Detection of a novel serotonin receptor subtype (5-HT$_{1E}$) in human brain: interaction with a GTP-binding protein', J. Neurochem. 53, 465-471.
43. Sumner, M.J. and Humphrey, P.P.A. (1989) '5-HT-1D binding sites in porcine brain can be subdivided by GR-43175', Br. J. Pharmacol. 98, 29-31.
44. Boulenguez, P., Chauveau, J., Segu, L., Morel, A., Lanoir, J., and Delaage, M. (1991) 'A new 5-hydroxy-indole derivative with preferential affinity for 5-HT$_{1B}$ binding sites', Eur. J. Pharmacol. 194, 91-98.
45. Palacios, J.M., Waeber, C., Bruinvels, A.T., and Hoyer, D. (1992) 'Direct visualization of serotonin1D receptors in the human brain using a new iodinated ligand', Molec. Brain Res. 13, 175-179.
46. Bruinvels, A.T., Landwehrmeyer, B., Waeber, C., Palacios, J.M., and Hoyer, D. (1991) 'Homogeneous 5-HT$_{1D}$ recognition sites in the human substantia nigra identified with a new iodinated radioligand', Eur. J. Pharmacol. 202,

89-91.
47. Bruinvels, A.T., Lery, H., Nozulak, J., Palacios, J.M., and Hoyer, D. (1992) '5-HT_{1D} binding sites in various species: similar pharmacological profile in calf, guinea pig, dog, monkey and human brain membranes', Naunyn Schmiedeberg's Arch. Pharmacol. 346, 243-248.
48. Bruinvels, A.T., Palacios, J.M., and Hoyer, D. (1993) 'Autoradiographic characterization and localization of 5-HT_{1D} compared to 5-HT_{1B} binding sites in rat brain', Naunyn Schmiedeberg's Arch. Pharmacol., in press.
49. Xiong, W.C. and Nelson, D.L. (1989) 'Characterization of a [^{3}H]-5-hydroxytryptamine binding site in rabbit caudate nucleus that differs from the 5-HT_{1A}, 5-HT_{1B}, 5-HT_{1C} and 5-HT_{1D} subtypes', Life Sci. 45, 1433-1442.
50. Hoyer, D., Lery, H., Waeber, C., Bruinvels, A.T., Nozulak, J., and Palacios, J.M. (1992) '5-HT_{1R}' or 5-HT_{1D} binding? Evidence for 5-HT_{1D} binding in rabbit brain', Naunyn Schmiedeberg's Arch. Pharmacol. 346, 249-254.
51. Schoeffter, P. and Hoyer, D. (1989) '5-Hydroxytryptamine $5HT_{1B}$ and 5-HT_{1D} receptors mediating inhibition of adenylate cyclase activity. Pharmacological comparison with special reference to the effects of yohimbine, rauwolscine and some ß-adrenoceptor antagonists', Naunyn Schmiedeberg's Arch. Pharmacol. 340, 285-292.
52. Mahle, C.D., Nowak, H.P., Mattson, R.J., Hurt, S.D., and Yocca, F.D. (1991) '[^{3}H]5-Carboxamidotryptamine labels multiple high affinity 5-HT_{1D}-like sites in guinea pig brain', Eur. J. Pharmacol. 205, 323-324.
53. Wilkinson, L.O. and Middlemiss, D.N. (1992) 'Metitepine distinguishes two receptors mediating inhibition of [^{3}H]5-hydroxytryptamine release in guinea pig hippocampus', Naunyn Schmiedeberg's Arch. Pharmacol. 345, 696-699.
54. Molderings, G.J., Fink, K., Schlicker, E., and Göthert, M. (1987) 'Inhibition of noradrenaline release in the rat vena cava via presynaptic 5-HT_{1B} receptors', Naunyn Schmiedeberg's Arch. Pharmacol. 336, 245-250.
55. Molderings, G.J., Werner, K., Likungu, J., and Gothert, M. (1990) 'Inhibition of noradrenalin release from the sympathetic nerves of the human saphenous vein via presynaptic 5-HT receptors similar to the 5-HT(1D) subtype', Naunyn Schmeideberg's Arch. Pharmacol. 342, 371-377.
56. Saito, K., Maskowitz, S., and Moskowitz, M.A. (1988) 'Ergot alkaloids block neurogenic extravasation in dura mater. Proposed action in vascular headache', Ann. Neurol. 24, 732-737.
57. Buzzi, M.G. and Moskowitz, M.A. (1990) 'The antimigraine drug sumatriptan GR-43175 selectively blocks neurogenic plasma extravasation from blood vessels in dura mater', Br. J. Pharmacol. 99, 202-206.
58. Buzzi, M.G., Carter, W.B., Shimizu, T., Heath, H., and Moskowitz, M.A. (1991a) 'Dihydroergotamine and sumatriptan attenuate the increase in plasma CGRP levels within rat superior sagittal sinus during electrical trigeminal ganglion stimulation', Neuropharmacol. 5, 5-18.

59. Buzzi, M.G., Dimitriadou, V., Theoharides, T.C., and Moskowitz, M.A. (1992) '5-hydroxytryptamine receptor agonists for the abortive treatment of vascular headaches block mast cell, endothelial and platelet activation within the rat dura mater after trigeminal stimulation', Brain Res. 583, 137-149.
60. Nozaki, K., Moskowitz, M.A., and Boccalini, P. (1992) 'CP-93,129, sumatriptan, dihydroergotamine block c-FOS expression within rat trigeminal nucleus caudalis caused by chemical stimulation of the meninges', Br. J. Pharmacol. 106, 409-415.
61. Schoeffter, P. and Hoyer, D. (1990) '5-hydroxytryptamine (5-HT) induced endothelium-dependent relaxation of pig coronary arteries is mediated by 5-HT receptors similar to the 5-HT$_{1D}$ receptor subtype', J. Pharmacol. Exp. Therap. 252, 387-395.
62. Gupta, P. (1992) 'An endothelial 5-HT receptor that mediates relaxation in guinea pig isolated jugular vein resembles the 5-HT$_{1D}$ subtype', Br. J. Pharmacol. 106, 703-709.
63. Sahin-Erdemli, I., Hoyer, D., Stoll, A., Seiler, M.P., and Schoeffter, P. (1991) '5-HT1-like receptors mediate 5-hydroxytryptamine-induced contraction of guinea-pig isolated iliac artery', Br. J. Pharmacol. 102, 386-390.
64. Schoeffter, P. and Sahin-Erdemli, I. (1992) 'Further characterization of the 5-hydroxytryptamine 5-HT$_1$-like receptor mediating contraction of the guinea pig iliac artery', Eur. J. Pharmacol. 219, 295-301.
65. Van Charldorp, K.J., Tulp, M.T.M., Hendriksen, B., Mons, H., Couwenberg, P., and Wouters, W. (1990) '5-HT-receptors in porcine basilar arteries closely resemble 5-HT(1D)-binding sites', Eur. J. Pharmacol. 183, 1106-1107.
66. Hamel, E. and Bouchard, E. (1991) 'Contractile 5-HT-1 receptors in human isolated pial aterioles correlation with 5-HT-1D binding sites', Br. J. Pharmacol. 102, 227-233.
67. Martin, G.R., Prentice, D.J., and MacLennan, S.J. (1991) 'The 5-HT receptor in rabbit saphenous vein', Pharmacological identity with the 5-HT$_{1D}$ recognition site?' Fund. Clin. Pharmacol. 54, 417.
68. Bax, W.A., Van Heuven-Nolsen, D., Bos, E., Simoons, M., and Saxena, P.R. (1992) '5-hydroxytryptamine-induced contractions of the human isolated saphenous vein: involvement of 5-HT$_2$ and 5-HT$_{1D}$-like receptors, and a comparison with grafted veins', Naunyn Schmiedeberg's Arch. Pharmacol. 345, 500-508.
69. Higgins, G.A., Jordon, C.C., and Skingle, M. (1991) 'Evidence that the unilateral activation of 5-HT(1D) receptor in the substantia nigra of the guinea-pig elicits contralateral rotation', Br. J. Pharmacol. 102, 305-310.

5-HT$_3$ RECEPTORS

G.J. KILPATRICK and H. ROGERS
Glaxo Group Research Ltd.
Ware, Hertfordshire SG12 0DP
UK

ABSTRACT. The 5-HT$_3$ receptor directly gates an ion-channel. This makes it unique among the receptors for 5-HT. This unique nature is reflected in some of the characteristics of this receptor. In this review we give an overview of the characteristics of the 5-HT$_3$ receptor focusing on the more interesting features.

1. Introduction

The 5-HT$_3$ receptor is unique among the receptors for 5-HT and other biogenic amines because it is not a single-subunit G-protein linked receptor. Rather, this receptor is a multi-subunit ligand-gated ion channel like the nicotinic acetylcholine receptor and GABA$_A$ receptor. The pace of research into the 5-HT$_3$ receptor has accelerated greatly over the last few years and much of the stimulus for this has come from the therapeutic utility of the selective 5-HT$_3$ receptor antagonists. Such compounds are now used as anti-emetics in cancer therapy and postoperative nausea and vomiting [1,2]. In addition, behavioral studies in animals and early clinical trials indicate that 5-HT$_3$ receptor antagonists may also be useful in the treatment of psychiatric disorders, such as cognitive dysfunction, anxiety, psychoses, and drug dependence [3]. In this review it is our intention to give an overview of the characteristics of the 5-HT$_3$ receptor, paying particular attention to some of the more interesting features in recent advances that have been made.

2. Pharmacology

We now have available to us a wide range of potent and selective 5-HT$_3$ receptor antagonists including MDL 72222, tropisetron, and ondansetron. Several thorough reviews have been written on the characteristics of the 5-HT$_3$ antagonists [3, 4]; because of this we do not intend to go into the details of these compounds here. However, some of the characteristics observed with the agonists that interact with the 5-HT$_3$ receptor are worthy of discussion. The range of selective and potent agonists available to us is more limited than that for the antagonists. Compounds such as 2-methyl-5-HT and the most potent agent, meta-chlorophenylbiguanide (mCPBG) are available and their study has revealed some unusual features of the 5-HT$_3$ receptor. Firstly, 5-HT$_3$ agonists are more potent as inhibitors of radioligand binding to the 5-HT$_3$ receptor than as agonists in functional assays (table 1 [5]). This is most pronounced for mCPBG which is some 100 times more potent to inhibit binding than it is functionally. The reasons for this discrepancy are not clear, but it is possible that the binding studies recognize a desensitized state of the receptor that is not important functionally. A second unusual feature of the agonists in the binding studies is reflected in Hill coefficients significantly above unity. This has been observed with several radioligands in brain tissue [6,7] but is less obvious in binding to peripheral tissues such as the rat vagus nerve and the ileum (table 2 [8,9]). The high Hill coefficients may reflect positive co-operativity in the binding of agonists to the receptor as has been observed functionally. The reason for the variation in Hill coefficients between tissues is less clear but it may reflect heterogeneity in the receptor (see also the later discussion on 5-HT$_3$ receptor subtypes).

TABLE 1. Agonist potencies to inhibit [^{3}H]GR67330 binding to rat entorhinal cortex and to depolarize the rat isolated vagus nerve.

Compound	[^{3}H]GR67330 binding IC$_{50}$(nM)	Rat isolated vagus nerve EC$_{50}$(nM)
5-HT	130 ± 26	460 ± 40
2-methyl-5-HT	207 ± 30	3200 ± 80
Phenylbiguanide	141 ± 9	900 ± 120
mCPBG	1.47 ± 0.07	50 ± 10

Results are mean ± sem of $\rangle$ 3 experiments [5].

It is often thought that the 5-HT$_3$ agonists have little use *in vivo* because of the likely rapid clearance from the body and because their hydrophilicity would mean that they would penetrate the brain poorly. However, recent studies that we have conducted would indicate that this may not be the case. *Ex vivo* binding experiments reveal that, surprisingly, 2-methyl-5-HT will penetrate the brain but, in particular, mCPBG seems to be potent in this respect (figure 1). Limited behavioral studies have been conducted with mCPBG but more are clearly warranted since the full utility of the 5-HT$_3$ agonists in this area may not have been fully recognized.

TABLE 2. Hill coefficients for 5-HT_3 receptor agonist inhibition of [^{3}H]GR67330 binding to rat entorhinal cortex and ileum.

Compound	Hill Number	
	Rat cortex	Rat ileum
5-HT	1.51 ± 0.07	1.30 ± 0.02
2-methyl-5-HT	1.71 ± 0.04	1.23 ± 0.08
Phenylbiguanide	1.74 ± 0.04	1.14 ± 0.28
mCPBG	1.67 ± 0.23	1.15 ± 0.05

Results are the mean ± sem ⟩ 3 experiments [9].

The subject of heterogeneity of the 5-HT_3 receptor has been an area of debate for several years. Fozard [10] and Richardson et al. [11] were the first to address this issue but the range of compounds and preparations available to them were somewhat limited. A more recent study by Butler et al. [12] has examined the potency of three agonists and eleven antagonists in several 5-HT_3 receptor containing preparations. The conclusion of this study and that of others [13,14] is that there is clear heterogeneity in the 5-HT_3 receptor but at present this can only be observed when comparing between species.

However, it seems likely that in the near future subtypes of the 5-HT_3 receptor resulting from assembly of different subunits in different ways are likely to be identified, as with other members of the multi-subunit ligand-gated ion channel family.

3. Distribution

The 5-HT_3 receptor was long described as a 'peripheral neuronal receptor' because effects had only been observed in peripheral neurones. This description is now clearly incorrect. It is still likely that this is an exclusively neuronal receptor but the availability of radioligands had enabled the identification of 5-HT_3 receptors in the brain [15]. The distribution has been examined using homogenate binding and autoradiographic techniques [16]. It is clear that the density of 5-HT_3 receptors in the brain is lower than that for many other 5-HT receptors but high levels have been identified particularly in the hind brain. The densest levels are observed in the nucleus tractus solitarius, an area of the brain involved, amongst other things, in the vomiting reflex and indeed the antiemetic effects of 5-HT_3 receptor antagonists may be partially mediated in this area [17]. The forebrain distribution varies somewhat with species but the highest levels are routinely detected in limbic areas including the hippocampus and nucleus accumbens. In peripheral tissues the highest levels of receptors, as identified using radioligand binding, can be identified in tissues such as the ileum and the vagus nerve [8,9].

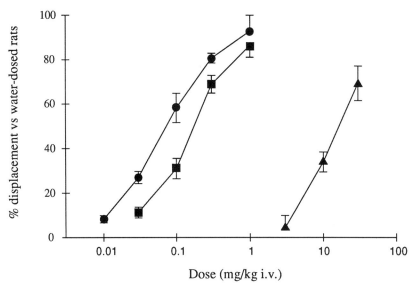

COMPOUND	t 1/2 min	ID_{50} (mg kg^{-1} i.v.)
▲ 2-methyl-5-HT	14	16
● mCPBG	50	0.08
■ GR38032	55	0.16

Figure 1. Inhibition of *ex vivo* [^{3}H]GR65630 binding. Rats (n=4-8/group) were dosed IV with drug and killed with an overdose of pentobarbitone 5 minutes later. The entorhinal cortex was removed from each rat and the specific binding of ^{3}H-GR65630 determined. Results are expressed as the mean percentage inhibition of binding relative to water-dosed rats sem. The t½ values were calculated from time vs displacement profile for a single dose of each compound and the ID_{50} values from the dose-effect curves.

4. Electrophysiology

5-HT$_3$ receptors mediate some of the excitatory effects of 5-HT. The responses observed have often been described as nicotine-like, ie. a rapid response followed by a rapid desensitization. Indeed, the electrophysiological work has been very important in the identification of the 5-HT$_3$ receptor as a ligand-gated ion channel [18]. Immortal cell lines, including NG108-15, N1E-115, and NCB20 cells, have proved a useful source of 5-HT$_3$ receptors for electrophysiological study and indeed many papers have been published examining 5-HT$_3$ receptors in these cells [19]. In terms of body tissues, responses mediated by 5-HT$_3$ receptors have been identified in peripheral neurones including the guinea pig myenteric plexus and nodose ganglion and, more recently, in rat brain tissue.

The effects observed in brain tissue are worthy of particular mention. Yakel and Jackson [20] were the first to identify 5-HT$_3$-mediated responses in brain tissue when they conducted studies in dissociated hippocampal cells in primary culture. More recently Ropert and Guy [21] and Sugita et al. [22] have observed 5-HT$_3$ mediated responses in slices of rat hippocampus and amygdala, respectively. The study of Ropert and Guy described 5-HT$_3$ receptors on hippocampal GABA interneurones where the application of a 5-HT$_3$ agonist caused depolarization of these cells and the release of GABA which could be measured as IPSPs in hippocampal pyramidal cells. Sugita and colleagues also saw an excitatory response. Examining amygdaloid neurones, in the presence of antagonists of glutamate and GABA receptors, they were able to observe an EPSP that was blocked by selective 5-HT$_3$ antagonists and mimicked by 5-HT. 5-HT$_3$ receptor-mediated responses have also been observed using *in vivo* electrophysiology. For example, Ashby et al. [23] have shown that 5-HT$_3$ agonists inhibit the firing of cells in the medial prefrontal cortex. The responses observed by Ashby and colleagues were slow in onset and inhibitory and thus not consistent with the rapid excitatory effects expected of the opening of the ligand-gated ion channel. However, it is possible that this response is mediated indirectly, i.e. by the stimulation of inhibition of the release of another transmitter.

Further electrophysiological studies are clearly required to elucidate the full role of 5-HT$_3$ receptors particularly in the brain. Nevertheless, significant advances have been made over the last few years.

5. Biochemistry/Molecular Biology

At a neurochemical level the main effects that have been observed to be mediated by 5-HT$_3$ receptors are the modulation of transmitter release. What is surprising however is the number of transmitters whose release is modulated by 5-HT$_3$ receptors, including dopamine, CCK, noradrenaline, acetylcholine, and 5-HT itself [24]. Indeed this is even more surprising when one considers that most of these types of release experiments require prolonged (minutes) exposure to the agonist. Since the 5-HT$_3$ receptor desensitizes very quickly, such effects must be mediated largely by residual responses which remain after desensitization. It is likely that more effects and indeed more profound effects may be observed if more rapid ways of analyzing transmitter release are employed.

The molecular characterization of the 5-HT$_3$ receptor has been reviewed in detail recently [25]; therefore, we will not go into great detail here. The receptor has been solubilized from several sources and more recently purified, but perhaps the greatest advance in this field came at the end of 1991 when Maricq and colleagues

[26] reported the cloning of a subunit of the $5\text{-}HT_3$ receptor. This subunit contains the 5-HT binding site which has similar pharmacology to the native receptor, as assessed using radioligand binding. The homo-oligomer is able to form a functional ion channel in xenopus oocytes which shares many of the characteristics of the native receptor, i.e. a rapidly opening and closing ion channel that conducts cations. One unusual feature of this cloned receptor was an apparent voltage-dependent magnesium and calcium blockade, akin to the voltage-dependent magnesium blockade of the NMDA receptor. This has not been observed in studies with the native receptor and one wonders whether this may be because we are dealing with a single subunit in this case and the native receptor has other subunits which are required for full functioning. In fact other members of the ligand-gated ion channel receptor family have multiple subunits that can be assembled in different ways to produce pharmacologically and functionally distinct receptors. It therefore seems unlikely that the $5\text{-}HT_3$ receptor has only one type. Undoubtedly the study of Maricq will stimulate the search for other subunits which may lead to the identification of receptor subtypes that have not been detected using conventional pharmacological means.

6. Conclusion

Research over the last few years has revealed the unique nature of the $5\text{-}HT_3$ receptor and spawned the development of selective $5\text{-}HT_3$ receptor antagonists which clearly have therapeutic utility. Nevertheless further study is still required to elucidate the full physiological and pathophysiological role of the $5\text{-}HT_3$ receptor. Of particular interest is whether subtypes of the $5\text{-}HT_3$ receptor may exist and therefore whether it may be possible to identify further novel therapeutic agents.

7. References

1. Bunce, K., Tyers, M.B., and Beranek, P. (1991) 'Clinical evaluation of $5\text{-}HT_3$ receptor antagonists as anti-emetics', Trends in Pharmacol. Sci. 12, 46-48.
2. Leeser, J. and Lip, H. (1991) 'Prevention of postoperative nausea and vomiting using ondansetron, a new selective $5\text{-}HT_3$ receptor antagonist', Anaesth. Analg. 76, 751-755.
3. Costall, B., Naylor, R.J., and Tyers, M.B. (1990) 'The psychopharmacology of $5\text{-}HT_3$ receptors', Pharmac. Ther. 47, 181-202.
4. Kilpatrick, G.J., Bunce, K.T., and Tyers, M.B. (1990) '$5\text{-}HT_3$ receptors', Med. Res. Rev. 10, 441-475.
5. Kilpatrick, G.J., Butler, A., Burridge, J., and Oxford, A.W. (1990) '1-(m-chlorophenyl)-biguanide, a potent high affinity $5\text{-}HT_3$ receptor agonist', Eur. J. Pharmacol. 182, 193-197.
6. Barnes, J.M., Barnes, N.M., Costall, B., Jagger, S.M., Naylor, R.J.,

Robertson, D.W., and Roe, S.Y. (1992) 'Agonist interactions with 5-HT$_3$ receptor recognition sites in the rat entorhinal cortex labelled by structurally diverse radioligands', Br. J. Pharmacol. 105, 500-504.
7. Kilpatrick, G.J., Butler, A., Hagan, R.M., Jones, B.J., and Tyers, M.B. (1990) '[3H]GR67330 a very high affinity ligand for 5-HT$_3$ receptors', Naunyn-Schmeideberg's Arch. Pharmacol. 342, 22-30.
8. Kilpatrick, G.J., Barnes, N.M., Cheng, C.H.K., Costall, B., Naylor, R.J., and Tyers, M.B. (1991) 'The pharmacological characterization of 5-HT$_3$ receptor binding sites in rabbit ileum: comparison with those in rat ileum and brain', Neurochem. Int. 19, 389-396.
9. Kilpatrick, G.J., Jones, B.J., and Tyers, M.B. (1989) 'Binding of the 5-HT$_3$ ligand, [3H]GR65630, to rat area postrema, vagus nerve and the brains of several species', Eur. J. Pharmacol. 159, 157-164.
10. Fozard, J.R. (1983) 'Differences between receptors for 5-hydroxytryptamine on autonomic neurones revealed by Nor(-)cocaine', J Auton. Pharmac. 3, 21-16.
11. Richardson, B.P., Engel, G., Donatsch, P., and Stadler, P.A. (1985) 'Identification of serotonin M-receptor subtypes and their specific blockade by a new class of drugs' Nature 316, 126-131.
12. Butler, A., Elswood, C.J., Burridge, J., Ireland, S.J., Bunce, K.T., Kilpatrick, G.J., and Tyers, M.B. (1990) 'The pharmacological characterization of 5-HT$_3$ receptors in three isolated preparations derived from guinea-pig tissues', Br.J. Pharmacol. 101, 591-598.
13. Malone, H.M., Peters, J.A., and Lambert, J.J. (1991) 'Physiological and pharmacological properties of 5-HT$_3$ receptors - A patch clamp study', Neuropeptides 19 (Suppl.), 25-30.
14. Newberry, N.R., Cheshire, S.H., and Gilbert, M.J. (1991) 'Evidence that the 5-HT$_3$ receptors of the rat, mouse and guinea-pig superior cervical ganglion may be different', Br. J. Pharmacol. 102, 615-620.
15. Kilpatrick, G.J., Jones, B.J., and Tyers, M.B. (1987) 'Identification and distribution of 5-HT$_3$ receptors in rat brain using radioligand binding', Nature 330, 746-748.
16. Laporte, A.M., Kidd, E.J., Vergé, P., Gozlan, H., and Hamon, M. (1992) 'Autoradiographic mapping of central 5-HT$_3$ receptors', in M. Hamon (ed.), Central and Peripheral 5-HT$_3$ Receptors, Academic Press, New York, pp. 157-188.
17. Higgins, G.A., Kilpatrick, G.J., Bunce, K.T., Jones, B.J., and Tyers, M.B. (1989) '5-HT$_3$ receptor antagonists injected into the area postrema inhibit cisplatin-induced emesis in the ferret', Br. J. Pharmacol. 97, 247-255.
18. Derkach, V., Surprenant, A., and North, R.A. (1989) '5-HT$_3$ receptors are membrane ion channels', Nature 339, 706-709.
19. Peters, J.A. and Lambert, J.J. (1989) 'Electrophysiology of 5-HT$_3$ receptors in neuronal cell lines', Trends in Pharmacol. Sci. 10, 172-175.

20. Yakel, J.L., Trussell, L.O., and Jackson, M.B. (1988) 'Three serotonin responses in cultured mouse hippocampal and striatal neurons', J. Neurosci. 8, 1273-1285.
21. Ropert, N. and Guy, N. (1991) 'Serotonin facilitates GABAergic transmission in the CA1 region of rat hippocampus in vitro', J. Physiol. 441, 121-136.
22. Sugita, S., Shen, K.Z., and North, R.A. (1992) '5-hydroxytryptamine is a fast excitatory transmitter at 5-HT_3 receptors in rat amygdala', Neuron. 8, 199-203.
23. Ashby, C.R., Edwards, E., Harkins, K., and Wang, R.Y. (1989) 'Characterization of 5-HT_3 receptors in the medial prefrontal cortex: a microiontophoretic study', Eur. J. Pharmacol. 173, 193-196.
24. Kilpatrick, G.J. and Tyers, M.B. (1992) 'The pharmacological properties and functional roles of central 5-HT_3 receptors', in M. Hamon, (ed.), Central and Peripheral 5-HT_3 Receptors, Academic Press, New York, pp. 33-57.
25. McKernon, R.M. (1992) 'Biochemical properties of the 5-HT_3 receptor', in M. Hamon, (ed), Central and Peripheral 5-HT_3 Receptors, Academic Press, New York, pp. 89-102.
26. Maricq, A.V., Peterson, A.S., Brake, A.J., Myers, R.M., and Julius, D. (1991) 'Primary structure and functional expression of the 5-HT_3 receptor, a serotonin-gated ion channel', Science 254, 432-437.

5-HT$_4$ RECEPTOR: CURRENT STATUS

DAVID E. CLARKE and *JOEL BOCKAERT
Institute of Pharmacology
Syntex Research
Palo Alto, California 94304
USA, and
*Centre CNRS-INSERM de Pharmacologie-Endocrinologie
34094, Montpellier Cedex 5
France

ABSTRACT. The 5-HT$_4$ receptor is reviewed with regard to historical development, distribution, function, pharmacology, and therapeutic significance. Recent information on agonists and antagonists is included, along with mechanistic data relating to 5-HT$_4$ receptor desensitization.

1. Introduction

The 5-HT$_4$ receptor was first described by Bockaert and co-workers in mouse and guinea pig brain [1; see 2-4 for background] followed by its definition in guinea pig ileum [5,6; see 7-9 for background], human heart [10,11], and porcine heart [12-14; see 15-17 for background]. The 5-HT$_4$ receptor is resistant to blockade by antagonists at 5-HT$_1$, 5-HT$_2$, and 5-HT$_3$ receptors and exhibits a unique profile of agonism by substituted benzamides (e.g. cisapride, renzapride) and benzimidazolones (e.g. BIMU 1 and BIMU 8), as well as by 5-HT and certain related idoleamines (e.g. 5-methoxytryptamine; 5-MOT).

An account of early findings on the 5-HT$_4$ receptor [18] and more recent reviews [19,20] have been published.

2. Distribution, Function, and Significance of the 5-HT$_4$ Receptor

The 5-HT$_4$ receptor has now been identified in a wide variety of tissues and species (table 1). Functionally, the receptor is coupled, via G$_S$ to adenylyl cyclase and cyclic AMP production [1] and through a cyclic AMP-dependent protein kinase serves to close voltage-dependent potassium channels in mouse colliculi neurons [21; but see 22] and increase calcium current through voltage-sensitive calcium channels in heart [11,23].

In neurons, closure of potassium channels by 5-HT$_4$ receptor activation would be expected to lead to depolarization, followed by calcium entry and neurotransmitter release. However, direct release of neurotransmitter by 5-HT_4 agonism has been demonstrated only for acetylcholine (ACh) in guinea pig ileum [24,25]; a tissue in which 5-HT_4 receptor-mediated increases in cyclic AMP have yet to be demonstrated. Evidence for involvement of ACh in the CNS is tentative and indirect [26,27] and, to date, no specific neurochemical substrate to 5-HT_4 receptor ligands. In this regard, it is worth noting that a cyclic AMP-linked 5-HT receptor system operates in *Aplysia* for long-term memory storage [28]. Recently, this system has been linked with neuronal growth and synapse formation [29]. The mechanism appears related to endocytotic activation and redistribution of membrane components, including cell adhesion molecules. Thus, it is reasonable to speculate that 5-HT_4 receptor activation may evoke long-term changes in neuronal plasticity, and research which measures immediate, short-term responses may overlook important time-dependent events.

In the heart (atrial tissue), increased calcium conductance adequately explains the inotropic response to 5-HT_4 receptor agonism [11,23] but whether calcium mediates the tachycardiac response has yet to be resolved. Interestingly, the cardiac 5-HT_4 receptor in man and pig appears restricted to atria, with little or no evidence for its presence in ventricular muscle [30,31]. 5-HT_4-induced increases in force of ventricular contraction may therefore be secondary to rate [30] and, in a clinical setting, express an unwanted side effect. It has been noted that intravenous infusion of cisapride in healthy volunteers evokes tachycardia [32] and that exaggerated responses to 5-HT_4 agonists may result in isolated strips of atria from patients treated chronically with beta-adrenoceptor antagonists [33]. Thus, the development of 5-HT_4 receptor agonists as putative inotropic agents is without rational foundation and their use would be contra-indicated in heart failure patients. On the other hand, 5-HT_4-induced tachycardia encourages speculation that 5-HT_4 antagonists may function as novel antiarrhythmic drugs.

There is now abundant evidence that 5-HT_4 receptor agonism facilitates peristalsis in the gastrointestinal (GI) tract of guinea pig [34-36]. Both longitudinal [34] and circular [37] muscle contraction occurs via ACh release. The actual size of ACh release is still uncertain but enteric synapses [38] and cholinergic motor neurons innervating smooth muscle directly have been implicated [39]. In addition, a role in atropine-resistant peristalsis has been observed [40] and potentiation of noncholinergic neurotransmission claimed [41]. Atropine-resistant peristalsis is mediated by substance P [42] and it is possible that the 5-HT_4 receptor, as with the enteric 5-HT_3 receptor [7], can induce the release of substance P. Definitive proof is awaited.

Whereas 5-HT_4 receptor activation increases motility in isolated segments of guinea pig intestine [34-36], inhibitory effects on smooth muscle of the alimentary tract of man and rat appear to predominate (table 1). For example, the rat esophageal 5-HT_4 receptor is located non-neuronally, on smooth muscle cells [43,44], and activation of the receptor results in relaxation due to cyclic AMP production [45,46]. Thus the 5-HT$_4$ receptor appears to be "plugged-in" to enteric neuro-muscular circuits in a species-dependent manner, perhaps indicative of a modulatory, rather than an essential role, in GI function. As such the 5-HT$_4$ receptor may be an attractive target for the treatment of irritable bowel disease (IBS), in which both hyper- and hypomotility states exist. *In vivo*, 5-HT$_4$ agonists are clearly prokinetic in both dog and man [47-29]. However, whether antagonists reduce hypermotility conditions in IBS or colonic spasm has yet to be determined.

TABLE 1. 5-HT$_4$ Receptor: Distribution and Function[1]

Species	Tissue	Mechanism	Transduction Effect
Guinea pig	Hippocampus	↑ cAMP	-
Rat	Hippocampus	-	↓ K$^+$ conductance, slow depolarization
Mouse	Colliculi neurons	↑ cAMP	↓ K$^+$ conductance
Rat	CNS	-	↑ EEG-energy
Rat	Vagus nerve	-	Depolarization
Guinea pig	Myenteric neurons	-	↑ Fast EPSP
Guinea pig	Ileum	-	Concentration, peristalsis
Guinea pig	Colon	-	Concentration
Guinea pig[2]	Ileum	-	↑ Short circuit current
Rat	Colon	-	↑ Short circuit current
Rat	Ileum	-	Relaxation
Rat	Esophagus	↑ cAMP	Relaxation
Man	Colon	-	Circular muscle relaxation
Guinea pig	Atria	-	↑ Rate
Piglet	Atria	↑ cAMP	↑ Rate and force
Pig	Heart	-	↑ Rate (*in vivo*)
Monkey	Heart	-	↑ Rate (*in vivo*)
Man	Right atrial appendage	↑ cAMP	↑ Force
Man	Left atrium	-	↑ Force
Man[2]	Urinary bladder	-	↑ Contraction
Rhesus Monkey[2]	Urinary bladder	-	↓ Contraction
Frog	Adrenal	↑ cAMP	Steroid release
Man	Adrenal	↑ cAMP	Steroid release

[1]For references see text, Bockaert et al. [20] and Hoyer et al. [58]. [2]Putative 5-HT$_4$; awaits full characterization

Similarly, whether 5-HT$_4$ antagonists will be of value in the treatment of diarrhea, as may be speculated from the prosecretory action of 5-HT$_4$ receptor activation (see table 1), remains for resolution.

Other functions attributed to 5-HT$_4$ receptor activation are steroid release [50,51] and putative effects on the urinary bladder [52,53; table 1]. It may be pertinent to recall that steroids elevate mood and that behavioral activity seen with 5-HT$_4$ agonists in laboratory animals may result from or be complicated by this event.

3. Agonists and Antagonists at the 5-HT$_4$ Receptor

Agonists at the 5-HT$_4$ receptor have been reviewed recently [20] and are summarized in table 2. SDZ compound 9b and (S)-RG 12718 are recent disclosures.

TABLE 2. Agonists at the 5-HT$_4$ Receptor[1]

Class	Equipotent Concentration Ratios Relative to 5-HT (pEC$_{50}$ = 8.2)
Indoles	5-MOT (2) > α-M-5-HT > 5-CT (95) > Tryptamine (906) > 2-M-5-HT (very weak or inactive)
Benzamides	SC-53116[2] (1.4) > Cisapride (12) = (S)-Zacopride (12) > Renzapride (12) > Zacopride (76) > (S) - RG 12718 (?)[3] > Metoclopramide (128) > (R) - Zacopride (130)
Benzimidazolones	BIMU 8 (1.6) > BIMU 1 (2)
Quinolines	SDZ compound 9b (0.9)[4]

[1]Data from Baxter et al. [45 and 59] in rat esophagus (see also Bockaert et al., [20]). Not agonized by 8-OH-DPAT, isapirone, buspirone, 1-phenylbiguanide or RU 249696. [2]Flynn et al. [60]. [3]Martin et al. [68] and [4]Blum et al. [61] both studied in the field stimulated guinea pig ileum.

Distinct progress has been made with 5-HT$_4$ receptor antagonists since early days when tropisetron (ICS 205, 930) was the only available drug tool [18; table 3]. Of the newer compounds, GR 113808 is by far the most affine and selective. However, as with SB 203186, RS-23597-190, SDZ 205, 557, LY 297524, and LY 297582, GR 113808 is an ester readily susceptible to hydrolysis and exhibits only a short duration of action *in vivo*. In this regard, the greater metabolic stability of DAU 6285 and SC-53606 may be advantageous.

The high affinity of GR 113808 for the 5-HT$_4$ receptor represents a major breakthrough which allows, for the first time, labeling of the receptor [67]. Autoradiographic studies in guinea pig and rat brain show highest specific biding in olfactory tubercles, striatum, substantia nigra, and superior colliculus. The 5-HT$_4$

receptor is also detected in hippocampus (dorsal, medial, and ventral) but only a very low density of labeling was found in frontal cortex and cerebellum. As yet, no binding has been reported in peripheral tissues. Success concerning the genomic cloning of the 5-HT_4 receptor seems imminent but, to date, no such claim has been made. This event, plus GR 113808, will assist in the resolution of questions concerning putative subtypes of the 5-HT_4 receptor [see 20 for references].

TABLE 3. Antagonists at the 5-HT_4 Receptor[1]

Compound	pA_2 5-HT_4	pA_2 5-HT_3	pA_2 MUSCARINIC	Ratio 5-HT_3/5-HT_4
Tropisetron[2]	6.5	(8.0) 10	5.4	0.0003
DAU 6285[2]	7.0	(5.0) 6.3	< 5.0	5
SDZ 205,557[2]	7.1	(5.9) 6.9	< 4.0	1.6
SB Comp. 2[3]	7.2	BJR	---	---
SB 203186[4]	8.3	7.2	---	12
SC-53606[5]	7.7	6.8	---	8
LY 297524[6]	7.7	(5.2)	---	---
LY 297582[6]	7.7	(5.3)	---	---
RS-23597-190[7]	7.8	5.7	< 5.0	126
GR 113808[8]	9.5	6.0	< 5.0	3000

[1]Not blocked by 5-HT_1, 5-HT_2, or 5 HT_3 antagonists [1]. [2]See Bockaert et al. [20]. [3]Kaumann et al. [62]. [4]Kaumann et al. [63]. [5]Moormann et al. [65]. [6]Krushinski et al. [65]. [7]Eglen et al. [66]. [8]Grossman et al. [67]. The numbers in parentheses give pA_2 estimates in guinea pig ileum which is a low affinity site. BJR; potent inhibition of the von Bezold Jarisch reflex.

4. Desensitization of the 5-HT_4 Receptor

A major concern in developing agonists for chronic use is the question of receptor desensitization. Early work on the 5-HT_4 receptor revealed its susceptibility to desensitization, and, in fact, the process was used as a tool to identify better and discriminate the 5-HT_4 receptor, as desensitization of the receptor was highly homologous [54]. It now seems that desensitization of the 5-HT_4 receptor resembles, in part, that seen with beta-adrenoceptors [55]. The process is agonist-dependent but cyclic AMP-independent, and is likely to involve both phosphorylation by βARK (beta-adrenoceptor kinase) or another specific agonist-dependent receptor kinase [56] as well as receptor sequestration. Heparin and zinc, both inhibitors of βARK [57], reduce desensitization as does concanavalin A, a compound which also inhibits the sequestration of beta-adrenoceptors. As the magnitude of desensitization correlates positively with agonist potency it would seem wise to develop partial agonists for the $5\text{-}HT_4$ receptor. As such, these compounds may not only retain clinical efficacy upon prolonged exposure but may provide the required organ selectivity, due to tissue-dependent variations in $5\text{-}HT_4$ receptor number, coupling efficiency, and machinery for desensitization.

5. References

1. Dumuis, A., Bouhelal, R., Sebben, M., Cory, R., and Bockaert, J. (1988) 'A nonclassical 5-hydroxytryptamine receptor positively coupled with adenylate cyclase in the central nervous system,' Mol. Pharmacol. 34, 880-887.
2. Shenker, A., Maayani, S., Weinstein, H., and Green, J.P. (1987) 'Pharmacological characterization of two 5-hydroxytryptamine receptors coupled to adenylate cyclase in guinea pig hippocampal membranes,' Mol. Pharmacol. 31, 357-367.
3. Enjalbert, A., Hamon, M., Bourgoin, S., and Bockaert, J. (1978) 'Postsynaptic serotonin-sensitive adenylate cyclase in the central nervous system,' Mol. Pharmacol. 14, 11-23.
4. Dumuis, A., Bouhelal, R., Sebben, M., and Bockaert, J. (1988) 'A 5-HT receptor in the central nervous system, positively coupled with adenylate cyclase, is antagonized by ICS 205 930,' Eur. J. Pharmacol. 146, 187-188.
5. Craig, D.A. and Clarke, D.E. (1989) '5-Hydroxytryptamine and cholinergic mechanisms in guinea pig ileum,' Br. J. Pharmacol. 96, 246P.
6. Craig, D.A. and Clarke, D.E. (1990) 'Pharmacological characterization of a neuronal receptor for 5-hydroxytryptamine in guinea pig ileum with properties similar to the 5-hydroxytryptamine$_4$ receptor,' J. Pharmacol. Exp. Ther. 252, 1378-1386.
7. Buchheit, K.-H., Engle, G., Mutschler, E., and Richardson, B. (1985) 'Study of the contractile effect of 5-hydroxytryptamine (5-HT) in isolated longitudinal muscle strip from guinea pig ileum. Evidence for two distinct release mechanisms', Naunyn-Schmiedeberg's Arch. Pharmacol. 329, 36-41.
8. Sanger, G.J. (1987) 'Increased gut cholinergic activity and antagonism of 5-hydroxytryptamine M-receptors by BRL 24924: Potential clinical importance of BRL 24924', Br. J. Pharmacol. 91, 77-87.
9. Schuurkes, J.A.J., van Neuten, J.M., van Daele, P.G.H., Reyntjens, A.J., and Janssen, P.A.J. (1985) 'Motor stimulating properties of cisapride in isolated gastrointestinal preparations of the guinea pig', J. Pharmacol. Exp. Ther. 234, 775-783.
10. Kaumann, A.J., Murray, K.J., Brown, A.M., Sanders, L., and Brown, M.J. (1989) 'A receptor for 5-HT in human atrium', Br. J. Pharmacol. 98, 664P.
11. Kaumann, A.J., Sanders, L., Brown, A.M., Murray, K.J., and Brown, M.J.

(1990) 'A 5-hydroxytryptamine receptor in human atrium', Br. J. Pharmacol. 100, 879-885.

12. Kaumann, A.J. (1990) 'Piglet sinoatrial 5-HT receptors resemble human atrial 5-HT$_4$-like-receptors', Naunyn-Schmiedeberg's Arch. Pharmacol. 342, 619-622.

13. Villalón, C.M., den Boer, M.O., Heiligers, J.P.C., and Saxena, P.R. (1990) 'Mediation of 5-hydroxytryptamine-induced tachycardia in the pig by the putative 5-HT$_4$ receptor', Br. J. Pharmacol. 100, 665-667.

14. Villalón, C.M., den Boer, M.O., Heiligers, J.P.C., and Saxena, P.R. (1991) 'Further characterization, by use of tryptamine and benzamide derivatives, of the putative 5-HT$_4$ receptor mediating tachycardia in the pig', Br. J. Pharmacol. 102, 107-112.

15. Duncker, D.J., Saxena, P.R., and Verdouw, P.D. (1985) '5-Hydroxytryptamine causes tachycardia in pigs by acting on receptors unrelated to 5-HT$_1$, 5-HT$_2$, or M-type', Br. J. Pharmacol. 86, 596P.

16. Bom, A.H., Duncker, D.J., Saxena, P.R., and Verdouw, P.D. (1988) '5-Hydroxytryptamine-induced tachycardia in the pig: possible involvement of a new type of 5-hydroxytryptamine receptor', Br. J. Pharmacol. 93, 663-671.

17. Saxena, P.R. (1986) 'Nature of the 5-hydroxytryptamine receptors in mammalian heart', Prog. Pharmacol. 6, 173-185.

18. Clarke, D.E., Craig, D.A., and Fozard, J.R. (1989) 'The 5-HT$_4$ receptor: naughty, but nice', Trends Pharmacol. Sci. 10, 385-386.

19. Turconi, M., Schiantarelli, P., Borsini, F., Rizzi, C.A., Ladinsky, H., and Donetti, A. (1991) 'Azabicycloalkyl benzimidazolones: interaction with serotonergic 5-HT$_3$ and 5-HT$_4$ receptors and potential therapeutic implications', Drugs of the Future 16, 1011-1026.

20. Bockaert, J., Fozard, J.R., Dumuis, A., and Clarke, D.E. (1992) 'The 5-HT$_4$ receptor: a place in the sun', Trends Pharmacol. Sci. 13, 141-145.

21. Fagni, L., Dumuis, A., Sebben, M., and Bockaert, J. (1992) 'The 5-HT$_4$ receptor subtype inhibits K$^+$ current in colliculi neurones via activation of a cyclic AMP-dependent protein kinase', Br. J. Pharmacol. 105, 973-979.

22. Andrade, R. and Nicoll, R.A. (1987) 'Pharmacologically distinct actions of serotonin on single pyramidal neurones of the rat hippocampus recorded in vitro', J. Physiol. 394, 99-124.

23. Quadid, H., Sequin, J., Dumuis, A., Bockaert, J., and Nargeot, J. (1992) 'Serotonin increases calcium current in human atrial myocytes via the newly described 5-hydroxytryptamine$_4$ receptors', Mol. Pharmacol. 41, 346-351.

24. Kilbinger, H., Gebauer, A., and Ladinsky, H. (1992) 'Effects of the 5-HT$_4$ receptor agonist BIMU8 on release of acetylcholine (ACh) and 5-hydroxytryptamine (5-HT) from guinea pig ileum', Naunyn-Schmiedeberg's Arch. Pharmacol. (suppl.) 345, 3.

25. Kilbinger, H. and Wolf, D. (1992) 'Effect of 5-HT$_4$ receptor stimulation on basal and electrically evoked release of acetycholine from guinea-pig

myenteric plexus', Naunyn-Schmiedeberg's Arch. Pharmacol. 345, 270-275.
26. Boddeke, H.W.G.M. and Kalkman, H.O. (1990) 'Zacopride and BRL 24924 induce an increase in EEG-energy in rats', Br. J. Pharmacol. 101, 281-284.
27. Boddeke, H.W.G.M. and Kalkman, H.O. (1992) 'Agonist effects at putative central 5-HT$_4$ receptors in rat hippocampus by R(+)- and S(-)-zacopride; no evidence for stereo-selectivity', Neurosci. Let. 134, 261-263.
28. Goelet, P., Castellucci, V.F., Schacher, S., and Kandel, E.R. (1986) 'The long and the short of long-term memory--a molecular framework', Nature (Lond.) 322, 419-422.
29. Bailey, C.H., Chen, M., Keller, F., and Kandel, E.R. (1992) 'Serotonin-mediated endocytosis of apCAM: An early step of learning-related synaptic growth in *Aplysia*', Science 246, 645-649.
30. Schoemaker, R.G., Du, X.Y., Bax, W.A., and Saxena, P.R. (1992) '5-Hydroxytryptamine increases contractile force in porcine right atrium but not left ventricle', Eur. J. Pharmacol., in press.
31. Schoemaker, R.G., Du, X.Y., Bax, W.A., and Saxena, P.R. (1992) '5-Hydroxytryptamine stimulates human isolated atrium but not ventricle', Naunyn-Schmiedeberg's Arch. Pharmacol., in press.
32. Bateman, D.N. (1986) 'The action of cisapride on gastric emptying and pharmacodynamics and pharmacokinetics of oral diazepam', Eur. J. Clin. Pharmacol. 30, 205-208.
33. Kaumann, A.J. (1991) 'Regulation of cardiac β-adrenoceptors by antagonists', in S.A. Abadi (ed.), Adrencoeptors: Structure, Mechanisms, Function, Birkhäuser Verlag, Basel, pp. 221-230.
34. Craig, D.A. and Clarke, D.E. (1991) 'Peristalsis evoked by 5-HT and renzapride: evidence for putative 5-HT$_4$ receptor activation', Br. J. Pharmacol. 102, 563-564.
35. Buchheit, K.-H., and Buhl, T. (1991) 'Prokinetic benzamides stimulate peristaltic activity in the isolated guinea-pig ileum by activation of 5-HT$_4$ receptors', Eur. J. Pharmacol. 205, 203-208.
36. Rizzi, C.A., Coccini, T., Onori, L., Manzo, L., and Tonini, M., 'Benzimidazolone derivatives: a new class of 5-hydroxytryptamine$_4$ receptor agonists with prokinetic and acetylcholine releasing properties in the guinea pig ileum', J. Pharmacol. Exp. Ther. 261, 412-419.
37. Tonini, M., Stefano, C.M., Onori, L., Coccini, T., Manzo, L., and Rizzi, C.A. (1992) '5-Hydroxytryptamine$_4$ receptor agonists facilitate cholinergic transmission in the circular muscle of guinea-pig ileum: antagonism by tropisetron and DAU6285', Life Sci. 50, 173-178.
38. Tonini, M., Galligan, J.J., and North, R.A. (1989) 'Effects of cispride on cholinergic neurotransmission and propulsive motility in the guinea pig ileum', Gastroenterology 96, 1257-1264.
39. Tonini, M., Rizzi, C.A., Manzo, L., and Onori, L. (1991) 'Novel enteric 5-HT$_4$ receptors and gastrointestinal prokinetic action', Pharmacol. Res. 24, 5-

14.
40. Clarke, D.E., Baxter, G.S., Young, H., and Craig, D.A. (1991) 'Pharmacological Properties of the Putative 5-HT$_4$ Receptor in Guinea-Pig Ileum and Rat Oesophagus: Role in Peristalsis', in J.R. Fozard and P. R. Saxena (eds.), Serotonin: Molecular Biology, Receptors and Functional Effects, Birkhäuser Verlag, Basel, pp. 232-242.
41. King, B.F. and Sanger, G.J. (1992), 'Facilitation of noncholinergic neurotransission by renzapride (BRL 24924) in circular muscle of guinea-pig ileum', Proc. Br. Pharmacol. Soc., London meeting, in press.
42. Suzuki, N. and Gomi, Y. (1992) 'Effects of CP-96, 345, a novel non-peptide antagonist of NK$_1$ receptor, on the peristalsis of isolated guinea pig ileum', Jap. J. Pharmacol. 58, 473-477.
43. Bieger, D. and Triggle, C. (1985) 'Pharmacological properties of mechanical responses of the rat oesophageal muscularis mucosae to vagal and field stimulation', Br. J. Pharmacol. 84, 93-160.
44. Baxter, G.S., Craig, D.A., and Clarke, D.E. (1991) '5-Hydroxytryptamine$_4$ receptors mediate relaxation of the rat oesophageal tunica muscularis mucosae', Naunyn-Schmiedeberg's Arch. Pharmacol. 343, 439-446.
45. Ford, A.P.D.W., Baxter, G.S., Eglen, R.M., and Clarke, D.E. (1992) '5-Hydroxytryptamine stimulates cyclic AMP formation in the tunica muscularis mucosae of the rat oesophagus via 5-HT$_4$ receptors', Eur. J. Pharmacol. 211, 117-120.
46. Moummi, C., Yang, D.C., and Gullikson, G.W. (1992) '5-HT$_4$ receptor activation induces relaxation and associated cAMP generation in rat esophagus', Eur. J. Pharmacol. 216, 47-52.
47. Yoshida, N., Mizumoto, A., Iwanaga, Y., and Itoh, Z. (1991) 'Effects of 5-hydroxytryptamine 3 receptor antagonists on gastrointestinal motor activity in conscious dogs', J. Pharmacol. Exp. Ther. 256, 272-278.
48. Gullikson, G.W., Loeffler, R.F., and Viriña, M.A. (1991) 'Relationship of serotonin-3 receptor antagonist activity to gastric emptying and motor-stimulating actions of prokinetic drugs in dogs', J. Pharmacol. Exp. Ther. 258, 103-110.
49. Hawkey, C.J. (1991) 'The place of cisapride in therapeutics: An interim verdict', Aliment. Pharmacol. Ther. 5, 351-356.
50. Idres, S., Delarue, C., Lefebvre, H., and Vaudry, H. (1991) 'Benzamide derivatives provide evidence for the involvement of a 5-HT$_4$ receptor type in the mechanism of action of serotonin in frog adrenocortical cells', Mol. Brain Res. 10, 251-258.
51. Lefebvre, H., Contesse, V., Delarve, C., Fevilloley, M., Hery, F., Grise, P., Raynaud, G., Verhofstad, A.A.J., Wolfe, L.M., and Vaudry, H. (1992) 'Serotonin-induced stimulation of cortisol secretion from human adrenocortical tissue is mediated through activation of a serotonin-4 receptor subtype', Neurosci. 47, 999-1007.

52. Waikar, M.V., Ford, A.P.D.W., Hedge, S.S., and Clarke, D.E. (1992) 'DAU 6285: A probe for the 5-HT_4 receptor in rat oesophagus and monkey bladder', Proc. Brit. Pharmacol. Soc., Dublin meeting, in press.
53. Ford, A.P.D.W., Waikar, M.V., and Clarke, D.E. (1992) '5-HT_4 receptor agonism inhibits neuronally-mediated responses in monkey urinary bladder', 2nd International Symposium on Serotonin from Cell Biology to Pharmacology and Therapeutics, Abst. Book, p. 53.
54. Craig, D.A., Eglen, R.M., Walsh, L.K.M., Perkins, L.A., Whiting, R.L., and Clarke, D.E. (1990) '5-Methoxytryptamine and 2-methyl-5-hydroxytryptamine-induced desensitization as a discriminative tool for the 5-HT_3 and putative 5-HT_4 receptors in guinea pig ileum', Naunyn-Schmiedeberg's Arch. Pharmacol. 342, 9-16.
55. Roth, N.S., Campbell, P.T., Caron, M.G., Lefkowitz, R.J., and Lohse, M. (1991) 'Comparative roles of desensitization of β-adrenergic receptor kinase and the cyclic AMP-dependent protein kinase', Proc. Natl. Acad. Sci. USA 88, 6201-6204.
56. Ansanay, H., Sebben, M., Bockaert, J., and Dumuis, A. (1992) 'Characterization of homologous 5-HT_4 receptor desensitization in colliculi neurons', Mol. Pharmacol., in press.
57. Benovic, J.L., Mayor Jr., F., Staniszewski, C., Lefkowitz, R.J., and Caron, M.G. (1987) 'Purification and characterization of the β-adrenergic receptor kinase,' J. Biol. Chem. 262, 9026-9032.
58. Hoyer, D., Fozard, J.R., Saxena, P.R., Mylecharane, E.J., Clarke, D.E., Martin, G.R., and Humphrey, P.P.A (1993) 'A modern classification of receptors for 5-hydroxytryptamine (serotonin)', Pharmacol. Revs., submitted.
59. Baxter, G.S. and Clarke, D.E. (1992) 'Benzimidazolone derivatives act as 5-HT_4 receptor ligands in rat oesophagus', Eur. J. Pharmacol. 212, 225-229.
60. Flynn, D.L., Zabrowski, D.L., Becker, D.P., Nosal, R., Villamil, C.I., Gullikson, G.W., Moummi, C., and Yang, D.C. (1992) 'SC-53116: The First Selective agonist at the newly identified serotonin 5-HT_4 receptor subtype', J. Med. Chem. 35, 1486-1489.
61. Blum, E., Buchheit, K.H., Reuscher, I.I.H., Gamse, R., Kloeppner, E., Meigel, H., Papageorgiou, C., Waclchli, R., and Revesz, L. (1992) 'Design and synthesis of novel ligands for the 5-HT_3 and the 5-HT_4 receptor', Bioorganic Med. Chem. Lett. 2, 461-466.
62. Kaumann, A.J., King, F.D., and Young R.C. (1992) 'Indazole as an indole bioisotere: 5-HT_4 receptor antagonism', Bioorganic Med. Chem. Lett. 2, 419-420.
63. Kaumann, A.J. Medhurst, A., Boyland, P., Vimal, M., and Young R.C. (1992) 'SB 203186, A potent selective 5-HT_4-receptor antagonist', 2nd International Symposium on Serotonin from Cell Biology to Pharmacology and Therapeutics, Abst. Book, p. 44.
64. Moormann, A.E., Yang, D.C., Gullikson, G.W., and Flynn, D.L. (1992) 'New

imidazopyridines which act as serotonin 5-HT$_3$ antagonists and/or 5-HT$_4$ antagonist', Amer. Chem. Soc., Division Med. Chem., 204th National meeting, August, 1992, Washington, D.C.
65. Krushinski, J.H., Susemichel, A., Robertson, D.W., and Cohen, M.L. (1992) 'Interaction of metoclopramide analogues with 5-HT$_4$ receptors', Amer. Chem. Soc., Division Med. Chem., 203rd National meeting, April, 1992, San Francisco.
66. Eglen, R.M., Bonhaus, D.W., Clarke, R., Hegde, S.S., Leung, E., and Whiting, R.L., (1992) 'RS-23597-190: a potent and selective 5-HT$_4$ receptor antagonist', Proc. Brit. Pharmacol. Soc., London meeting, in press.
67. Grossman, C.J., Kilpatrick, G.J., Bunce, K.T., Oxford, A.W., Gale, J.D., Whitehead, J.F., and Humphrey, P.P.A. (1992). 'Development of a radioligand binding assay for the 5-HT$_4$ receptor: Use of a novel antagonist', 2nd International Symposium on Serotonin from Cell Biology to Pharmacology and Therapeutics, Abst. Book, p. 31.
68. Martin, G.E., Davis, M.A., Pendley, C.E., and Fitzpatrick, L.R. (1992) 'Serotonin-4 receptor agonist activity of the prokinetic agent ($\pm$)-RG 12718,' Pharmacologist 34, 151.

SEROTONERGIC RECEPTORS: FROM LIGANDS TO SEQUENCE

P.A.J. JANSSEN and H. MOEREELS
Janssen Research Foundation
B-2340 Beerse
Belgium

ABSTRACT. With the help of molecular modelling and computational chemistry a number of residues are identified in the $5HT_2$ and $5\text{-}HT_{1D}$ receptor protein sequences as being important for the binding of serotonin. In the $5\text{-}HT_2$ receptor the following residues are involved: Trp^{151}, Asp^{155} and Phe^{158} in transmembrane domain 3 (TM-3), Ser^{239} and $Phe^{240,243,244}$ in TM-5. In the $5\text{-}HT_{1D}$ receptor: Trp^{114}, Asp^{118}, and Cys^{121} in TM-3, Thr^{202}, Cys^{203}, Phe^{206} and Tyr^{207} in TM-5. The obtained models revealed a new concept, the so-called "aromatic box," which is formed by aromatic residues that constitute the walls and parcel up serotonin. In this box a Cys residue can be the substitute for Phe. For the $5\text{-}HT_2$ receptor it is also shown that Asn^{92}, Asp^{120}, Ser^{372}, Asn^{376}, and Tyr^{380} can glue TM-1, 2, and 7 together by a network of hydrogen bonds. Apart form Ser^{372} these residues are conserved among neurotransmitter receptors and are proposed to be important for the receptor architecture.

1.   Introduction

Serotonin receptors belong to the large superfamily of G-protein (guanine nucleotide binding protein)-coupled receptors. After being triggered by the appropriate hormone these receptors transduce a signal via interaction with a particular G-protein that activates enzymatic reactions generating second messengers. The hallmark of these receptors is the tentative presence of seven hydrophobic domains packed in a bundle of helices spanning the plasma membrane. Many hundreds of protein sequences for these receptors have now been cloned. All the properties and characteristics of proteins, ranging from folding to function, are encoded in their sequences. Although this is only the primary structure information it enables several investigators to identify and analyze the function of, in principle, every residue in the sequence or strings of consecutive residues. Several successes have indeed been reported. As a result, our knowledge about the tentative importance and or function of specific residues or particular receptor domains is continuously growing. It is clear that any residue in the sequence may be important for numerous functions that have not been detected yet. However, in a first approach a rather simple anticipation is made. Some residues could be necessary for binding a ligand whereas others govern the receptor architecture. In this contribution we report the identification of residues that are good candidates for being involved in the binding of serotonin to the 5-HT_2 and 5-HT_{1D} receptors. The obtained serotonin binding model reveals a new concept, the so-called "aromatic box," which is formed by aromatic residues that constitute the walls and parcel up serotonin. Apart from these residues other tentative residues are identified for the receptor architecture. These are selected by testing the hypothesis that some of the fully conserved residues, among neurotransmitter receptors, can be important for the architecture. For the G-protein-coupled receptors no detailed three-dimensional information from experiment is available as yet. In the study of their structure it is therefore appropriate to use molecular modelling and computational chemistry as our tools of choice. Although the photon receptor bacteriorhodopsin translocates protons and is not G-protein coupled, its rough three-dimensional structure is available and is used as a source of inspiration.

2. Methods

The molecular modelling and computations are performed using the BIOSYM, Insight and Discover software, release 2.00 (BIOSYM TECHNOLOGIES, San Diego, CA 92121, USA), running on an IRIS Silicon Graphics 4D/220GTX computer. For sequence alignment the CGEMA is used, running on the same hardware [1]. The starting molecular complexes were minimized with the conjugate gradient method until the rms energy derivative was less than 0.01 Kcal/mole.Å. No distance constraints between atoms are imposed and the cut-off distance for the van der Waals and electrostatic interactions was set to a value of 8 Å, and a constant dielectric $\epsilon = 1$ was used. Standard Discover atomic charges were employed. Asp^{155} (in 5-HT_2), Asp^{118} (in 5-HT_{1D}), and serotonin are charged but the complete complex is neutral. Instead of the harmonic bond stretching term, a Morse type potential form and the bond-angle cross terms option was used.

3. Results

3.1 SELECTION OF SOME KEY RESIDUES

In the model describing the binding of catecholamines to adrenergic receptors three residues play a key role [2]. The protonated nitrogen of the catecholamine interacts with the conserved Asp residue in TM-3, whereas the meta- and parahydroxyl groups form hydrogen bonds with the hydroxyl side chains of respectively, Ser^{204} and Ser^{207} (TM-5). In serotonin, however, only one hydroxyl group is present. From an alignment between adrenergic and serotonin receptors it is obvious that only Ser^{204} of the β_2 receptor is conserved among the serotonin receptors which show a Ser or Thr in that position. Thus, Ser^{239} in the 5-HT_2 and Thr^{202} in the 5-HT_{1D} receptors are tentative candidates to interact with the serotonin hydroxyl group by forming a hydrogen bond. The Asp residue in TM-3 is fully conserved among neurotransmitter receptors and at position 155 in the 5-HT$_2$ receptor. This Asp residue is the obvious candidate for interaction with the protonated nitrogen of serotonin. The alignment of different neurotransmitter receptors revealed several fully conserved residues. These residues are indicated in figure 1.

According to the three-dimensional model of bacteriorhodopsin, the seven TM domains are located consecutively in an anti-clockwise manner, when looking from outside the cell, with TM-1, 2, and 7 packed close together [4]. As a consequence the following conserved residues, characterized by their potential to form hydrogen bonds, are selected as candidates to glue TM-1, 2, and 7 together: Asn^{92} in TM-1, Asp^{120} in TM-2, and Ser^{373}, Asn^{376}, Tyr^{380} in TM-7. In the following sections the selection of tentative residues made for serotonin binding and for architecture will be examined further by making use of molecular modelling and computational chemistry.

Figure 1. The seven TM domains of the 5-HT$_2$ receptor [3]. Residues in bold are conserved among neurotransmitter receptors.

3.2 THE BINDING SEROTONIN

At first, serotonin was constructed and energy minimized. Then TM-3 and TM-5 of the 5-HT$_2$ receptor were constructed as regular α-helices and energy minimized. Using the minimized structures a complex was constructed between serotonin and the helices of TM-3 and TM-5. The serotonin molecule was positioned in a spatial location according to the following requirements: the OH-group of serotonin being in a tentative position to form a hydrogen bond with the OH-group of SER^{239} in TM-5, concomitantly the protonated nitrogen of serotonin in a position to interact with the COO$^-$ side chain of Asp^{155} in TM-3 while the spatial arrangement of both TM-3 and TM-5 is in agreement with that observed in bacteriorhodopsin. This complex was energy minimized and the result is shown in figure 2. The same procedure was used to construct the complex between serotonin and the helices of TM-3 and TM-5 of the 5-HT$_{1D}$ receptor. The energy minimized complex is shown in figure 3. In both minimized complexes the serotonin OH-group forms a proper hydrogen bond with Ser[239] or Thr[202]. Serotonin functions as the donor while Ser or Thr are concomitantly forming a hydrogen bond with a carbonyl oxygen atom located in the helix backbone which, for the sake of clarity, is now shown in the figures. The protonated nitrogen of serotonin forms a strong interaction with the COO$^-$ side chain of Asp[155] or Asp[118]. The complexes however reveal a number of additional interesting interactions. Figure 2 clearly shows how serotonin is closely packed in an aromatic box which is formed by aromatic residues that constitute the walls and literally parcel up serotonin. At the bottom of this box, the indole-NH of serotonin is forming a typical hydrogen bond with the π-system of Phe[244]. Phe[243], at the right wall of the box, shows a stacking interaction with the serotonin indole ring. At the back, Phe[240] forms an edge-to-edge interaction with the indole ring, while Phe[158] is involved in an edge-to-face interaction at the left. Finally, the cover of the box is formed by Trp[151]. An interesting aspect is, that during the energy minimization, this Trp residue moved about 15 Å to its final position. Although Phe[240] and Phe[158] are both replaced by a Cys residue in the 5-HT$_{1D}$ receptor, the minimized complex looks very close to that obtained for the 5-HT$_2$ receptor. In fact, these Cys residues are involved in typical sulphur-aromatic interactions, and therefore are good substitutes for the Phe residues present in the 5-HT$_2$ receptor. This in fact means that the stabilizing interactions between the serotonin indole part and receptor residues can be obtained by both Phe or Cys residues.

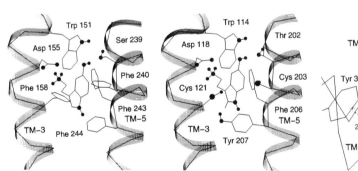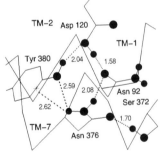

Figure 2. Minimized complex between serotonin, TM-3, and TM-5 of the 5-HT$_2$ receptor. Large spheres: N and O atoms; small spheres: H atoms.

Figure 3. Minimized complex between serotonin, TM-3, and TM-5 of the 5-HT$_{1D}$ receptor. Large spheres: N, O and S atoms; small spheres: H atoms.

Figure 4. Minimized complex of TM-1,2, and 7. Large spheres: N and O atoms; small spheres: H atoms.

3.3. THE GLUE BETWEEN TM-1, 2, AND 7 IN THE 5-HT$_2$ RECEPTOR

Regular α-helices were constructed for TM-1, 2, and 7, and energy minimized. In the minimized TM-7 helix Ser^{372} (not Ser^{373}), Asn ^{376} and Tyr^{380} protrude at the same side of the helix and are involved in a chain of hydrogen bonds. Ser^{372} donates a hydrogen bond to the side chain carbonyl oxygen of Asn^{376} while the side chain NH of the latter residue donates a hydrogen to the side chain oxygen of Tyr^{380}. With the minimized helices a complex was constructed according to the spatial arrangement as observed in bacteriorhodopsin. In this complex Asn^{92} (TM-1), Asp^{120} (TM-2), and Ser^{372}, Asn^{376}, Tyr^{380} (TM-7) all point to the inside area between the three helices. This complex was energy minimized and the result is shown in figure 4. The complex is characterized by a network of hydrogen bonds. The NH of Asn^{376} is no longer forming a proper hydrogen bond with the OH of Tyr^{380} but forms a weak hydrogen bond with the aromatic π-system at a ring edge. The hydrogen bond between Ser^{372} and Asn^{376} remained intact. In addition, hydrogen bonds are observed between Tyr^{380} and Asp^{120}, Asp^{120} and Asn^{92}, and Asn^{92} and Asn^{376}. Although Ser^{373} was initially selected as a tentative candidate it is not involved in any interactions except for a hydrogen bond towards its helix backbone. Nevertheless, the minimized complex shows clearly that four out of the five selected residues can contribute in the packing of TM-1, 2, and 7 and thus be considered as being important for the receptor architecture.

4. Discussion

A serious caveat in any molecular modelling experiment is that the final result depends on the starting structure used. However, molecular modelling and computational chemistry techniques should be considered as an experiment that can contribute in the study of receptor structures. It should be stressed that the main goal of this contribution is not the unravelling of the "exact" receptor structure and serotonin binding place but as an examination of a hypothesis. Our simple hypothesis about residues being important for ligand binding or for receptor architecture is indeed supported by some experimental results obtained for different receptors of the family. In the β$_2$ and muscarinic acetylcholine receptors, for example, the Asp residue in TM-3 is shown to be crucial for ligand binding. The fully conserved Pro residue present in TM-7 however, is involved in the processing event of receptor folding [5]. The binding of serotonin in an aromatic box should be considered as a new concept which could hold for the binding of other agonists to their appropriate receptors. The more so as Phe^{243} is fully conserved among neurotransmitter receptors, with known sequence, and in position 244 at the bottom of the box we find either a Phe or Tyr residue. It should also be noticed that in the Trp^{151} position an aromatic residue like Phe, Trp, or Tyr is always present in each neurotransmitter receptor. The human D$_4$ receptor forms the only exception and shows a Leu. In general, we postulate that the conserved Asp residue in TM-3 and one or two Ser or Thr residues in TM-5 function as the pegs for the agonist ligand by supplying the initial driving force. It is clear that these forces, at any time, can be compensated or compensate, by or for, interaction with water or different ions that are present. The major binding of the ligand is governed by the interactions with aromatic rings. As shown for the 5-HT_{1D} receptor the side chain of a Cys residue can substitute for an aromatic ring. The aromatic interactions with sulphur are indeed observed X-ray data of 36 different protein structures [6]. It is also interesting to note that after completion of the modelling a recent report came to our attention showing that in the 5-HT_{1A} receptor the $Ser^{198} \rightarrow Ala^{198}$ mutant has a decreased ligand affinity whereas the affinity for the $Thr^{199} \rightarrow Ala^{199}$ (Ser^{239} in 5-HT_2) mutant was too weak to be measured [7].

The molecular complex between TM-1, 2, and 7 does not support Ser^{373} as a candidate for being important in the receptor architecture but reveals Ser^{372}. Among the neurotransmitter receptors, with known sequence, an alignment (now shown) indicates the presence of a Ser, Asn, or Cys residue in this position. As Asn or Cys have both the potential to be involved in hydrogen bonding it is assumed that the model identifying the glue residues, between TM-1, 2, and 7, may hold for all of the neurotransmitter receptors. The function of Ser^{373} remains still unclear in our present model. From preliminary modelling, energy minimization, and molecular dynamics performed on the complete 7 α-helical bundle it is seen however that Ser^{373} is a good candidate to interact with the conserved Phe^{332} or Trp^{336} residues in TM-6. This observation however needs to be further explored. In another series of additional modelling and computation it is observed that the conserved Pro^{377} in TM-7, needed for folding in the $β_2$ receptor, can have a role in governing the hydrogen bond network, due to the fact that it can serve as a hinge in the helix. Although this needs further to be examined, the same property could be important for the conserved Pro^{246} in TM-5. The hypothesis about "architecture residues" is further supported by the $Asp^{79} \rightarrow Asn^{79}$ (Asp^{120} in 5-HT_2) mutant of the $α_2$-adrenoceptor [8]. This study shows that the carboxylic acid side chain can bind cations and contribute to a conformational state of the receptor that functions in receptor-G-protein interaction. The mutant receptor however cannot achieve the conformation necessary to activate G-proteins. Although Asn has the potential to form hydrogen bonds which probably will not violate the hydrogen bond network, it cannot serve as an anion to bind cations and therefore will not induce conformational changes.

5. Conclusion

In this report it is shown that a number of residues present in the 5-HT_2 and 5-HT_{1D} receptor sequences, can be selected for the binding of a ligand. The selection is supported with success by making use of molecular modelling and computational chemistry. These techniques prove to be valuable tools in the study of receptor structures. They further reveal a new concept for the binding of serotonin to the

5-HT$_2$ and 5-HT$_{1D}$ receptors which probably holds also for other transmitters and their receptors. The so called "aromatic box" which is formed by aromatic residues that constitute the walls parcels up serotonin. In the 5-HT$_2$ receptor the following residues are involved: Trp[151], Asp[155], and Phe[158] present in TM-3, Ser[239] and Phe[240,243,244] in TM-5. In the 5-HT$_{1D}$ receptor Phe[244] is replaced by a Tyr whereas Phe[158,240] are replaced by a Cys residue. The obtained results prove that Cys can be a substitute for a Phe residue in its interaction with the indole ring of serotonin. For the 5-HT$_2$ receptor is it shown that Asn[92], Asp[120], Ser[372], Asn[376], and Tyr[380] can glue TM-1, 2, and 7 together by a network of hydrogen bonds. Apart from Ser[372] all these residues are conserved among neurotransmitter receptors and proposed to be important for receptor architecture. The results obtained should be considered as the outcome of an experiment that explores a hypothesis. Further experiments are in progress to test the quality of the results presented here.

6. Acknowledgements

Dr. J. E. Leysen is acknowledged for stimulating discussions and Dr. L. Koymans for a critical review of the manuscript.

7. References

1. Moereels, H., De Bie, L., and Tollenaere, J.P. (1990) 'CGEMA and VGAP: a Colour Graphics Editor for Multiple Alignment using a Variable GAP penalty. Application to the muscarinic acetylcholine receptor', J. Comp. Aided Mol. Des. 4, 131-145.
2. Strader, C.D., Candelore, M.R., Hill, W.S., Sigal, I.S., and Dixon, R.A.F. (1989) 'Identification of two serine residues involved in agonist activation of the β-adrenergic receptor', J. Biol. Chem. 264, 13572-13578.
3. Saltzman, A.G., Morse, B., Whitman, M.M., Ivanshchenko, Y., Jaye, M., and Felder, S. (1991) 'Cloning of the human serotonin 5-HT$_2$ and 5-HT$_{1C}$ receptors subtypes', Biochem. Biophys. Res. Commun. 181, 1469-1478.
4. Henderson, R., Baldwin, J.M., Ceska, T.A., Zemlin, F., Beckmann, E., and Downing, K.H. (1990) 'Model for the structure of bacteriorhodopsin based on high-resolution electron cryo-microscopy', J. Mol. Biol. 213, 899-929.
5. Strader, C.D., Sigal, I.S., Register, R.B., Candelore, M.R., Rands, E., and Dixon, R.A.F. (1987) 'Identification of residues required for ligand binding to the β-adrenergic receptor', Proc. Natl. Acad. Sci. USA 84, 4384-4388.
6. Reid, K.S.C., Lindley, P.F., and Thornton, J.M. (1985) 'Sulphur-aromatic interactions in proteins', FEBS Lett. 190, 209-213.
7. Ho, B.Y., Karschin, A., Branchek, T., Davidson, N., and Lester, H.A. (1992) 'The role of conserved aspartate and serine residues in ligand binding and in function of the 5-HT$_{1A}$ receptor: a site-directed mutation study', FEBS Lett. 312, 259-262.

8. Surprenant, A., Horstman, D.A., Akbarali, H., and Limbird, L.E. (1992) 'A point mutation of the α_2-adrenoceptor that blocks coupling to potassium but not calcium currents', Science 257, 977-980.

POSSIBLE MECHANISMS OF ACTION OF DRUGS EFFECTIVE AGAINST ACUTE MIGRAINE ATTACKS

PRAMOD SAXENA
Department of Pharmacology
Dutch Migraine Research Group
Faculty of Medicine and Health Sciences
Erasmus University Rotterdam
Post Box 1738
3000 DR Rotterdam
The Netherlands

ABSTRACT. The acute headache phase of migraine is chiefly characterized by a vasodilatation in extracerebral (both intra- and extracranial) large arteries and arteriovenous anastomoses. In addition, there seems to be some evidence of release of neuropeptides from perivascular sensory nerves. The drugs active in aborting acute migraine attacks (ergotamine, dihydroergotamine, and sumatriptan) have a potent vasoconstrictor action on the above vessels and can also reduce plasma extravasation following trigeminal ganglion stimulation. Though a presynaptic action of these antimigraine drugs blocking the release of neuropeptides may be involved, the main mechanism of action seems to be vasoconstriction of the dilated extracerebral vessels in migraine.

1. Pathophysiological findings in migraine

More than 50 years ago Graham and Wolff [1] reported that during the headache phase of migraine increased pulsations of the ipsilateral temporal artery are noticed, and that both pulsations and headache can be abolished by manual pressure on the temporal artery or administration of ergotamine. These authors therefore hypothesized that migraine headache is due to a dilatation mainly within the extracerebral vascular bed. Subsequently, Heyck [2] reported an increased oxygen saturation in the jugular venous blood on the headache side, possibly due to dilatation of arteriovenous anastomoses.

More recent investigations have shown that the relief of headache by manual compression of the superficial temporal artery, as demonstrated by Graham and Wolff [1], was noticed in only one in three patients [3]. The use of transcranial Duplex-Doppler to measure flow velocity has revealed evidence of vasodilatation in the common and internal carotid and the anterior, middle, and posterior cerebral arteries during attacks of migraine without aura; however, no clear lateralization or correlation with the side of the headache was found [4]. Dilatation of the middle cerebral artery, as indicated by increased blood flow velocity without cerebral blood flow changes, during the migraine headache was confirmed by Friberg et al. [5]. Although their patients suffered both from common and classic migraine and the side of velocity increase correlated with the headache side. On the other hand, Zwetsloot et al. [6] reported unchanged blood flow velocities in the major cerebral arteries, except for a bilateral dilatation of the common carotid artery.

Biochemically, during migraine headache a decrease in blood 5-hydroxytryptamine (5-HT) [7] and an increase in the sensory neuropeptide CGRP [8] have been demonstrated. This last may reflect involvement of perivascular trigeminal nerves in the migraine syndrome, although whether as a cause or as a result of the pain is not certain. Nonetheless, the release of sensory peptides can be experimentally induced by stimulation of the trigeminal ganglion [9] or sensory afferents in the superior sagittal sinus [10]. Curiously, such a neurogenic mechanism, involving release of sensory peptides from the trigeminal nerve, implies involvement of the vasculature, since CGRP is a prominent vasodilator agent.

Another important discovery favoring a neurovascular etiology is that stimulation of different brain stem regions appears to have marked influence on cranial hemodynamics, giving vasculature on one side of the head only [11]. 5-HT may be involved in this cerebral control of the cranial circulation, since the 5-HT-containing dorsal and median raphe nuclei innervate cerebral and pial blood vessels [12] and stimulation of these regions increases both cerebral and extracranial blood flow [11].

2. Pharmacological findings in migraine

2.1 ERGOTAMINE

Ergotamine possesses a potent and long-lasting vasoconstrictor activity on a variety of blood vessels, including the coronaries [13]. However, in *in vivo* experiments ergotamine may cause a selective vasoconstriction within the external carotid arterial bed of the dog [14] as well as humans [15]. In anesthetized animals, this selectivity extends further, as least with lower concentrations, to the arteriovenous anastomotic (non-nutrient) fraction of the vasculature since the drug has little influence on tissue perfusion, including in the brain [16](table 1). There is a good correlation between the reduction in arteriovenous shunting and the increase in the difference in oxygen content between arterial and jugular venous blood. It should be noted that the vasoconstrictor action on arteriovenous anastomoses does not involve either α-adrenoceptors or $5-HT_2$ receptors and seems to be mediated partly by $5-HT_1$ and partly by novel unknown receptors [16,17]. The arteriovenous anastomoses are located in skin, ears, eyes, tongue, and dura mater [18,19], but those in the dura mater are not affected by ergotamine [20].

Ergotamine also inhibits extravasation of plasma in the dura mater following 'neurogenic inflammation' due to stimulation of trigeminal ganglion in the rat [21]. Since ergotamine and dihydroergotamine, but not angiotensin II or phenylephrine, inhibited neurogenic and capsaicin-induced edema, the authors concluded that the effect of the two ergot derivatives is independent of vasoconstriction. However, these results were not backed up by simultaneous measurement of dural (or carotid) blood flow or arterial blood pressure and, therefore, the effectiveness of angiotensin II and phenylephrine to induce dural vasoconstriction or even systemic vasoconstriction is uncertain.

TABLE 1. Effect of 5-HT and antimigraine drugs on arterioles and arteriovenous anastomoses (AVAs) in the porcine carotid artery bed.

Drugs	Arterioles	AVAs	Receptor
5-HT	----	+++ +	5-HT$_1$[1]
Ergotamine	0	+++ +	Partly 5-HT$_1$, Partly unknown
Dihydroergotamine	0	+++ +	Partly 5-HT$_1$, Partly unknown
Sumatriptan	0	+++ +	5-HT$_1$
Methysergide	-	+++	5-HT$_1$

[1], The 5-HT$_1$ receptors mediating vasodilatation and vasoconstriction not identical; -, dilatation; +, contraction; 0, no effect.

2.2 DIHYDROERGOTAMINE

The basic pharmacologic features of dihydroergotamine resemble those of ergotamine but dihydroergotamine exerts less vasoconstriction and has a stronger α-adrenoceptor antagonist property. The drug is more effective in constricting capacitance vessels than resistance vessels [13]. Similar to ergotamine, it decreases arteriovenous anastomotic blood flow [16,17], but not in dura mater [20] (table 1).

2.3 SUMATRIPTAN

The 5-HT$_1$ receptor agonist sumatriptan was developed with the premise that vasoconstriction within the carotid bed is responsible for the therapeutic action of ergot alkaloids and that this can be achieved more selectively via stimulation of 'special' 5-HT receptors [22].

Sumatriptan causes contraction of different isolated blood vessels: dog saphenous vein and middle cerebral artery and primate basilar artery and human basilar artery [23]. *In vivo,* sumatriptan selectively increases carotid resistance in the dog, cat, and pig; in the latter two species the increase in carotid resistance has been shown to be confined to its arteriovenous anastomotic fraction [23-25] (table 1). Unlike ergotamine, this effect of sumatriptan is mediated selectively by a subtype of 5-HT$_1$ receptors [24], but like ergotamine it also does not affect dural shunting [20]. Compared with the ergot alkaloids, the peripheral vasoconstrictor effects of sumatriptan [26], including on the human isolated coronary artery (Bax et al., unpublished), are weaker.

Sumatriptan has also been reported to inhibit the extravasation of plasma from dural vessels following trigeminal ganglion stimulation or local capsaicin application in the rat [27], but it is not known whether this effect is not secondary to the selective vasoconstriction of the cephalic blood vessels.

Sumatriptan does not readily cross the blood-brain-barrier [28] nor does it have any analgesic activity in conventional tests [23].

3. Mechanism of action of acutely acting drugs

Both vascular [22,29] and neural [30] effects have been implicated to explain the therapeutic action of sumatriptan (figure 1). Several studies have provided evidence

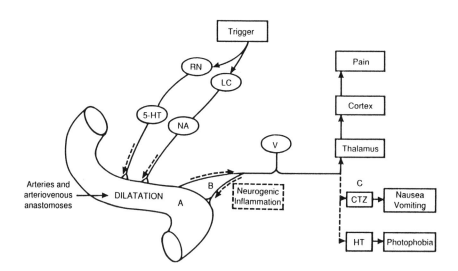

Figure 1. Possible chain of events during migraine. Changes in locus ceruleus (LC) and/or raphe (RN) nuclei activity can induce cephalic (dural and scalp) vasodilatation, which may stimulate perivascular sensory afferents of the fifth cranial nerve (V) to cause headache and associated symptoms. Ergot alkaloids and sumatriptan appear to abort migraine attacks chiefly by constricting the dilated cephalic vessels (A), but may also have inhibitory influence at the perivascular nerve terminals (B) and chemoreceptor trigger zone (C). From Saxena and Ferrari [22].

suggestive of ipsilateral dilatation of extracerebral cranial large arteries (not arterioles) and/or arteriovenous anastomoses during the headache phase of migraine [1,2,5]. Large arterial and, particularly, arteriovenous anastomotic dilatation will lower diastolic pressure and thus increase pulsations within the carotid artery concerned. The resulting increase in vascular pulsations may then act on 'stretch receptors' in the vessel wall to initiate pain and other symptoms of migraine. The proponents of primarily vascular action suggest that sumatriptan, which does not readily cross the blood-brain barrier [28], constrict the dilated blood vessels to abort a migraine attack (figure 1). Apart from the potent and selective action on carotid vasculature, as is also noticed with ergot derivatives [14,16,17], evidence for a constrictor effect of sumatriptan on cranial arteries during a migraine attack is also available [5, 31].

On the other hand, the proponents of the neural effects of acutely acting drugs believe that they act on presynaptic $5-HT_1$-like receptors located on sensory nerve endings to inhibit the process of 'neurogenic' inflammation responsible for migraine pain. Thus, the decrease in CGRP levels in migraine patients [8] and experimental animals during trigeminal ganglion stimulation [9] by sumatriptan is thought to be a reflection of the presynaptic action. However, such a direct action of sumatriptan on sensory nerve fibers has not been found in isolated arteries [32,33] and, therefore, the reduction of CGRP release may simply be due to vasoconstriction and consequent pain relief.

4. References

1. Graham J.R. and Wolff H.G. (1938) 'Mechanism of migraine headache and action of ergotamine tartrate', Arch. Neurol. Psychiat. 39, 737-763.
2. Heyck H. (1969). Pathogenesis of migraine. Res. Clin. Stud. Headache 2, 1-28.
3. Drummond P.D. and Lance J.W. (1988) 'Contribution of the extracranial circulation to the pathophysiology of headache', in J. Olesen and L. Edvinsson (eds.), Basic Mechanisms of Headache, Elsevier Science Publishers, Amsterdam, pp. 321-288.
4. Thie A., Fuhlendorf A., Spitzer K., and Kunze K. (1990) 'Transcranial Doppler evaluation of common and classic migraine. Part II Ultrasonic features during attacks', Headache 30, 209-215.
5. Friberg L., Olesen J., Iversen H.K., and Sperling B. (1991) 'Migraine pain associated with middle cerebral artery dilatation: reversal by sumatriptan', Lancet 338, 13-17.
6. Zwetsloot C.P., Caekebeke J.F.V., Jansen J.C., Odink J., and Ferrari M.D. (1991) 'Blood flow velocity changes in migraine attacks- a transcranial Doppler study', Cephalalgia 11, 103-107.
7. Saxena P.R. (1990) '5-Hydroxytryptamine and migraine', in P.R.Saxena, D.I. Wallis, W. Wouters and P. Bevan (eds.), Cardiovascular pharmacology of 5-

hydroxytryptamine. Prospective therapeutic applications, Kluwer Academic Publishers, Dordrecht, pp. 407-416.
8. Goadsby P.J., Edvinsson L., and Ekman R. (1990) 'Vasoactive peptide release in the extracerebral circulation of humans during migraine headache', Ann. Neurol. 28, 183-187.
9. Buzzi M.G., Carter W.B., Shimizu T., Heath H., and Moskowitz M.A. (1991) 'Dihydroergotamine and sumatriptan attenuate levels of CGRP in plasma in rat superior sagittal sinus during electrical stimulation of the trigeminal ganglion', Neuropharmacology 30, 1193-1200.
10. Zagami A.S., Goadsby P.J., and Edvinsson L. (1990) 'Stimulation of the superior sagittal sinus in the cat causes release of vasoactive peptides', Neuropeptides 16, 69-75.
11. Lance J.W., Lambert G.A., Goadsby P.J., and Zagami A.S. (1989) '5-Hydroxytryptamine and its putative etiological involvement in migraine', Cephalalgia 9 (Suppl. 9), 7-13.
12. Edvinsson L., Deguerce A., Duverger D., MacKenzie E.T., and Scatton B. (1983) 'Central serotonergic nerves project to the pial vessels of the brain', Nature 306, 55-57.
13. Müller-Schweinitzer E. and Weidmann H. (1978) 'Basic pharmacological properties', in B. Berde and H.O. Schild (eds.), Ergot alkaloids and related compounds. Handbook of experimental pharmacology, Springer-Verlag, Berlin, pp. 87-232.
14. Saxena P.R. and de Vlaam-Schluter G.M. (1974) 'Role of some biogenic substances in migraine and relevant mechanism in antimigraine action of ergotamine. Studies in an experimental model for migraine', Headache 13, 142-163.
15. Puzich R., Girke W., Heidrich H., and Rischke M. (1983) 'Dopplersonographische Untersuchungen der extrakraniellen Hirngefäse bei Migräne-Patienten nach Gabe von Ergotamintartrat', Dtsch. Med. Wochenschr. 108, 457-461.
16. Saxena P.R. (1987) 'Arteriovenous anastomoses and veins in migraine research', in J.N. Blau (eds.), Migraine, Clinical, Therapeutic, Conceptual and Research Aspects, Chapman and Hall Medical, London, pp. 581-596.
17. Den Boer M.O., Heiligers J.P.C., and Saxena P.R. (1991) 'Carotid vascular effects of ergotamine and dihydroergotamine in the pig: no exclusive mediation via 5-HT_1-like receptors', Br. J. Pharmacol. 104, 183-189.
18. Kerber C.W. and Newton T.H. (1973) 'The macro and microvasculature of the dura mater', Neuroradiology 6, 175-179.
19. Saxena P.R. and Verdouw P.D. (1982) 'Redistribution by 5-hydroxytryptamine of carotid arterial blood at the expense of arteriovenous blood flow', J. Physiol. (Lond). 332, 501-520.
20. Den Boer M.O., Somers J.A.E., and Saxena P.R. (1992) 'The porcine dura mater: arteriovenous anastomotic shunting is not affected by the

antimigraine drugs sumatriptan, ergotamine and dihydroergotamine', Br. J. Pharmacol. 107, 577-583.
21. Saito K., Markowitz S., and Moskowitz M.A. (1988) 'Ergot alkaloids block neurogenic extravasation in dura mater: Proposed action in vascular headaches', Ann. Neurol. 24, 732-737.
22. Saxena P.R. and Ferrari M.D. (1992) 'From serotonin receptor classification to the antimigraine drug sumatriptan', Cephalalgia 12, 187-196.
23. Humphrey P.P.A., Apperly E., Feniuk W., and Perren M.J. (1990) 'A rational approach to identifying a fundamentally new drug for the treatment of migraine', in P.R. Saxena, D.I. Wallis, W. Wouters and P. Bevan (eds.), Cardiovascular Pharmacology of 5-Hydroxytryptamine, Prospective Therapeutic Applications, Kluwer Academic Publishers, Dordrecht, pp. 417-428.
24. Den Boer M.O., Villalón C.M., Heiligers J.P.C., Humphrey P.P.A., and Saxena P.R. (1991) 'Role of 5-HT_1-like receptors in the reduction of porcine cranial arteriovenous anastomotic shunting by sumatriptan', Br. J. Pharmacol. 102, 323-330.
25. Perren M.J., Feniuk W., and Humphrey P.P.A. (1989) 'The selective closure of feline arteriovenous anastomoses (AVAs), by GR43175', Cephalalgia 9 (Suppl. 9), 41-46.
26. Nielsen T.H. and Tfelt-Hansen P. (1989) 'Lack of effect of GR43175 on peripheral arteries in man', Cephalalgia 9 (Suppl. 9), 93-95.
27. Buzzi M.G. and Moskowitz M.A. (1990) 'The antimigraine drug, sumatriptan (GR43175), selectively block neurogenic plasma extravasation from blood vessels in dura mater', Br. J. Pharmacol. 99, 202-206.
28. Sleight A.J., Cervenka A., and Peroutka S.J. (1990) '*In vivo* effects of sumatriptan (GR 43175), on extracellular levels of 5-HT in the guinea pig', Neuropharmacology 29, 511-513.
29. Humphrey P.P.A. and Feniuk W. (1991) 'Mode of action of the anti-migraine drug sumatriptan', Trends Pharmacol. Sco. 12, 444-446.
30. Moskowitz M.A. (1992) 'Neurogenic versus vascular mechanisms of action of sumatriptan and ergot alkaloids in migraine', Trends Pharmacol. Sci. 13, 307-311.
31. Caekebeke J.F.V., Ferrari M.D., Zwetsloot C.P., Jansen J., and Saxena P.R. (1992) 'Antimigraine drug sumatriptan increases blood flow velocity in large cerebral arteries during migraine attacks', Neurology 42, 1522-1526.
32. Butler A., Worton S., and Connor H.E. (1992) 'Sumatriptan does not attenuate sensory nerve-mediated relaxations of guinea-pig isolated pulmonary artery', Br. J. Pharmacol. 105, 273P.
33. Waldron G.J., O'Shaughnessy C.T., and Connor H.E. (1992) 'Sumatriptan does not modify capsaicin-induced relaxation of guinea-pig isolated basilar artery', Br. J. Pharmacol. 105, 272P.

SB 203186, A POTENT 5-HT$_4$ ANTAGONIST FOR CARDIAC 5-HT$_4$ RECEPTORS

ALBERTO J. KAUMANN*, ANDREW D. MEDHURST, PENNY BOYLAND, MYTHILY VIMAL, and RODNEY C. YOUNG
*Clinical Pharmacology Unit
Cambridge University, Addenbrooke's Hospital
Cambridge CB2 2QQ and
SmithKline Beecham Pharmaceuticals
The Frythe
Welwyn, Hertforshire AL6 9AR
UK

ABSTRACT. To select an antagonist as a tool for 5-HT$_4$ receptor blockade we investigated the affinity of three compounds for 5-HT$_4$ receptors and 5-HT$_3$ receptors. SB 203186 ([1-piperidinyl] ethyl 1H-indole 3-carboxylate) was a more potent antagonist (pK$_B$ = 8.26) than tropisetron (pk$_B$ = 6.89) and SDZ 205-557 ([2-diethylaminoethyl] 2-methoxy-4-amino-5-chloro-benzoate) (pK$_B$ = 7.35) of 5-HT-evoked tachycardia on piglet sinoatrial 5-HT$_4$ receptors. The three compounds also bound to rat brain 5-HT$_3$ receptors labelled with [^{3}H]-BRL 43694 (granisetron). SB 203186 was 12 times 5-HT$_4$-selective, SDZ 205-557 1.6 times 5-HT$_4$-selective, and tropisetron 275 times 5-HT$_3$-selective. SB 203186 is useful to characterize 5-HT$_4$ receptors.

1. Introduction

5-HT$_4$ receptors mediate a variety of effects in the central nervous system, gut, and heart [1]. There is a need for selective blocking agents of 5-HT$_4$ receptors. After the introduction of 5-HT$_3$ receptor-selective tropisetron (ICS 205-930) as a weak surmountable blocker of 5-HT$_4$ receptors [2-4] identified SDZ 205-557 ([2-diethylaminoethyl] 2-methoxy-4-amino-5-chloro-benzoate) as a selective 5-HT$_4$ receptor antagonist. We now present data showing that SB 203186 ([1-piperidinyl] ethyl 1H-indole 3-carboxylate) is a 5-HT$_4$ receptor antagonist more potent and more selective (with respect to 5-HT$_3$ receptors than SDZ 205-557. The structures of the 3 compounds are shown in figure 1.

SB 203186

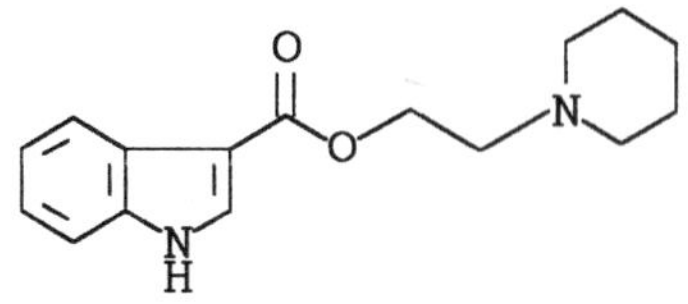

SDZ 205557

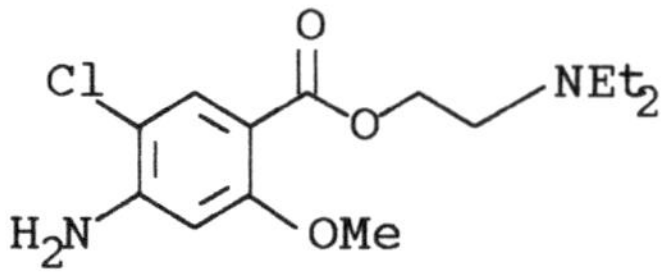

TROPISETRON

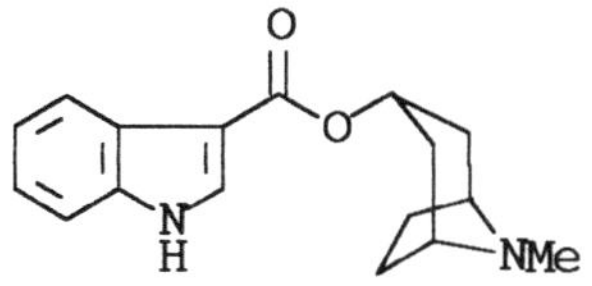

Figure 1. Formulae of 5-HT$_4$ receptor antagonists.

2. Methods

We estimated the affinity of SB 203186, SDZ 205-557, and tropisetron for 5-HT$_4$ receptors of piglet atrium [5] and 5-HT$_3$ receptors of rat brain. The antagonism of 5-HT-evoked tachycardia was investigated on spontaneously beating right atria (a model for human atrial 5-HT$_4$ receptors) [6,7], in the presence of ascorbate (0.2 mM), cocaine (6 μM), and ($\pm$)-propranolol (400 nM) as described by Kaumann [5]. A single concentration-effect curve to 5-HT was determined in the absence or presence of antagonist (preincubated for at least 1 hour). 5-HT concentration-ratios, caused by the antagonists, were calculated and apparent pK$_B$ values estimated, also as described by Kaumann [5].

The affinity of the antagonists for 5-HT$_3$ receptors was estimated from the inhibition of binding of [^{3}H]-BRL 43694 (granisetron) to previously frozen crude entorhinal cortical and hippocampal tissue obtained from male Hooded Listar rats. The tissues were homogenized (Ultra Turrax T25; 10 s at 20,500 rpm) in 40 volumes of ice-cold Hepes buffer (50 mM, pH 7.5). Homogenates were allowed to stand at room temperature for 15 minutes and placed on ice until use. The binding assay consisted of 0.1 ml [^{3}H]-BRL 43694 (0.3 nM, specific activity 61 Ci/mmole; NEN Dupont), 0.1 ml of competing drug and 0.6 ml of tissue homogenate containing approximately 0.8 mg protein as measured by the method of Bradford [8]. At least 8 concentrations (2 per log unit) of competing compound were used. The incubation was at 23°C for 30 minutes and terminated by filtering rapidly through

GF/B filters (presoaked for 30 minutes in 0.1% polyethyleneimine) using a 48 well Brandel cell harvester. The filters were rinsed immediately with 3 x 5 ml of ice-cold Hepes buffer and radioactivity measured by liquid scintillation spectrometry. Specific binding was defined as the excess over blank in the presence of metoclopramide (100 μM).

3. Results

The three antagonists of table 1 were devoid of stimulant or depressant effects but caused surmountable antagonism of the positive chronotropic effects of 5-HT. Comparison of apparent pK_B values revealed that SB 203186 was 23 and 8 times more potent that tropisetron and SDZ 205-557, respectively, as blocker of piglet sinoatrial 5-HT$_4$ receptors. The three compounds of table 1 competed with [^{3}H]-BRL 43694 for binding to rat brain 5-HT$_3$ receptors. The affinity of tropisetron for

TABLE 1. Comparison of the affinity estimates of SB 203186, SDZ 205-557, and tropisetron for 5-HT$_3$ and 5-HT$_4$ receptors.

Piglet Right Atrium (5-HT$_4$)			
Antagonist	pK_B^a, M	n	Schild slope
SB 203186	8.26 ± 0.09	10	0.96 (10-3000 nM)
SDZ 205-557	7.35 ± 0.11	12	0.58 (0.1-10 μM)
Tropisetron	6.89 ± 0.12	7	1.09 (0.3-3.0 μM)
Rat Brain (5-HT$_3$)			
Antagonist	pK_D, M^b	n	Slope
SB 203186	7.19 ± 0.04	3	1.12
SDZ 205-557	7.18, 7.10	2	1.09, 1.03
Tropisetron	9.33 ± 0.06	3	1.01

^{a}: K_B values were calculated by forcing unit slope through Schild-plot data; ^{b}: Each binding assay was run in triplicate; n: Number of animals; Data are mean ± sem.

5-HT$_3$ receptors was around 200 times higher than the affinities of both SB 203186 and SDZ 205-557. Comparison of affinity estimates for piglet sinoatrial 5-HT$_4$ receptors and rat brain 5-HT$_3$ receptors, SDZ 205-557 is only marginally selective for 5-HT$_4$ receptors (1.6 times), and tropisetron is 275 times selective for 5-HT$_3$ receptors.

4. Discussion

Our results show that SB 203186 is a potent competitive antagonist of the positive chronotropic effects of 5-HT mediated through piglet sinoatrial 5-HT$_4$ receptors. The blocking potency of SB 203186 is 8 and 23 times higher than that of SDZ 205-557 and tropisetron, respectively. The high blocking potency of SB 203186 has recently been verified *in vivo* in anesthetized Yucatan minipigs; SB 203186 was 20 times more potent than tropisetron as antagonist of 5-HT-evoked tachycardia [9]. Furthermore, on isolated human atrial 5-HT$_4$ receptors SB 203186 was 100 times more potent than tropisetron as antagonist of 5-HT-induced positive inotropic effects [9].

Comparison of the affinities of SB 203186 for piglet sinoatrial 5-HT$_4$ receptors and rat brain 5-HT$_3$ receptors revealed a 12-fold selectivity for 5-HT$_4$ receptors. In agreement with other workers [10], tropisetron was 275 times selective for 5-HT$_3$ receptors. Also in agreement with Eglen et al. [10] we found that SDZ 205-557 was only marginally selective for 5-HT$_4$ receptors, which is at variance with the results of Buchheit et al. [3,4] who using guinea pig ileum showed a 30-fold selectivity for 5-HT$_4$ receptors. The reason for the discrepancy is that guinea pigs possess a 5-HT$_3$ receptor homologue that has systematically lower affinities for ligands than 5-HT$_3$ receptors for other species [11]. The higher affinity and selectivity of SB 203186 for 5-HT$_4$ receptors, compared to SDZ 205-557, make SB 203186 a better tool for further characterization of 5-HT$_4$ receptors.

5. References

1. Bockaert, J., Fozard, J., Dumuis, A., and Clarke, D.E. (1992) 'The 5-HT$_4$ receptor: a place in the sun', Trends Pharmacol. Sci. 13, 141-145.
2. Dumuis, A., Bouhelal, R., Sebben, M., Cory, R., and Bockaert, J. (1988) 'A nonclassical 5-hydroxytryptamine receptor positively coupled with adenylate cyclase in the central nervous system', Mol. Pharmacol. 34, 880-887.
3. Buchheit, K.-H., Gamse, R., and Pfannkuche, H.-J. (1991) 'SDZ 205-557, a selective antagonist at 5-HT$_4$ receptors in the isolated guinea pig ileum', Eur. J. Pharmacol. 200, 373-374.
4. Buchheit, K.-H., Gamse, R., and Pfannkuche, H.-J. (1992) 'SDZ 205-557, a selective, surmountable antagonist for 5-HT$_4$ receptors in the isolated guinea pig ileum', Naunyn Schmiedeberg's Arch. Pharmacol. 345, 387-393.
5. Kaumann, A.J. (1990) 'Piglet sinoatrial 5-HT receptors resemble human atrial 5-HT$_4$-like receptors', Naunyn Schmiedeberg's Arch. Pharmacol. 342, 619-622.
6. Kaumann, A.J., Sanders, L., Brown, A.M., Murray, K.J., and Brown, M.J. (1990) 'A 5-hydroxytryptamine receptor in human atrium', Br. J. Pharmacol. 100, 879-885.
7. Kaumann, A.J., Sanders, L., Brown, A.M., Murray, K.J., and Brown, M.J.

(1991) 'A 5-HT$_4$-like receptor in human atrium', Naunyn Schmiedeberg's Arch. Pharmacol. 344, 150-159.

8. Bradford, M. (1976) 'A rapid and sensitive method for the quantitation of microgram quantities of protein utilizing the principle of protein-dye binding', Anal. Biochem. 72, 248-254.

9. Parker, S.G., Hamburger, S., Taylor, E.M., and Kaumann, A.J. (1993) 'SB 203186, a potent 5-HT$_4$ receptor antagonist, in porcine sinoatrial node and human and porcine atrium', Br. J. Pharmacol., in press.

10. Eglen, R.M., Alvares, R., Johnson, L.G., Leung, E., and Wong, E.H.G. (1993) 'The action of SDZ 205-557 at 5-hydroxytryptamine (5-HT$_3$ and 5-HT$_4$) receptors', Br. J. Pharmacol. 108, 376-382.

11. Kilpatrick, G. and Tyers, M.B. (1992) 'Interspecies variants of the 5-HT$_3$ receptor', Biochem. Soc. Trans. 20, 118-123.

ROLE OF THE LATERAL TEGMENTAL FIELD IN THE SYMPATHOLYTIC ACTION OF 5-HT$_{1A}$ AGONISTS

ROBERT B. McCALL and MARK E. CLEMENT
The Upjohn Company
Kalamazoo, Michigan 49001
USA

ABSTRACT. This study was designed to examine the role of the lateral tegmental field (LTF) in the sympatholytic effect of 5-HT$_{1A}$ agonists such as 8-OH-DPAT. Three types of sympathetic-related neurons were identified in the lateral tegmental field of anesthetized cats based on their response to baroreceptor activation (i.e. sympathoexcitatory, sympathoinhibitory, or baroreceptor unresponsive). Intravenous 8-OH-DPAT inhibited sympathoexcitatory neurons with a high degree of correlation to inhibition of sympathetic activity, and excited sympathoinhibitory neurons in a dose-dependent manner. Iontophoretic application of 8-OH-DPAT and 5-HT onto sympathoexcitatory neurons caused inhibition of spontaneous activity. Iontophoretic application of 8-OH-DPAT to sympathoinhibitory neurons had variable effects although 5-HT consistently caused excitation. Microinjection of kainic acid into the LTF blocked the sympatholytic effects of 8-OH-DPAT but not that produced by clonidine. These data indicate that 8-OH-DPAT acts in the LTF to inhibit sympathetic nerve activity and therefore lower arterial blood pressure.

1. Introduction

5-HT$_{1A}$ receptor agonists such as 8-OH-DPAT cause a centrally mediated reduction in sympathetic nerve discharge [1,2]. Several studies have attempted to identify the site and mechanism by which 5-HT$_{1A}$ agonists produce their sympatholytic effects. Midcollicular transection failed to affect the sympatholytic action of 8-OH-DPAT indicating the drug inhibits sympathetic activity at a site in the lower brain stem or spinal cord [3]. It is unlikely that the sympatholytic effect of 8-OH-DPAT results from an inhibitory action on medullary 5-HT neurons which project to autonomic area of the brainstem and spinal cord. In this regard, there is a lack of correlation between the inhibition of medullary 5-HT neuronal firing and the decrease in sympathetic nerve activity in response to systemic administration of 8-OH-DPAT [4]. Furthermore, extensive lesions of the midline medulla and pons fail to alter the sympathetic inhibition produced by 8-OH-DPAT [4]. Thus, 5-HT$_{1A}$ agonist-induced inhibition of 5-HT cell firing does not account for the sympatholytic effects of 5-HT_{1A} agonists. Similarly a spinal action is unlikely since microiontophoretic application of 8-OH-DPAT fails to affect the firing of sympathetic preganglionic neurons located in the intermediolateral cell column [3].

Microinjection or local application of 8-OH-DPAT into the rostral ventrolateral medulla (RVLM) decreases blood pressure, heart rate, and sympathetic nerve discharge (SND) in rats, dogs, and cats [5-7]. The action of 8-OH-DPAT in this area has been further localized by examination of individual neurons thought to be associated with cardiovascular regulation [3]. These studies illustrate a remarkable correlation between the inhibition of sympathoexcitatory neurons in the RVLM and the decrease in SND in response to intravenous doses of 8-OH-DPAT. These results suggest that 8-OH-DPAT may exert its sympatholytic effect in part by inhibiting neurons in the RVLM responsible for the generation of sympathetic tone. Microiontophoresis of 8-OH-DPAT onto these neurons in the cat had no effect on spontaneous activity however, suggesting that the receptors for 8-OH-DPAT are located on distal dendrites or on neurons located antecedent to RVLM sympathoexcitatory neurons [3].

Evidence suggests that the lateral tegmental field (LTF) of the medulla is a source of basal sympathetic activity which lies antecedent of the RVLM [8,9]. The LTF contains both sympathoexcitatory and sympathoinhibitory neurons. Sympathoexcitatory neurons in the LTF have been shown to project to the RVLM [8]. Based on the fact that sympathoexcitatory neurons in the LTF fulfill the criteria of being a putative source of basal sympathetic activity which lies antecedent to the RVLM, and on the lack of a direct effect of 8-OH-DPAT on RVLM sympathoexcitatory neurons, we investigated the role of the LTF in the sympatholytic action of 8-OH-DPAT. Specifically, we sought to determine the effects of intravenous and iontophoretic 8-OH-DPAT on the spontaneous activity of sympathetic-related neurons in the LTF. In addition, we determined the effect of LTF kainic acid lesions on the sympatholytic effects of 8-OH-DPAT. The data indicate that neurons in the LTF play an important role in the sympatholytic effect of 5-HT_{1A} agonists.

2. Material and Methods

Cats were administered a preoperative dose of ketamine HCl (11 mg/kg) and anesthetized by an intravenous injection of α-chloralose (80 mg/kg) or diallybarbiturate sodium (60 mg/kg), urethane (240 mg/kg), and monoethylurea (240 mg/kg). A femoral artery and vein were cannulated to measure arterial blood pressure and to permit intravenous drug administration, respectively. A Fogarty embolectomy catheter was inserted in the opposite femoral artery to permit occlusion of the descending aorta. The dorsal aspect of the medulla was exposed by removal of portions of the overlying occipital bone and cerebellum. Following surgery, animals were immobilized with gallamine triethiodide (4 mg/kg, IV) and artificially respired.

The obex was used as a surface landmark for placement of the recording electrode. Neurons were recorded in an area previously described by Barman and Gebber [8,9]. Unitary discharges of sympathetic-related neurons in intravenous studies were recorded using extracellular techniques with a tungsten microelectrode. A platinum reference electrode was placed on the frontal bone. Single-unit action potentials were amplified using capacity-coupled preamplification and displayed on an oscilloscope. In other experiments 5-barrel glass microelectrodes containing monofilament glass fibers were used to simultaneously record the extracellular potentials from single neurons and to microiontophoretically apply drugs at the recording site. A retaining current of 15 nA was applied between ejecting periods. Sympathetic related neurons in the LTF were identified using previously described criteria [3,8].

Peripheral sympathetic nerve activity was recorded from the central end of the sectioned left postganglionic inferior cardiac nerve. Potentials were recorded monophasically under mineral oil with a bipolar platinum electrode. A band pass of 1-1000 Hz was used to display the synchronized discharges of the whole sympathetic nerve in the form of slow waves.

Microinjections were made in an area in which we found sympathetically related neurons in the LTF (see below). Pipette tip diameters were approximately 40 microns. Microinjections of 1.25-2.5 μM (50 to 100 nL) of glutamic acid, NMDA, or kainic acid were made by advancing the glass microelectrode using a David Kopf Microdrive connected to the plunger of a 1 μL Hamilton syringe. Pontamine sky blue was incorporated into the drug solutions for later identification of injection sites. Injection times lasted approximately 5 minutes per site. Evoked potentials in cardiac nerve discharge produced by electrical (0.5 ms square wave, 5 volt) stimulation of the RVLM were used to determine that the descending fibers from the RVLM remained intact.

The brainstem was removed at the end of each experiment. Frozen sagittal sections of 20 μm thickness were cut with a cryostat microtome and stained with nuclear fast red. Lesion and stimulation sites were identified in these sections according to the stereotaxic atlas of Berman [10].

3. Results

Sympathetic neurons recorded from the LTF were characterized by a slow (i.e. 2.5 Hz), irregular spontaneous firing rate (2.5 ± 0.2 Hz) which exhibited a temporal relationship to inferior cardiac nerve discharge and to the cardiac cycle. Units were classified as either sympathoexcitatory if an increase in blood pressure resulted in a decrease in firing rate or as sympathoinhibitory if firing rate increased during baroreceptor activation. A third group of sympathetic-related neurons was identified which did not respond to changes in blood pressure. These three types of neurons were found dispersed within the same general location, i.e anterior 1.5-

3.0 mm from the obex, lateral 2.5-3.5 mm from the midline, and ventral 2.0-4.0 mm from the dorsal surface of the medulla.

In order to examine the sympatholytic effects of 8-OH-DPAT in the LTF we simultaneously recorded activity from the inferior cardiac nerve and sympathetic-related neurons in the LTF. Increasing intravenous doses of 8-OH-DPAT resulted in a dose-related inhibition of sympathoexcitatory neuronal firing and inferior cardiac sympathetic nerve discharge and an increase in the firing of LTF sympathoinhibitory neurons. Figure 1 illustrates data from twelve experiments and reveals a high degree of correlation between overall LTF sympathoexcitatory neuronal inhibition and the sympathetic inhibition elicited by intravenous 8-OH-DPAT. The dose of 8-OH-DPAT required to produce a 50% inhibition of LTF sympathoexcitatory unit activity and sympathetic activity was 6 μg/kg and a 90% reduction in both parameters was achieved at 30 μg/kg. These data suggest that LTF sympathoexcitatory neurons are an integral part of the sympatholytic action of 8-OH-DPAT. In contrast LTF sympathoinhibitory neurons exhibited an inverse relationship to the inhibition of SND produced by intravenous 8-OH-DPAT.

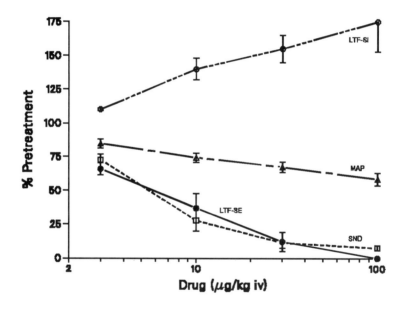

Figure 1. Cumulative dose response curve illustrating effects of 8-OH-DPAT on firing rate of sympathoexcitatory (LTF-SE) and sympathoinhibitory (LTF-SI) neurons, sympathetic nerve discharge (SND) and mean arterial blood pressure (MAP).

To more thoroughly examine the effects of 8-OH-DPAT on individual neurons in the LTF, we microiontophoretically applied 8-OH-DPAT and 5-HT directly onto sympathetic-related neurons in this area. Figure 2 represents a rate histogram depicting the inhibitory action of 8-OH-DPAT and 5-HT on a LTF sympathoexcitatory neuron (n=7). Ejection currents of 30 nA 8-OH-DPAT and 5-HT were equally effective in reducing the spontaneous firing rate of this neuron. Inhibitory effects of 8-OH-DPAT lasted 3 to 4 times longer the inhibitory effect of 5-HT. Three LTF-SE units were found which did not respond to microiontophoretic 5-HT or 8-OH-DPAT.

We also examined the effects of iontophoretic 8-OH-DPAT and 5-HT on the sympathoinhibitory neurons found in the LTF. The majority of sympathoinhibitory neurons exhibited an excitatory response to 5-HT, although one neuron was inhibited and one unit had no response. LTF sympathoinhibitory neurons were typically unresponsive to direct application of 8-OH-DPAT.

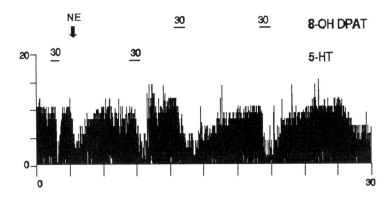

Figure 2. Effects of microiontophoretically applied 8-OH-DPAT and 5-HT on firing of LTF sympathoexcitatory neuron. NE is intravenous norepinephrine. Ordinate cell discharges/10 seconds. Abscissa is in minutes.

In order to further investigate the importance of the LTF in the sympatholytic action of 5-HT_{1A} agonists, we studied the effects of LTF kainic acid microinjections on the sympathoinhibitory effect of 8-OH-DPAT. In this regard, kainic acid has been reported to destroy cell bodies while leaving fibers of passage intact [11]. Incremental intravenous doses of 8-OH-DPAT or clonidine were given thirty minutes following bilateral microinjection of 50 to 100 nL kainic acid (1.25-2.5 μM) into the LTF. Figure 3 illustrates the effects of bilateral microinjections of kainic acid into the LTF, and the subsequent administration of IV 8-OH-DPAT. As can be seen from this example, 8-OH-DPAT had no effect on sympathetic activity or blood pressure at doses which normally result in marked hypotensive effects in control animals. The cumulative result of eleven of these experiments is summarized in figure 4. This figure represents a dose response curve of IV 8-OH-DPAT in kainic acid microinjected and sham animals which received microinjections of saline. This figure reveals the marked difference in the response of the two groups, with 8-OH-DPAT causing only slight reductions in blood pressure and heart rate, and a slight increase in sympathetic activity in kainic-acid-injected animals, compared to the usual reductions of all three parameters in sham animals. In contrast to 8-OH-DPAT, clonidine inhibited sympathetic activity in kainic-acid-lesioned animals. Histological examination of the microinjection sites revealed the extent of the diffusion of kainic acid to range from 300 to 500 microns from the point of injection, with diffusion being slightly greater along the dorsal-ventral axis. The extent of the injection as indicated by the dispersion of pontamine sky blue dye was described by a roughly circular area approximately 400 μm in diameter. The symmetrical nature of the injection site was reflected along the rostro-caudal axis.

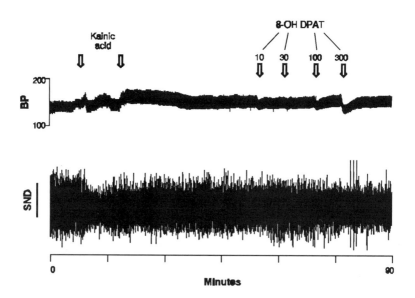

Figure 3. Effect of bilateral kainic acid microinjections (50 nl) into the LTF on the sympatholytic effect of intravenous 8-OH-DPAT (10-300 μg/kg). BP is blood pressure. SND is sympathetic nerve discharge.

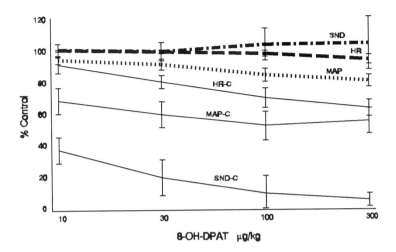

Figure 4. Cumulative dose response curve illustrating effects of 8-OH-DPAT on mean arterial blood pressure (MAP), heart rate (HR) and sympathetic nerve discharge (SND) in control (C) and LTF lesioned animals.

4. Discussion

The present study provides evidence that the LTF plays an important role in the sympatholytic action of 5-HT_{1A} agonists. We found that intravenous 8-OH-DPAT produced a dose-dependent inhibition of sympathoexcitatory neurons identified in the lateral tegmental field. Reduction in inferior cardiac nerve activity exhibited a direct relationship to unit inhibition. Furthermore, direct application of 8-OH-DPAT and 5-HT by microiontophoresis inhibited the discharge frequency of the majority of these neurons, suggesting that 5-HT_{1A} receptors mediating the sympatholytic effects of 8-OH-DPAT reside on these neurons. Sympathoexcitatory neurons in the LTF project to the vicinity of sympathoexcitatory neurons in the RVLM [8]. The response of both RVLM and LTF sympathoexcitatory neurons to intravenous 8-OH-DPAT is nearly identical. Both neuronal types are inhibited and the inhibition is tightly correlated with the shutoff of inferior cardiac sympathetic nerve activity. Iontophoretic 8-OH-DPAT inhibits LTF, but not RVLM, sympathoexcitatory neurons [3]. These data suggest that 8-OH-DPAT induced inhibition of RVLM sympathoexcitatory neurons is secondary to inhibition of LTF sympathoexcitatory neurons.

This conclusion is further supported by our kainic acid experiments. We found that bilateral microinjections of kainic acid into the LTF of the cat medulla abolishes the sympatholytic action of intravenous 8-OH-DPAT. Lesions in the current series of experiments were restricted to the LTF corresponding to the dorsomedial pressor area and including the region in which sympathetically related neurons have been identified by our laboratory and others [8,9]. The integrity of the RVLM and the associated descending sympathoexcitatory pathway remained intact as evidenced by the evoked sympathetic nerve potentials which could be evoked by microstimulation of the RVLM. The sympatholytic effects of clonidine was not affected by kainic acid LTF lesions. Thus destruction of the neurons in the area of the LTF blocks the usual 8-OH-DPAT induced inhibition of sympathetic activity and decrease of blood pressure, indicating that integrity this area is crucial to the sympatholytic mechanism of 8-OH-DPAT.

In conclusion sympathoexcitatory neurons of the lateral tegmental field appear to be involved in the sympatholytic effects of 8-OH-DPAT and may represent the primary site of action thereof. In this regard, these units are sensitive to 8-OH-DPAT and 5-HT. In particular, sympathoexcitatory neurons in the LTF, which send projections to and influence RVLM sympathoexcitatory neurons, are directly sensitive to the inhibitory actions of 8-OH-DPAT. Finally, discrete lesions of neuronal cell bodies in the LTF prevent the sympatholytic effect of 8-OH-DPAT.

5. References

1. McCall, R.B., Patel, B.N., and Harris, L.T. (1987) 'Effects of serotonin$_1$ and serotonin$_2$ receptor agonists and antagonists on blood pressure, heart rate and sympathetic nerve activity', J. Pharmacol. Exp. Ther. 242, 1152-1159.
2. Ramage, A.G. and Fozard, J.R. (1987) 'Evidence that the putative 5-HT$_{1A}$ receptor agonists 8-OH-DPAT and ipsapirone have a central hypotensive action that differs from that of clonidine in anaesthetized cats', Eur. J. Pharmacol., 138, 179-191.
3. Clement, M.E. and McCall, R.B. (1990) 'Studies on the site and mechanism of the sympatholytic action of 8-OH-DPAT', Brain Research 525, 232-241.
4. McCall R.B., Clement M.E., and Harris L.T. (1989) 'Studies on the mechanism of the sympatholytic effect of 8-OH-DPAT: lack of correlation between inhibition of serotonin neuronal firing and sympathetic activity', Brain Research 501, 73-83.
5. Gillis, R.A., Hill, K.J., Kirby, J.S., Quest, J.A., Hamosh, P., Norman, W.P., and Kellar, K.J. (1989) 'Effect of activation of central nervous system serotonin 1A receptors on cardiorespiratory function', J. Pharmacol. Exp. Ther. 248, 851-857.
6. Laubie, M., Drouillat, M., Dabire, H., Cherqui, C., and Schmitt, H. (1989) 'Ventrolateral medullary pressor area: site of hypotensive and sympatho-inhibitory effects of 8-OH-DPAT in anesthetized dogs', Eur. J. Pharmacol., 160, 385-394.
7. Lovick, T.A. (1989) 'Systemic and regional hemodynamic responses to microinjection of 5-HT agonists in rostral ventrolateral medulla in the rat',

Neurosci. Lett, 107, 157-161.
8. Barman S.M. and Gebber G.L. (1987) 'Lateral tegmental field neurons of cat medulla: a source of basal activity of ventral medullospinal sympathoexcitatory neurons', J. Neurophysiol. 57, 1410-1424.
9. Gebber G.L. and Barman S.M. (1985) 'Lateral tegmental field neurons of cat medulla: a potential source of basal sympathetic discharge', J. Neurophysiol., 54, 1498-1512.
10. Berman, A.L. (1968) The Brain Stem of the Cat, University of Wisconsin Press, Madison.
11. McGeer, P.L. and McGeer, E.G. (1982) 'Kainic acid: The neurotoxic breakthrough', CRC Crit. Rev. Toxicology, 1-26.

URAPIDIL: THE ROLE OF 5-HT$_{1A}$ AND α-ADRENERGIC RECEPTORS IN BLOOD PRESSURE REDUCTION

N. KOLASSA, K.D. BELLER, R. BOER, H. BOSS, and K.H. SANDERS
Byk Gulden
Postfach 10 03 10
D-7750 Konstanz
Germany

ABSTRACT. The CNS-mediated hypotensive response to urapidil is attributed to 5-HT$_{1A}$ receptor stimulation in anesthetized cats. This effect significantly participates in the blood pressure reduction following peripheral drug administration besides inhibition of vascular α1-adrenoceptors. In conscious dogs, the α1-antagonist effect prevents the appearance of a 5-HT syndrome, which is seen after 'pure' 5-HT$_{1A}$ agonists, and contributes to the good tolerability of urapidil. It is concluded that the well-balanced properties of the hybrid molecule of urapidil make it an efficient and well-tolerable antihypertensive drug.

1. Introduction

Urapidil is an antihypertensive drug which was characterized at first as a peripheral α1-adrenoceptor antagonist with additional central hypotensive activity unrelated to inhibition of α1-adrenoceptors [1]. The notion that the central effect might be due to stimulation of α2-adrenoceptors was ruled out: urapidil does not stimulate α2-adrenoceptors but in fact inhibits α2-adrenoceptor-mediated effects on, for example, vascular system and transmitter release [2,3,4]. Moreover, the central hypotensive action of urapidil is insensitive to pretreatment with the α2-antagonist yohimbine [4].

When Ramage [5] investigated the influence of urapidil on preganglionic sympathetic nerve activity in anesthetized cats, he pointed out (i) differences between urapidil and 'pure' α1-adrenoceptor antagonists and (ii) similarities of urapidil to 5-HT$_{1A}$ receptor agonists. Thereafter, Fozard and Mir [6] showed that the potency of urapidil as a partial agonist at 5-HT$_{1A}$ receptors in several isolated tissues is similar to its potency as an α1-adrenoceptor antagonist.

In the following we describe experiments which identified the central hypotensive effect of urapidil as the consequence of 5-HT$_{1A}$ receptor stimulation in the hindbrain of anesthetized cats. In addition, studies in conscious dogs suggest that the balance between the 5-HT$_{1A}$ agonist and α1-antagonist properties of urapidil is the key to the prevention of a 5-HT behavioral syndrome.

2. Materials and Methods

Receptor binding experiments were performed with pig cortical membranes in the presence of a fixed concentration (0.1 to 0.3 nM) of (^{3}H)8-OH-DPAT (8-hydroxy-2-[di-n-propylamino]tetralin) and increasing concentrations of competing drugs. Nonspecific binding was determined in the presence of 10 μM serotonin. Furthermore, experiments were performed with rat cortical membranes in the presence of a fixed concentration of (^{3}H)prazosin (0.2 to 0.5 nM) and increasing concentrations of competing drugs. Nonspecific binding was determined in the presence of 1 μM prazosin or 10 μM phentolamine. Bound and free radioactivity were separated by filtration and the radioactivity retained by the filter was measured by liquid scintillation counting. Data were analyzed by the ligand program from Elsevier Biosoft (Amsterdam).

Normotensive mongrel cats of either sex were anesthetized with a mixture of α-glucochloralose and ethylurethane (80 and 120 mg/kg i.p.). After tracheotomy and left-sided thoracotomy, the animals were artificially ventilated with room air. A catheter was introduced into the left subclavian artery and pushed upstream until its tip was situated just distal to the ostium of the vertebral artery [7]. A femoral artery and vein were cannulated for blood pressure recording and for IV administration of drugs, respectively. Drugs were injected during 1 minute into either the femoral vein or the vertebral artery in a volume of 0.2 ml; pretreatment was given 15 minutes before the test drug. Maximal decreases in mean arterial blood pressure were noted.

Six normotensive, male beagle dogs were used in a cross-over design. The animals were trained to stand in a frame with a loose-fitting dog sling. Arterial pressure was measured every 5 minutes by an oscillometric cuff method with a Dinamap®-1846 (Critikon, Tampa, Florida). After a 30-minute run-in period, drugs were injected subcutaneously in the neck region and post-treatment blood pressure values were averaged for the following 60 minutes. The appearance of tremor, salivation, and hyperventilation was noted and the severity of these behavioral changes was rated as absent, low, medium, or high (score 0,1,2,3).

3. Results

Urapidil reduces blood pressure at lower doses when injected into the vertebral artery than after IV administration in anesthetized cats (compare figures 1 and 2) [4,8]. Such a difference in the potency of an antihypertensive drug is taken as evidence for a CNS-mediated action on blood pressure regulation, since the vertebral artery in cats delivers the drug to the brain stem [7]. When the 5-HT$_{1A}$ antagonists, spiroxatrine, or (-)pindolol are given into the vertebral artery 15 minutes before administration of urapidil via the same route, the hypotensive effect of urapidil is reduced in relation to the dose (figure 1). The results support the assumption that the central hypotensive effect of urapidil is due to stimulation of 5-HT$_{1A}$ receptors in the hindbrain.

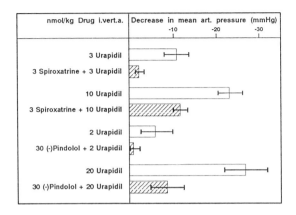

Figure 1. Influence of pretreatment with spiroxatrine or (-)pindolol on the hypotensive effect of urapidil; drugs were administered via the vertebral artery. Means ± sem, 6 cats; data from [9,10].

A further aim was to demonstrate that the central hypotensive action of urapidil is also operative under the conditions of peripheral drug administration. As can be seen in figure 2, the central administration of the 5-HT$_{1A}$ antagonist spiroxatrine significantly attenuates the blood pressure reduction following IV injection of urapidil.

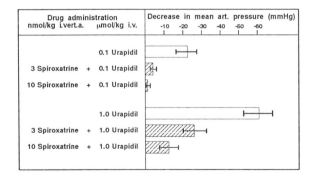

Figure 2. Influence of pretreatment with spiroxatrine via the vertebral artery on the hypotensive effect of urapidil given IV. Means ± sem, 6 cats; data from [8].

TABLE 1. Drug affinities (pK$_i$, -log mol/l) for 5-HT$_{1A}$ and α1-adrenergic receptor binding sites.

Drug	5-HT$_{1A}$ site	α1 site	(α1A:α1B)
Prazosin	5.38	9.98	
Urapidil	7.74	7.07	
5-Methyl-urapidil	9.47	7.94	(9.11:7.17)
Flesinoxan	8.95	6.68	
8-OH-DPAT	9.72	5.5	

Drugs with different affinities for 5-HT$_{1A}$ receptors and α1-adrenoceptors (table 1) were given subcutaneously to conscious dogs. The doses were adjusted in order to elicit similar reductions in blood pressure (left panel in figure 3). Under these conditions, no behavioral changes were noted with the 'pure' α1-antagonist prazosin nor with the hybrid α1-antagonist and 5-HT$_{1A}$ agonist urapidil. When the affinity for 5-HT$_{1A}$ receptors exceeds that for α1-adrenoceptors by more than a factor of 10 as with 5-methyl-urapidil, flesinoxan, and 8-OH-DPAT, side effects appear corresponding to a 5-HT behavioral syndrome in dogs (right panel of figure 3).

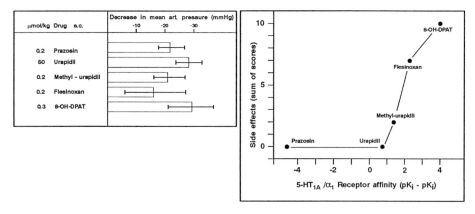

Figure 3. Hypotensive effect of various drugs given subcutaneously; means ± sem, 6 dogs (left panel). Appearance of side effects in relation to the affinity ratio for 5-HT$_{1A}$ and α1-adrenergic receptors (right panel).

4. Discussion

The selective affinity of the α1-adrenoceptor antagonist urapidil for 5-HT$_{1A}$ binding sites has been repeatedly shown [6,12,13,14]. In addition, the 5-methyl analogue of urapidil exhibits an even higher binding affinity and has been found to represent a valuable ligand for 5-HT$_{1A}$ recognition sites [14,15].

Following the identification of urapidil as 5-HT$_{1A}$ receptor agonist', the central hypotensive effect was attributed to this property by use of various 5-HT$_{1A}$ antagonists. Experiments designed to localize the effect within the brain stem gave the following results. Blood pressure reduction is observed with topical application of urapidil or 5-methyl-urapidil to the intermediate area of the ventral surface of the medulla in cats [13]. Microinjection studies revealed that the hypotensive effect is elicited within the medial part of the area of the B1/B3 cell group corresponding to the region of the nucleus raphe magnus and pallidus in rats [16,17].

In cats, microinjection of urapidil or 5-methyl-urapidil into the medulla proved the subretrofacial nucleus (glutamate-sensitive site) to represent the dominant site for blood pressure reduction; minor responses are elicited also from the rostral part of the nucleus reticularis lateralis (bicuculline-sensitive site) and the parapyramidal areas [18]. Injection of 5-HT$_{1A}$ receptor antagonists into these medullary areas significantly attenuate the hypotensive response to peripherally injected urapidil or 5-methyl-urapidil [18]. Thus, a major contribution of the central action to the blood pressure lowering effect of systematic urapidil can be assumed.

The stimulation of 5-HT$_{1A}$ receptors in the medulla leads to a decrease in sympathetic nerve activity. Experiments have shown that this reduction in sympathetic tone can be elicited by urapidil in naive cats and also after full blockade of α1-adrenoceptors with prazosin. The administration of the 5-HT$_{1A}$ antagonist spiperone reverses the effect of urapidil on sympathetic nerve activity and on blood pressure [19].

The α1-adrenoceptor antagonist effect of urapidil not only contributes to the antihypertensive effect; yet it may also be important for the lack of specific side effects resulting from 5-HT$_{1A}$ receptor stimulation. 'Pure' 5-HT$_{1A}$ receptor agonists produce a 5-HT behavioral syndrome in rats [20] and dogs [21,22], which can be largely inhibited by pretreatment with an α1-adrenoceptor antagonist [12,20]. Thus, a fine balance between 5-HT$_{1A}$ receptor stimulation and α1-adrenoceptor inhibition apparently represents a valid antihypertensive principle.

5. References

1. Schoetensack, W., Bruckschen, E.G., and Zech, K. (1983) 'Urapidil', in A. Scriabine (ed.), New Drugs Annual: Cardiovascular Drugs, Raven Press, New York, pp. 19-48.
2. Jackisch, R., Kasakov, L., Feuerstein, T.J., and Hertting, G. (1987) 'Effects of urapidil on neurotransmitter release in CNS tissue *in vitro*', Arch. int.

Pharmacodyn. Ther. 285, 5-24.
3. Sanders, K.H., Kilian, U., Kolassa, N., and Schoetensack, W. (1985) 'Influence of urapidil on α- and β-adrenoceptors in pithed rats', J. Auton. Pharmac. 5, 307-316.
4. Van Zwieten, P.A., de Jonge, A., Wilffert, B., Timmermans, P.B.M.W.M., Beckeringh, J.J., and Thoolen, M.J.M.C. (1985) 'Cardiovascular effects and interaction with adrenoceptors of urapidil', Arch. int. Pharmacodyn. Ther. 276, 180-201.
5. Ramage, A.G. (1986) 'A comparison of the effects of doxazosin and alfuzosin with those of urapidil on preganglionic sympathetic nerve activity in anesthetized cats', Eur. J. Pharmacol. 129, 307-314.
6. Fozard, J.R. and Mir, A.K. (1987) 'Are 5-HT receptors involved in the antihypertensive effects of urapidil', Br. J. Pharmacol. 90, 24P.
7. Van Zwieten, P.A. (1975) 'Antihypertensive drugs with a central action', Progr. Pharmacol. 1, 1-63.
8. Sanders, K.H., Beller, K.-D., Eltze, M., and Kolassa, N. (1989) 'Urapidil and some analogs with high affinities for serotonin-1A and α1-adrenoceptor binding sites show potent hypotensive activity upon central administration', Curr. Opinion Cardiol. 4 (Suppl. 4), S49-S55.
9. Sanders, K.H., Beller, K.-D., Bischler, P., and Kolassa, N. (1988) 'Interaction of urapidil with brain serotonin-1A receptors increases the blood pressure reduction due to peripheral alpha-adrenoceptor inhibition', J. Hypertension 6 (Suppl. 2), S65-S68.
10. Sanders, K.H., Beller, K.-D., and Kolassa, N. (1990) 'Involvement of 5-HT$_{1A}$ receptors in blood pressure reduction by 8-OH-DPAT and urapidil in cats', J. Cardiovasc. Pharmacol. 15 (Suppl. 7), S86-S93.
11. Kolassa, N., Beller, K.-D., and Sanders, K.H. (1989) 'Involvement of brain 5-HT$_{1A}$ receptors in the hypotensive response to urapidil', Am. J. Cardiol. 64, 7D-10D.
12. Beller, K.-D., Boer, R., Sanders, K.H., and Walter, B. (1990) 'Is blockade of α1-adrenoceptors favorable in hypotension induced by stimulation of serotonin-1A receptors in conscious dogs?', Drugs 40, 38-41.
13. Gillis, R.A., Kellar, K.J., Quest, J.A., Namath, I.J., Martino-Barrows, A., Hill, K., Gatti, P.J., and Dretchen, K. (1988) 'Experimental studies on the neurocardiovascular effects of urapidil', Drugs 35 (Suppl. 6), 20-33.
14. Gross, G., Schüttler, K., Xin, X., and Hanft, G. (1990) 'Urapidil analogues are potent ligands of the 5-HT$_{1A}$ receptor', J. Cardiovasc. Pharmacol. 15 (Suppl. 7), S8-S16.
15. Laporte, A.-M., Schechter, L.E., Bolanos, F.J., Verg, D., Hamon, M., and Gozlan, H. (1991) '(^{3}H)5-Methyl-urapidil labels 5-HT$_{1A}$ receptors and α1-adrenoceptors in the rat CNS. *In vitro* binding and autoradiographic studies', Eur. J. Pharmacol. 198, 59-67.
16. Valenta, B., Kotai, E., Weisz, E., and Singer, E.A. (1990) 'Influence of

urapidil and 8-OH-DPAT on brain 5-HT turnover and blood pressure in rats', J. Cardiovasc. Pharmacol. 15 (Suppl. 7), S68-S74.

17. Valenta, B. and Singer, E.A. (1990) 'Hypotensive effects of 8-hydroxy-2-(di-n-propylamino)tetralin and 5-methylurapidil following stereotaxic microinjection into the ventral medulla of the rat', Br. J. Pharmacol. 99, 713-716.

18. Mandal, A.K., Kellar, K.J., and Gillis, R.A. (1991) 'The role of serotonin-1A receptor activation and alpha-1 adrenoceptor blockade in the hypotensive effect of 5-methyl-urapidil', J. Pharmacol. Exp. Ther. 257, 861-869.

19 Ramage, A.G. (1991) 'The mechanism of the sympathoinhibitory action of urapidil: role of 5-HT$_{1A}$ receptors', Br. J. Pharmacol. 102, 998-1002.

20. Tricklebank, M.D., Forler, C., and Fozard, J.R. (1984) 'The involvement of subtypes of the 5-HT$_1$ receptor and of catecholaminergic systems in the behavioral response to 8-hydroxy-2-(di-n-propylamino)tetralin in the rat', Eur. J. Pharmacol. 106, 271-282.

21. Di Francesco, G.F., Petty, M.A., and Fozard, J.R. (1988) 'Anti-hypertensive effects of 8-hydroxy-2-(di-n-propylamino)tetralin (8-OH-DPAT) in conscious dogs', Eur. J. Pharmacol. 147, 287-290.

22. Grohs, J.G., Fischer, G., and Raberger, G. (1990) 'Cardiovascular effects of flesinoxan in anesthetized and conscious dogs', Naunyn-Schmiedeberg's Arch. Pharmacol. 341, 472-475.

CLINICAL ASPECTS OF ANTIEMESIS: ROLE OF THE 5-HT$_3$ ANTAGONISTS

STEVEN M. GRUNBERG
Division of Medical Oncology
University of Southern California Comprehensive Cancer Center
2025 Zonal Avenue
Los Angeles, California 90033
USA

ABSTRACT. Development of effective antiemetics has depended upon understanding of the emetic reflex arc and identification of significant neurotransmitter receptors within that reflex arc. Appreciation of the role of histaminergic, cholinergic, and dopaminergic receptors led to development of the first generation of antiemetics. The increased efficacy of high-dose metoclopramide (an antidopaminergic agent) was later found to be due at least in part to additional blockade of serotonergic (5-HT$_3$) neurotransmitter receptors. Pure 5-HT$_3$ antagonists now allow effective antiemesis without antidopaminergic toxicity. However recent studies concerning dose-response effects and schedule effects with 5-HT$_3$ receptor blockers suggest that additional factors must also be identified to fully understand the physiology and pharmacology of chemotherapy-induced emesis and antiemetic activity.

1. Introduction

Nausea and vomiting are the toxicities of chemotherapy most feared by patients. This is somewhat paradoxical since nausea and vomiting are physiologic rather than pathologic processes. Vomiting is the body's natural defense against the ingestion of toxic substances. It is therefore not surprising to find that this basic defense mechanism is coordinated through some of the most basic areas of the central nervous system.

Emesis is essentially controlled by a reflex arc with multiple afferent limbs transferring information to the efferent limb through associative areas in the brainstem. The vomiting center is an integrated neural network in the brainstem near the area of the tractis nucleus solitarius. It is intimately associated with other basic control mechanisms including those for swallowing, gag, respiration, and arousal. From the vomiting center efferent neural impulses are sent to various muscle groups including those of the thoracic musculature, the gut, the diaphragm, and the abdominal musculature. When appropriate efferent messages are transmitted, the highly coordinated act of vomiting takes place. The emetic reflex arc differs from other reflex arcs in that the efferent limb receives input from a number of different sources (multiple afferent pathways). One major afferent pathway is the humoral pathway. Emetogenic substances cause mediators to be released which travel through the blood stream and cerebrospinal fluid to the fourth ventricle where the chemoreceptor trigger zone (sites in the area postrema of the brainstem that border the fourth ventricle) can be activated. Neuronal messages from the chemoreceptor trigger zone travel to the vomiting center to induce initiation of the vomiting response. A second afferent pathway is the peripheral. Through this pathway, ingested substances can activate nerve endings in the gut wall which then transmit neuronal messages through the vagus nerve directly to the vomiting center without a major contribution from the chemoreceptor trigger zone. A third afferent pathway is the cerebrocortical pathway by which learned responses (those associated with sights, smells, or memories) can activate the vomiting center even in the absence of the specific physical stimulus of chemotherapy (anticipatory vomiting), A fourth afferent pathway, which does not appear to have a major role in chemotherapy-induced emesis, is the vestibular pathway by which messages can be sent through the eighth cranial nerve to the vomiting center (motion sickness).

2. Traditional Antiemetic Agents

Antiemesis is therefore a matter of effectively blocking the relevant pathways for particular emetic stimuli. Various antiemetic agents have indeed been classified according to their neuroanatomic site of action. The reason for this anatomically-oriented differential activity becomes clear when one appreciates that different areas of the nervous system contain higher concentrations of different neurotransmitter receptors and that activity of antiemetics at the various sites can be correlated to their ability to block these local receptors. It was long appreciated that histaminergic, cholinergic, and dopaminergic (D_2) receptors were present in various sites within the emetic reflex arc. Since the chemoreceptor trigger zone has a particularly high concentration of D_2 receptors, families of antidopaminergic agents (phenothiazines, butyrophenones, substituted benzamides) constitute a large proportion of the traditional antiemetic agents. These agents in standard doses were indeed quite effective against mildly and moderately emetogenic chemotherapeutic agents such as 5-fluorouracil, cyclophosphamide, and adriamycin. However development of the severely emetogenic agent cisplatin in the late 1970s revealed the inadequacy of the antiemetics available at that time. This situation was markedly improved in 1981 when Gralla and colleagues [1] described the use of high-dose metoclopramide (metoclopramide 2 mg/kg IV every 2 hours x 5) which significantly reduced the median number of vomiting episodes after cisplatin (from 12 to 1) and provided complete protection (no vomiting episodes within 24 hours) for approximately 40% of patients receiving cisplatin. Antiemetic activity of high-dose metoclopramide could be further improved through concomitant use of a corticosteroid such as dexamethasone [2]. Unfortunately, use of such high-dose antidopaminergic agents could also be accompanied by significant antidopaminergic toxicity, particularly in younger patients.

3. Serotonin (5-HT$_3$) Antagonists

The next significant advance in antiemetic therapy resulted from the realization that other neurotransmitter receptors might also be of significance. Further antiemetic studies with high-dose metoclopramide had suggested that blockade of D$_2$ receptors might not explain all the observations. For example, other substituted benzamides that did not have significant antidopaminergic activity continued to have antiemetic activity. In addition, antiemetic activity of high-dose metoclopramide continued to increase even beyond maximal prokinetic doses (invalidating the assumption that neutralization of the retrograde peristalsis of vomiting might be the basis for the mechanism of action of metoclopramide). However it was also realized that high-dose metoclopramide antagonizes not only D$_2$ receptors but also certain serotonergic receptors (5-HT$_3$). These 5-HT$_3$ receptors are present in high concentrations in the chemoreceptor trigger zone and the periphery and have a significant independent role in the emetic reflex arc. Identification of the 5-HT$_3$ receptors also raised the possibility that specific 5-HT$_3$ receptor blockers could maintain antiemetic efficacy without antidopaminergic toxicity.

In 1987 the first clinical trials of the 5-HT$_3$ receptor antagonist ondansetron in the United States were performed at Memorial Sloan-Kettering Cancer Center and at the University of Southern California Cancer Center. In our study [3] 43 evaluable patients (33 men/10 women) receiving cisplatin (median dose 100 mg/m^2) were entered. These patients had a median age of 52 years and a median Karnofsky Performance Status of 80%. Thirty-eight of the patients had received no prior chemotherapy. Patients received ondansetron intravenously every 4 hours for 3 doses beginning one-half hour prior to chemotherapy. Doses ranged from 0.01 mg/kg x 3 to 0.48 mg/kg x 3. Although the lowest dose (0.01 mg/kg) was ineffective, significant antiemetic efficacy was seen at all higher dose levels such that an overall complete response rate (no vomiting episodes) of 44% and a major response rate (2 or fewer vomiting episodes) of 81% were noted. Nausea was also significantly controlled with 26% of patients having no nausea and 44% of patients having 2 hours or less of nausea. Toxicity was relatively mild and included headache (17 patients), lightheadedness (9 patients), elevated transaminase (8 patients), and diarrhea (7 patients). At the higher doses headaches increased to moderate intensity. However recent further analysis of ondansetron/cisplatin trials has suggested that elevation of transaminase may be due to both cisplatin and ondansetron [4], and that cisplatin-induced diarrhea may actually be partly counteracted in a dose-dependent manner by treatment with ondansetron [5]. We observed a significant hypersensitivity reaction in one patient which was successfully treated with diphenhydramine and corticosteroid. The toxicities of ondansetron appear to be common to this class of agents (5-HT$_3$ receptor antagonists) rather that to ondansetron per se since other 5-HT$_3$ antagonists have had similar toxicity profiles. Of particular note, in studies of several different 5-HT$_3$ antagonists, antidopaminergic toxicity has not been seen [6]. Ondansetron has demonstrated antiemetic activity not only against cisplatin but also against cyclophosphamide, doxorubicin, 5-fluorouracil, ifosfamide, melphalan, and radiotherapy-induced emesis, and comparative trials have suggested superiority to metoclopramide against both cisplatin and noncisplatin-induced emesis. Several trials have also combined ondansetron with dexamethasone. In a randomized double-blind trial, Roila et al. [7] demonstrated superiority of the ondansetron/dexamethasone combination to ondansetron alone in complete protection from vomiting, complete protection from nausea, and patient preference.

4. Future Directions

In spite of the excellent success of the 5-HT$_3$ antagonists, several observations have indicated that our understanding of these agents is not yet complete. Two areas of continuing interest are dose and schedule. Although the marked affinity of ondansetron for the 5-HT$_3$ receptor would indicate that only low doses are necessary, our Phase I study [3] suggested that a moderate dose-response effect might still exist. Complete protection from vomiting was seen in similar percentages of patients at dose levels of 0.06 mg/kg or higher. However comparison of drug exposure (AUC) to number of vomiting episodes suggested that higher doses of ondansetron would result in fewer failures (more that 3 vomiting episodes) [8]. This observation has now been confirmed in a large randomized double-blind trial comparing three different doses of ondansetron in patients receiving cisplatin. In this study Lane and colleagues [5] observed a decrease in failures from 25% to 16% when the ondansetron dose was increased from 0.15 mg/kg to 0.30 mg/kg x 3. Whether this observation of a further dose response represents dose-dependent access of the antiemetic to the eurotransmitter receptor site or antagonism of still another neurotransmitter receptor by very high-doses of ondansetron is not yet clear.

A second area of controversy has been the optimal schedule for administration of ondansetron. A schedule of three doses given at 4-hour intervals was originally adopted due to the relatively short half-life of ondansetron. However the high therapeutic index of this compound suggested that a large single dose would be equivalent to divided doses for antiemetic protection. In fact, the situation appears to be even more complicated. In a recent study Beck et al. [9] compared ondansetron 8 mg x 1 dose to ondansetron 0.15 mg/kg x 3 doses to ondansetron 32 mg x 1 dose in a total of 301 evaluable patients. When efficacy results were compared for both Complete Response and for Failure, the 0.15 mg/kg x 3 dose more closely resembled the 8 mg dose than the 32 mg dose, with the 32 mg dose showing the best results. Since the 0.15 mg/kg dose regimen is effectively a 10 mg x 3 regimen, it would appear that peak level (dependent upon the first dose of drug) is more important than cumulative dose. Pharmacologic implications of this observation concerning first dose (peak level) in a drug with such a short half-life remain to be elucidated.

Great advances have been made and many questions have been answered over the last 30 years in the field of antiemetics. However a perfect antiemetic has not yet been found. Questions of dose and schedule with our present agents as well as the identification of additional significant neurotransmitter receptors will lead to a greater understanding of the central nervous system as well as a better quality of life for patients receiving emetogenic chemotherapy.

5. References

1. Gralla, R.J., Itri, L.M., Pisko, S.E., Squillante, A.E., Kelsen, D.P., Braun, D.W., Bordin, L.A., Braun, T.J., and Young, C.W. (1981) 'Antiemetic efficacy of high-dose metoclopramide: randomized trials with placebo and prochlorperazine in patients with chemotherapy-induced nausea and vomiting', N. Engl. J. Med. 305, 905-909.
2. Grunberg, S.M., Akerley, W.L., Krailo, M.D., Johnson, K.B., Baker, C.R., and Cariffe, P.A. (1986) 'Comparison of metoclopramide and metoclopramide plus dexamethasone for complete protection from cisplatinum-induced emesis', Cancer Invest. 4, 379-385.
3. Grunberg, S.M., Stevenson, L.L., Russell, C.A., and McDermed, J.E. (1989) 'Dose ranging phase I study of the serotonin antagonist GR38032F for prevention of cisplatin-induced nausea and vomiting', J. Clin. Oncol. 7, 1137-1141.
4. Hesketh, P.J., Twaddell, T., and Finn, A. (1990) 'A possible role for cisplatin (DDP) in the transient hepatic enzyme elevations noted after ondansetron administration', Proc. Am. Soc. Clin. Oncol. 9, A1250.
5. Lane, M., Grunberg, S.M., Lester, E.P., Sridhar, K.S., and Sanderson, P.E. (1990) 'A double-blind comparison of three dose levels of IV ondansetron in the prevention of cisplatin-induced nausea and vomiting', Proc. Am. Soc. Clin. Oncol. 9, A1271.
6. Clark, R.A., Kris, M.D., Gralla, R.J., and Tyson, L.B. (1990) 'Serotonin antagonists demonstrate antiemetic effectiveness without extrapyramidal symptoms: analysis of studies with 3 new agents in 155 patients', Proc. Am. Soc. Clin. Oncol. 9, A1245.
7. Roila, F., Tonato, M., Cognetti, F., Cortesi, E., Favalli, G., Marangolo, M., Amadori, D., Bella, M.A., Gramazio, V., Donati, D., Ballatori, E., Del Favero, A. (1991) 'Prevention of cisplatin-induced emesis: a double-blind multicenter randomized crossover study comparing ondansetron and ondansetron plus dexamethasone', J. Clin. Oncol. 9, 675-678.

8. Grunberg, S.M., Groshen, S., Robinson, D.C., Stevenson, L.L., Sanderson, P. E. (1990) 'Correlation of anti-emetic efficacy and plasma levels of ondansetron', Eur. J. Cancer 26, 879-882.
9. Beck, T.M., Madajewicz, S., Navari, R.M., Pendergrass, K., Lester, E.P., House, K.W., and Bryson, J.C. (1992) 'A double-blind, stratified, randomized comparision of IV ondansetron administered as a multiple-dose regimen vs two single-dose regimens in the prevention of cisplatin-induced nausea and vomiting', Proc. Am. Soc. Clin. Oncol. 11, A1311.

THE ANTIEMETIC PROPERTIES OF 5-HT$_3$ RECEPTOR ANTAGONISTS

B. COSTALL, R.J. NAYLOR, and J.A. RUDD
Postgraduate Studies in Pharmacology
The School of Pharmacy
University of Bradford
Bradford BD7 1DP
UK

ABSTRACT. The ability of 5-HT$_3$ receptor antagonists such as ondansetron, tropisetron, and granisetron to prevent chemotherapy- and radiation-induced emesis in animal models has been confirmed in the clinic. In both animals and man the antiemetic effects have been attributed to a selective blockade of 5-HT$_3$ receptors located centrally in the area postrema and nucleus tractus solitarius and peripherally, on the afferent fibers of the vagus nerve in the gastrointestinal tract. The recent clinical demonstration of the efficacy of ondansetron in postoperative nausea and vomiting may also involve a 5-HT$_3$ receptor antagonism at central and peripheral sites. The breadth of action of ondansetron reveals the 5-HT$_3$ receptor antagonists as an important new therapy for the treatment of emesis.

1. Introduction

The present paper reviews the pharmacology of ondansetron and other 5-HT$_3$ receptor antagonists as antiemetic agents. The history of their development is reviewed elsewhere [1] and the present account is directed to an understanding of the mechanisms involved in the induction and antagonism of chemotherapy, radiation, and postoperative nausea and emesis.

2. Profile of Action of 5-HT$_3$ Receptor Antagonists to Inhibit Chemotherapy- and Radiation-Induced Emesis

Chemotherapeutic agents vary widely in their ability to induce emesis although severe nausea and emesis is an invariable concomitant to the use of an agent such as cisplatin [2]. This agent was chosen for use in animal studies as an emetogen of severe effect to detect the antiemetic activity of the 5-HT$_3$ receptor antagonists. The first studies using cisplatin in a ferret model showed that the 5-HT$_3$ receptor antagonists MDL72222 and tropisetron were highly effective to inhibit emesis. This was rapidly confirmed using ondansetron and granisetron which were one hundred times more potent than the reference agent metoclopramide [1]. Since the reporting of these initial studies, very many other 5-HT_3 receptor antagonists have been shown to be potent and efficacious agents to inhibit emesis induced by a wide range of chemotherapeutic agents (cyclophosphamide, actinomycin D, adriamycin, cycloheximide, doxorubicin, mustine, mechloroethamine, trimelalol) and radiation treatment [1,3]. 5-HT_3 receptor antagonists have also been shown to antagonize ipecacuanha-induced emesis in the ferret [4] and man [5]. The specific inhibitory action for such stimuli by the 5-HT_3 receptor antagonist is shown by their failure in high doses to antagonize emesis induced by apomorphine, morphine, copper sulphate, lisuride, protoveratrine A, histamine, pilocarpine, and motion sickness [6,7,8].

3. Sites of Action for Emesis Induction and its Antagonism

The synthesis of radiolabelled 5-HT_3 receptor antagonists allowed their use in ligand binding assays; 5-HT_3 receptors were found in high density in the area postrema and nucleus tractus solitarius and on the afferent nerve fibers of the vagus nerve [9,10]. This afforded both central and peripheral sites of action to inhibit emesis and evidence can be advanced to support effects at both sites. Thus the injection of zacopride, ondansetron, or MDL72222 into the cerebroventricular system or region of the area postrema can inhibit emesis induced by a peripheral treatment with cisplatin [11,12], and the ICV injection of cisplatin can induce emesis [11]. This is strong evidence to support a central effect for 5-HT_3 receptor antagonists to inhibit emesis and cisplatin has been shown to evoke the induction of c-fos protein in the ferret area postrema and nucleus solitary tract [13]. Such effects in the nucleus tractus solitarius were reduced by granisetron and vagotomy. Yet there is other evidence which can be interpreted to indicate a peripheral effect. Thus cisplatin treatment is reported to increase 5-HT turnover in the ferret ileum [14] and to release 5-HT from the isolated vascularly perfused small intestine of the cat and guinea pig [15,16]. Plasma 5-HT and urine 5-HIAA levels are reported to be raised in patients receiving chemotherapy (but not in the ferret) [17,18,19]. Lesions of the vagus nerve can also attenuate chemotherapy/radiation-induced sickness [20].

It has been hypothesized that the toxic effects of chemotherapy or radiation on the gastrointestinal mucosa may disrupt the enterochromaffin cells to cause a release of 5-HT to activate 5-HT_3 receptors on the afferent fibers of the vagus nerve. An activation of vagus cell firing via the input to the nucleus tractus solitarius or area postrema would trigger the emetic reflex to activate the effector nuclei of the "vomiting center." Thus the 5-HT_3 receptor antagonists could antagonize emesis at either or both the peripheral and central 5-HT_3 receptors (figure 1).

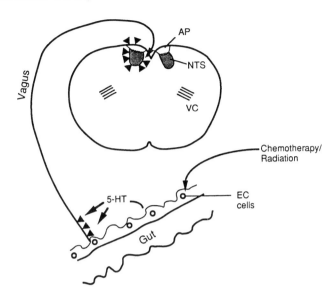

Figure 1. Diagramatic representation of the sites of action of 5-HT$_3$ receptor antagonists to inhibit chemotherapy- and radiation-induced emesis. (▲, 5-HT$_3$ receptors, located on the afferent vagus nerves in the gut and on the vagus terminals in the area postrema (AP) and nucleus tractus solitarius (NTS); vomiting centre (VC), enterochromaffin cells (EC).

4. The Actions of Ondansetron in Postoperative Nausea and Vomiting (PONV)

PONV is influenced by the nature of the operation and attendant procedures, i.e. the anesthetic used, the type of surgery, pre-, and postoperative medication, e.g. opiates, and movement of the patient and patient factors, e.g. sensitivity to emetic insult, sex, age, and body weight [21]. It could be envisaged that emetic stimuli could be channelled directly via central circuits to the "vomiting center" or indirectly via activation of vagus nerve firing. Thus abdominal surgery causing gut distension or by direct manipulation or by tissue damage could be envisaged to cause a release of 5-HT from the mucosal cells activating the 5-HT$_3$ receptors on the vagus nerves, triggering the emetic reflex. Anesthetic agents and opioid mechanisms could be envisaged to directly modify central emetic thresholds or induce emesis via the area postrema. In any event, ondansetron has been shown to be effective in the clinic to antagonize PONV to considerably extend its antiemetic potential [22]. The compound is at least as effective as existing therapies, fails to interact with anesthetic and other agents, and is without the side-effects of sedation, motor, or cognitive impairment. The antiemetic effects are probably achieved by blockade of the 5-HT$_3$ receptors in the central circuits and on the afferent vagus nerves (figure 2).

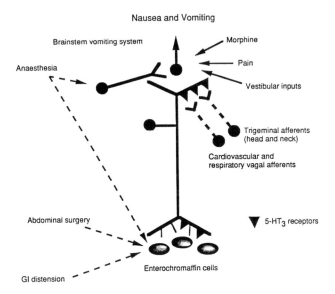

Figure 2. Diagramatic representation of the emetic stimuli contributing to postoperative nausea and vomiting. The 5-HT$_3$ receptors afford the sites of anti-emetic action of ondansetron.

5. Summary

There is an extensive literature indicating that 5-HT$_3$ receptor antagonists are the most potent and effective pharmacological agents to inhibit chemotherapy- and radiation-induced emesis in animals. This potential has been fully confirmed in the clinic and for ondansetron has been extended to the treatment of postoperative nausea and vomiting. The success of the pharmacological investigations of 5-HT$_3$ receptor antagonists is immediately apparent in the development of highly improved treatments in man. Perhaps in the longer term, an additional if not major benefit has been a renaissance of interest in emesis research.

6. References

1. Naylor, R.J. and Rudd, J.A. (1992) 'Mechanisms of chemotherapy-induced vomiting: involvement of 5-HT$_3$ receptors', in A.L. Bianchi, L. Grélot, A.D. Miller and G.L. King (eds.) Mechanisms and Control of Emesis, Colloque Inserm Vol. 223, John Libbey Eurotext Ltd., Montrouge, pp. 115-128.
2. Marty, M., Pouillart, P., School, S., et al. (1990) 'Comparison of the 5-hydroxytryptamine 3 (serotonin) antagonist ondansetron (GR38032F) with

high-dose metoclopramide in the control of cisplatin-induced emesis', New Eng. Med. 322, 816.
3. Miner, W.D., Sanger, G.J., and Turner, D.H. (1987) 'Evidence that 5-hydroxytryptamine3 receptors mediate cytotoxic drug and radiation-evoked emesis', Br. J. Cancer 56, 159.
4. Minton, N.A., Hlakk, A.L., Chilton, J.E., and Henry, J.A. (1991) 'Ipecacuanha-induced emesis: a potential human model for testing the antiemetic activity of 5-HT_3 receptor antagonists', Br. J. Pharmacol. 33, 221-222P.
5. Costall, B., Domeney, A.M., Naylor, R.J., Owera-Atepo, J.B., Rudd, J.A., and Tattersall, F.D. (1990) 'Fluphenazine, ICS205-930, and di-fenfluramine differentially antagonize drug-induced emesis in the ferret', Neuropharmacology 29, 453.
6. Costall, B. and Naylor, R.J. (1992) 'Neuropharmacology of emesis in relation to clinical response', Br. J. Cancer, in press.
7. Lucot, J.B. (1989) 'Blockade of 5-hydroxytryptamine3 receptors prevents cisplatin-induced but no motion or xylazine-induced emesis in the cat', Pharmacol. Biochem. Behav. 32, 207.
8. Stott, J.R.R., Barnes, G.R., Wright, R.J., and Ruddock, C.J.S. (1989) 'The effect on motion sickness and oculomotor function of GR38032F, a 5-HT_3 receptor antagonist with anti-emetic properties', Br. J. Clin. Pharmac. 27, 147-157.
9. Pratt, G.B., Bowery, N.G., Kilpatrick, G.J., et al. (1990) 'The distribution of 5-HT_3 receptors in mammalian hindbrain - a consensus', Trends in Pharmacol. Sci. 11, 135.
10. Kilpatrick, G.J., Jones, B.J., and Tyers, M.B. (1988) 'Binding of the 5-HT_3 ligand, [^{3}H]GR65630, to rat urea postrema, vagus nerve and the brains of several species', Eur. J. Pharmac. 159, 157.
11. Smith, W.L., Callaham, E.M., and Alphin, R.S. (1988) 'The emetic activity of centrally administered cisplatin in cats and its antagonism by zacopride' J. Pharm. Pharmac. 40, 142-144.
12. Higgins, G.A., Kilpatrick, G.J., Bunce, K.T., Jones, B.J. and Tyers, M.B. (1989) '5-HT_3 receptor antagonists injected into the area postrema inhibit cisplatin-induced emesis in the ferret', Br. J. Pharmac. 97, 247-257.
13. Reynolds, D.J.M., Barber, N.A., Grahame-Smith, D.G., and Leslie, R.A. (1991) 'Cisplatin evoked induction of c-fos protein in the brain stems of the ferret: the effect of cervical vagotomy and the anti-emetic 5-HT_3 receptor antagonist granisetron (BRL 43694)', Brain Res. 565, 231-236.
14. Gunning, S.J., Hagan, R., and Tyers, M.B. (1987) Cisplatin induces biochemical and histological changes in the small intestine of the ferret', Br. J. Pharmac. 90, 135P.
15. Minlano, S., Simon, C., and Grélot, L. (1991) '*In vitro* release and tissue levels of ileal serotonin after cisplatin-induced emesis in the cat', Clin. Auton.

Res. 1, 275-280.
16. Schworer, H., Rack, K., and Kilbinger, H. (1991) 'Cisplatin increases the release of 5-hydroxytryptamine (5-HT) from the isolated vascularly perfused small intestine of the guinea pig: involvement of 5-HT$_3$ receptors', Naunyn-Schmiedeberg's Arch. Pharmacol. 344, 143-149.
17. Barnes, N.M., Ge, J., Jones, W.G., Naylor, R.J., and Rudd, J.A. (1990) 'Cisplatin induced emesis: preliminary results indicative of changes in plasma levels of 5-hydroxytryptamine', Br. J. Cancer 62, 862-864.
18. Cubeddu, L.X., Hoffman, I.S., Fuenmayor, N.T., and Finn, A.L. (1990) 'Efficiency of ondansetron (GR38032F) and the role of serotonin in cisplatin-induced nausea and vomiting', New Eng. J. Med. 322, 810-817.
19. Rudd, J.A., Bunce, K.T., Cheng, C.H.K., and Naylor, R.J. (1992) 'The effect of cisplatin on plasma 5-HT, 5-HIAA and 5-HTP levels during emesis in the ferret', Br. J. Pharmac. 105, 274P.
20. Andrews, P.L.R., Bhandari, P., and Davis, C.J. (1992) 'Plasticity and modulation of the emetic reflex' In: A.L. Bianchi, L. Grélot, A.D. Miller and G.L. King, Mechanisms and control of emesis, Colloque Inserm Vol. 223, John Libbey Eurotext Ltd., Montrouge, pp. 275-284.
21. Palazzo, M.G.A. and Strunnin, L. (1984) 'Anaesthesia and emesis I: Etiology', Can. Anaesth. Soc. 31, 178-187.
22. Naylor, R.J. and Rudd, J.A. (1992) 'Ondansetron in the treatment of postoperative nausea and vomiting', Pharmaceutical Journal, June 6th, 749-750.

MECHANISMS OF THE EMETIC RESPONSE TO CHEMOTHERAPY AND OF THE ANTIEMETIC ACTION OF 5-HT$_3$ RECEPTOR ANTAGONISTS: CLINICAL STUDIES

LUIGI X. CUBEDDU and IRENE S. HOFFMANN
Department of Pharmacology,
Central University of Venezuela
Apartado Nueva Granada
Caracas
Venezuela

ABSTRACT. In this work, we evaluated the role of 5-HT and 5-HT$_3$ receptors on the emetic response to antineoplastic drugs, in human patients with cancer. 1) 5-HT$_3$ antagonists prevented vomiting induced by cisplatinum or cyclophosphamide-based chemotherapies. Emesis ensuing after high doses of ondansetron would be mediated by a non-5-HT$_3$ receptor mechanism (delayed emesis?). 2) Increases in plasma and on the urinary excretion of 5-hydroxyindoleacetic acid (5-HIAA) were produced by highly emetogenic treatments (high-dose cisplatinum or dacarbazine-based chemotherapies). For both drugs, the time courses for emesis were superimposed with those for the increases in 5-HIAA. With high-dose cisplatinum, the increases in 5-HIAA lasted 12 to 14 hours and no increases in 5-HIAA were observed during the period of delayed emesis. Platelet or plasma 5-HT were not modified by cisplatinum or cyclophosphamide. A direct relationship was observed between emetogenecity and serotonin-releasing properties of the drugs. The antiemetics, ondansetron, metoclopramide, and dexamethasone failed to modify the increases in 5-HIAA induced by high-dose cisplatinum. The emetic response and the increases in 5-HT metabolism induced by cisplatinum were not affected by the somatostatin analog (octreotide). These results suggest that 5-HT and 5-HT$_3$ receptors mediate the early, intense emetic response to chemotherapy. 5-HT appears to be released from the gut, since urinary 5-HIAA is a marker of gastrointestinal content and turnover of 5-HT and since platelet 5-HT does not contribute to the increases in 5-HIAA. Damage of the gut mucosa may be responsible for the release of serotonin induced by high-dose cisplatinum.

1. Introduction

The development of new chemical moieties derived from cocaine and/or metoclopramide, with selective antagonistic activity on 5-HT$_3$ receptors, led to extensive research on the role of serotonin in chemotherapy- and radiotherapy-induced nausea and vomiting [1]. Research conducted in experimental animals revealed that selective antagonists of 5-HT_3 receptors were very effective against retching and vomiting induced by strongly emetogenic antineoplastic drugs, such as high doses of cisplatinum and cyclophosphamide [1]. These observations were subsequently confirmed in clinical trials in human patients with cancer [2,3]. The observation in the ferret, that depletion of 5-HT stores with para-chlorophenylalanine, prevented vomiting induced by cisplatinum [4], provided additional support to the view that 5-HT mediates the emetic response to chemotherapeutic drugs. Based on these observations, we investigated the possible role of 5-HT in chemotherapy-induced nausea and vomiting in cancer patients. This was attempted by following the plasma and urine levels of 5-HT and of its main metabolite, 5-hydroxyindoleacetic acid (5-HIAA), following treatment with chemotherapeutic drugs. The temporal correlations between the development of vomiting and the increases in 5-HT and 5-HIAA were investigated. In addition, we explored: a) the existence of a direct relationship between emetogenecity and the magnitude of the changes in 5-HT and 5-HIAA; b) the mechanism of action of antiemetics: ondansetron, metoclopramide, and dexamethasone; c) the mechanisms by which chemotherapeutic drugs induce the release of 5-HT; and d) the role of 5-HT in delayed emesis, as well as on emesis induced during subsequent cycles of chemotherapy compared to the emesis induced by the first cycle of chemotherapy. The present work summarizes information recently published or accepted for publication [3,5-8].

2. Methods

Patients with histologically confirmed cancer, 18 years of age or older, who had not received previous chemotherapy were enrolled in the study. No study patients received any antiemetic medication within 24 hours prior to the first dose of study drug. Written informed consent was obtained from all patients, and the protocols were evaluated and accepted by the Institutional Review Board at participating institutions. The studies were conducted at the medical oncology divisions of the Luis Razetti, Padre Machado, and Domingo Luciani Hospitals of the city of Caracas.

Four types of chemotherapy regimens were studied: high-dose cisplatinum ($\geq$ 50 mg/m2), low-dose cisplatinum ($\leq$ 40 mg/m2), cyclophosphamide-based ($\geq$ 500 mg/m2) and dacarbazine- (250-300 mg/m2) based chemotherapies. Patients received only one of the strongly emetic regimens. The cyclophosphamide-based chemotherapy regimens did not contain cisplatin or dacarbazine. However, the cisplatinum or dacarbazine-based chemotherapies could include cyclophosphamide. Cisplatinum, cyclophosphamide, or dacarbazine were dissolved in 500 ml of 5% dextrose in 0.45% sodium chloride and administered as a 60-minute intravenous infusion. The primary agent was followed by administration of other chemotherapeutic drugs as required for treatment of the patients' neoplasia.

Blood samples for the determination of plasma and platelet 5-HT and 5-HIAA were drawn and immediately placed in chilled plastic tubes containing Na2EDTA (1 mg/ml) and sodium metabisulfite (2mg/ml). Urine samples were collected every two hours for a period of 8-12 hours, starting two hours prior to the initiation of chemotherapy. This was followed by a 12-16 hour urine sample, to complete the 24-hour collection period following chemotherapy. 5-HT and 5-HIAA were quantified by means of HPLC, as described by [4,6].

ANOVA and Duncan's multiple rank test were employed to compare differences between groups. A P value below 0.05 was considered to indicate statistical significance. The Mantel-Haenszel Test was employed to compare complete or major response between groups.

3. Results

3.1 EFFECTS OF HIGH-DOSE CISPLATINUM ON THE PLASMA AND URINE LEVELS OR 5-HT AND 5-H1AA

Nearly 98% of blood 5-HT (212 ± 27 ng/ml plasma) was found in the platelet-rich plasma fraction; whereas, free 5-HT (present in platelet-free plasma) averaged 3.2 ± 0.3 ng/ml plasma. The content of 5-HT per platelet averaged 0.33 ng/platelet. High-dose cisplatinum failed to modify the content of 5-HT per platelet, the number of platelets or the concentration of 5-HT per ml of plasma. Free 5-HT levels showed a tendency to increase from 3 to 7 hours after cisplatinum; however, the changes did not reach statistical significance. On the other hand, there were larger and sustained increases above baseline in the levels of 5-HIAA in plasma and urine following high-dose cisplatinum. At 3 and 5 hours after cisplatinum, the plasma 5-HIAA concentrations doubled those present at baseline. Subsequently, the plasma 5-HIAA levels declined, returning to the baseline levels 9 hours after cisplatinum administration. Consequently, the changes in the urinary excretion of 5-HIAA paralleled those of plasma 5-HIAA. The increases in urinary 5-HIAA were observed irrespectively of whether the results were expressed as the actual rate of excretion (micrograms of 5-HIAA/2 hours), or after correcting by urinary creatinine (nanograms of 5-HIAA/micrograms of creatinine). The increases in 5-HIAA were due to the cisplatinum, since in healthy volunteers (3M/3F), exposed to the similar hydration protocol to those of the cancer patients, showed no significant increases in the urinary excretion of 5-HIAA. No increases in urinary 5-HIAA were observed on the second, third, and fourth days following cisplatinum, times at which delayed nausea and emesis develop.

3.2 RELATIONSHIP BETWEEN EMETOGENECITY AND INCREASES IN 5-HT METABOLISM

The effects of regimens based on high-dose cisplatinum on 5-HT release and metabolism, were compared to regimens based on lower doses of cisplatinum, on cyclophosphamide, or on dacarbazine. Both low-dose cisplatinum and cyclophosphamide are considered of a moderate emetic activity, whereas high-dose cisplatinum and dacarbazine are strongly emetogenic agents. Dacarbazine-based chemotherapies (mean dose: 283 ± 22 mg/m2), similarly to high-dose cisplatinum-based regimens, produced marked increases in the urinary excretion of 5-HIAA. The increases in 5-HIAA paralleled the rapid onset of nausea and vomiting observed in the dacarbazine-treated patients. Despite antiemetic protection with metoclopramide the mean time to the onset of emesis was 2.5 ± 0.8 hours.

A much greater increase in the urinary excretion of 5-HIAA was observed with high-dose (84 ± 5 mg/m2) than with low-dose (31 ± 3 mg/m2) cisplatinum. For example, 4-6 hours after high-dose cisplatinum showed 5-HIAA levels that were 3-fold higher than those at baseline, whereas it was only 30% higher in the low-dose cisplatinum group. In addition, only half of the low-dose cisplatinum patients showed increases in 5-HIAA. Compared to high-dose cisplatinum and dacarbazine, cyclophosphamide produced small increases (30% above baseline) in the urinary excretion of 5-HIAA during the first 8 hours following its administration. Interestingly, the median time to emesis for cyclophosphamide-treated patients was 9.4 hours.

3.3 EFFECTS OF ONDANSETRON, METOCLOPRAMIDE, DEXAMETHASONE, AND OCTREOTIDE ON THE INCREASES IN 5-HT METABOLISM INDUCED BY HIGH-DOSE CISPLATINUM

The effects of three antiemetics, ondansetron, metoclopramide, and dexamethasone on the increases in the rate of excretion of 5-HIAA induced by high-dose cisplatinum-based chemotherapies were studied in two separate trials. In both studies, only chemotherapy-naive patients were studied. In the first study, the effects of ondansetron (0.15 mg/kg, IV e/4 hours, x 3) were compared to those of placebo; on the second trial the effects of metoclopramide (2 mg/kg IV, e/4 hours, x 2) were compared to those of dexamethasone (20 mg IV e/4 hours, x 2). Neither treatment had a significant effect on the absolute amount or the percentage of increase above baseline for 5-HIAA levels, induced by high-dose cisplatinum. For example, the percentage of increase above baseline for the period of 2 to 8 hours following high-dose cisplatinum, averaged 351 ± 39 for patients randomized to placebo and 410 ± 55 for those treated with ondansetron ($P > 0.1$). These similar amounts of 5-HT release and metabolism occurred despite the fact that ondansetron was very effective against cisplatinum-induced nausea and emesis. Consequently, the increases in 5-HIAA occurred irrespectively of the presence or absence of emesis. In the metocopramide-dexamethasone trial, the peak excretory rates for 5-HIAA were 2.5 and 2.9 times greater than baseline rates, respectively; in addition, a total of 1.1 ± 0.2 mg of 5-HIAA were released from 2 to 10 hours after cisplatinum in the metoclopramide group and 1.2 ± 0.3 mg were released in the dexamethasone-treated group ($P > 0.1$).

The long-acting somatostatin analog, octreotide, was given subcutaneously, at two different schedules: 1) 0.25 mg one hour before and one hour after the cisplatinum infusion and 2) 0.25 mg every 8 hours the day before and 0.5 mg one hour before and one hour after cisplatinum. These treatments failed to prevent the emetic response to cisplatinum as well as the increases in the urinary excretion of 5-HIAA induced by the chemotherapeutic agent. The greater percentage of increase above precisplatinum (baseline) levels was observed from 4 to 6 hours after the cisplatinum, both in the absence and presence of octreotide. This percentage of increase amounted 190 ± 25 % in the absence and 210 ± 40 in the presence of octreotide ($P > 0.1$).

3.4 RELATIONSHIP BETWEEN ANTIEMETIC EFFICACY AND CHEMOTHERAPY-INDUCED INCREASES IN 5-HT RELEASE AND METABOLISM

Since emetogenecity was related to increases in 5-HT release and metabolism, it appeared of interest to investigate if the clinical efficacy of drugs interfering with the action of 5-HT on 5-HT_3 receptors, was inversely related to amount of 5-HT released by the chemotherapeutic drug. For this purpose, the antiemetic efficacy of a fixed dose of ondansetron was compared against a highly and a moderately emetogenic chemotherapy regimen. The percentage of patients who experienced no vomiting (complete response) after ondansetron treatment was smaller after high-dose cisplatinum than after cyclophosphamide-based chemotherapies ($P < 0.01$).

Similar to ondansetron, a fixed-dose schedule of metoclopramide provided best antiemetic control in patients treated with low-dose than in patients receiving high-dose cisplatinum-based chemotherapies. In addition, better antiemetic coverage was obtained on the first cycle with high-dose than in subsequent cycles of chemotherapy. Complete response (no vomiting) with metoclopramide was achieved in 60% of patients with low-dose cisplatinum, 31% of patients on their first cycle with high-dose cisplatinum, and in 18% of patients on subsequent cycles with high-dose cisplatinum. Interestingly, higher peak levels and more sustained increases in 5-HIAA excretion were found after subsequent cycles compared to the first cycle of treatment with high-dose cisplatinum.

4. Discussion

The present set of studies provides further support to the role of 5-HT in chemotherapy-induced nausea and vomiting in human cancer patients [3,5-8]. A positive relationship between emetogenecity and 5-HT-releasing properties was observed; with high-dose cisplatinum and dacarbazine inducing the greatest release of 5-HT. In addition, the increases in 5-HT release occurred with a time course that paralleled the onset of nausea and vomiting, indicating the existence of a cause-and-effect relationship between 5-HT and the emetic response. In addition, our findings suggest that the amount and time course of 5-HT release determines the severity, the time of onset, and the pattern of emesis induced by the chemotherapeutic drug. If a cytotoxic drug induces a large release of 5-HT within a short period of time, an intense period of vomiting would be expected for this drug, associated with large increases in urinary excretion rate of 5-HIAA. However, if smaller amounts of 5-HT are released or even if the total amount of 5-HT released is comparable to that produced by high-dose cisplatinum, but the release occurs over a much longer period of time, only mild-to-moderate vomiting spreadout over the time during which 5-HT is being released, would develop. In addition, significant increases above baseline in urinary 5-HIAA excretion may not be observed since the amount of 5-HT released will be diluted over many hours.

Our observations also support the view that the most most likely source of the increase in plasma and urinary 5-HIAA, is the enterochromaffin and enteroendocrine cells of the gastrointestinal tract since: a) urinary 5-HIAA is a marker of gastrointestinal 5-HT content and turnover [9], b) 80% of the 5-HT is in the gastrointestinal tract (6-9 mg), 95% of which is within enterochromaffin cells [10], and c) neither high-dose cisplatinum nor cyclophosphamide affect the platelet 5-HT content [6]. Further support is given by the observation that cisplatinum increases the gastrointestinal turnover of 5-HT [11] and induces the release of 5-HT *in vitro* [12]. Consequently, it appears clear that at least the highly emetogenic chemotherapeutic drugs release 5-HT from the enterochromaffin cells of the gastrointestinal tract. Once released, 5-HT seems to act locally, since no evidence of passage into circulation was obtained [6]. A local intestinal action is further supported by the observation that visceral denervation in the ferret prevents cisplatinum-induced vomiting [13]. Finally, our findings indicate that cisplatinum induces 5-HT release via a somatostatin-insensitive mechanism, indicating as previously suggested [6,9,11], that 5-HT could be released because of the mucosal damage produced by the cytotoxics.

The present set of studies indicate that ondansetron and metoclopramide do not interfere with the release of 5-HT induced by high-dose cisplatinum. and that their antiemetic efficacy was inversely related to the amount of 5-HT released by the cytotoxic. Based on these findings and on their known actions, both ondansetron and metoclopramide should exert their antiemetic action by blocking the effects of the released 5-HT at the $5-HT_3$ sites. However, other additional mechanisms may not be disregarded [12]. Dexamethasone failed to prevent cisplatinum-induced 5-HT release and in addition, its antiemetic efficacy, contrary to that of metoclopramide, was independent on the amount of 5-HT released by the cytotoxic. Consequently, the mechanism of antiemetic action of dexamethasone is not due to inhibition of cytotoxic-induced 5-HT release. A central site of action has been proposed (for discussion see [10]). It should be emphasized that a central site of action for antagonists of 5-HT$_3$ receptors has also been proposed [14].

5. Acknowledgements

Supported by a grant from CDCH-UCV 07.10.2441/90. The authors are indebted to Dr. Nery Fuenmayor, Miss Indhira Garcia, and Mr. Juan Malave for their valuable assistance in the conduct of the study. The authors would also like to thank colleagues, personnel, and facilities of the Hospital Luis Razetti, Hospital Padre Machado, and Hospital Domingo Luciani where the patients were treated and the clinical studies conducted.

6. References

1. Andrews, R.L.R. (1992) 'Neuropharmacology of emesis induced by cytotoxic drugs and radiation', in E. Diaz Rubio and M. Martin (eds.), Antiemetic Therapy: Current Status and Future Prospects, Creaciones Elba, S.A., Madrid, Spain, pp. 17-38.
2. Leibundgut, U. and Lancranjan, I. (1986) 'First results with ICS 205-930 (5-HT, receptor antagonist', The Lancet (May) 1198.
3. Cubeddu, L.X., Hoffmann, I.S., Fuenmayor, N.T., and Finn, A.L. (1990) 'Efficacy of ondansetron (GR 38032F) and the role of serotonin in cisplatin induced nausea and emesis', N. Eng. J. Med. 322, 810-816.
4. Barnes, N.M., Barry, J.M., Costal, B., Naylor, R.J., and Tattersall, F.D. (1987) 'Antagonism by para-chlorophenylalanine of cisplatin-induced emesis', Br. J. Pharmacol. 92, 649P.
5. Cubeddu, L.X., Hoffmann, I.S., Fuenmayor, N.T., and Finn, A.L. (1990) 'Antagonism of serotonin S3 receptors with ondansetron prevents nausea and emesis induced by cyclophosphamide-containing chemotherapy regimens', J. Clin. Oncol. 8, 1721-1727.
6. Cubeddu, L.X., Hoffmann, I.S., Fuenmayor, N.T., and Malave, J.J. (1992) 'Changes in serotonin metabolism in cancer patients: its relationships to nausea and vomiting induced by chemotherapeutic drugs', Br. J. Cancer 66, 198-203.
7. Cubeddu, L.X. and Hoffmann, I.S. (1992) 'Participation of serotonin in early and delayed emesis induced by initial and subsequent cycles of cisplatinum-based chemotherapy: effects of antiemetics', J. Clin. Pharmacol., (in press)
8. Cubeddu, L.X. (1992) 'The role of serotonin in chemotherapy-induced emesis in cancer patients', in E. Diaz Rubio and M. Martin (eds.), Antiemetic Therapy: Current Status and Future Prospects, Creaciones Elba, S.A., Madrid, pp. 40-55.
9. Bertaccini, G. (1960) 'Tissue 5-hydroxytryptamine and urinary 5-

hydroxyindoleacetic acid after partial or total removal of the gastro-intestinal tract in the rat', J. Physiol. 153, 239-249.
10. Feldberg, W. and Toh, C.C. (1953) 'Distribution of 5-hydroxytryptamine (serotonin, enteramine) in the wall of the digestive tract', J. Physiol. (London) 119, 352-362.
11. Gunning, S.J., Hagan, R.M., and Tyers, M.B. (1987) 'Cisplatin induced biochemical and histological changes in the small intestine of the ferret', Br. J. Pharmacol. 90, 135P.
12. Schworer, H., Racke, K., and Kilbinger, H. (1991) 'Cisplatinum increases the release of 5-hydroxytryptamine (5-HT) from the isolated vascularly perfused small intestine of the guinea-pig: involvement of 5-HT3 receptors', Naunyn Schmiedeberg's Arch. Pharmacol. 344, 143-149.
13. Hawthorn, J., Ostler, K.J., and Andrews, P.L.R. (1988) 'The role of the abdominal visceral innervation and 5-hydroxytryptamine M-receptors in vomiting induced by the cytotoxic drugs cyclophosphamide and cisplatin in the ferret', Quarterly J. Exper. Physiol. 73, 7-21.
14. Higgins, G.A., Kilpatrick, G.J., Bunce, K.T., Jones, B.J., and Tyers, M.B. (1989) '5-HT3 receptor antagonists injected into the area postrema inhibit cisplatin-induced emesis in the ferret', Br. J. Pharmacol. 97, 247-255.

SEROTONIN AND SEROTONERGIC DRUGS IN EMESIS

U. WELLS, M. RAVENSCROFT, P. BHANDARI, and P.L.R. ANDREWS
Department of Physiology
St. George's Hospital Medical School
Cranmer Terrace
London SW17 0RE
UK

ABSTRACT. The involvement of the major subtypes of 5-HT receptor in emesis is briefly reviewed; 5-HT$_{1A}$ receptors have been implicated particularly in motion sickness, although evidence is emerging for a more generalized involvement in other types (e.g. copper sulphate, cisplatin); 5-HT$_2$ receptors may be involved in apomorphine and copper sulphate emesis but further studies are required; 5-HT$_3$ receptor antagonists are the most extensively studied and are in clinical use for the treatment of emesis induced by intragastric copper sulphate.

1. Introduction

The identification by Miner and Sanger [1] of the antiemetic effect of 5-hydroxytryptamine$_3$ receptor antagonists against emesis induced by the highly emetic cytotoxic drug cisplatin stimulated research into the involvement of 5-hydroxytryptamine in many types of emesis. It is of interest that prior to this publication, of the 'classical neurotransmitters (e.g. acetylcholine, noradrenaline, histamine), 5-HT was notable by its virtual absence from the emetic literature. This paper will briefly review the involvement of several types of 5-HT receptor in emesis and, in particular, 5-HT$_3$ receptors.

2. 5-HT$_1$ Receptors

Studies in the cat by Lucot and colleagues [2,3,4] have revealed that emesis induced by motion in a 'Ferris wheel' was reduced or blocked by a range of 5-HT$_{1A}$ receptor agonists. A study using four different agonists given systemically (8-OH-DPAT, flesinoxan, BMY 7378, buspirone) showed a dose-dependent inhibition of emesis with the rank order of potency being flesinoxan > DPAT > BMY 7378 > buspirone. Complete blockade of emesis occurred at a dose of flesinoxan of 100 μg/kg. Based on a comparison of the doses required for antiemesis and those for anxiolytic activity the authors suggested that in general the doses required for antiemesis were higher, although it must be noted that the anxiolytic study was undertaken in the rat in contrast to the antiemesis study in the cat. The authors argued that the 5-HT_{1A} receptors involved in the antiemetic effects were located postsynaptically, although their precise location in the motion-sickness pathway has not yet been identified.

The antiemetic effects of 5-HT_{1A} receptor activation appear to be more widespread than motion emesis as 8-OH-DPAT and buspirone suppress the emesis induced by the cytotoxic drug cisplatin, and the alpha$_2$-adrenergic receptor agonist xylazine in the cat [2,5]. 8-OH-DPAT also reduced the emesis induced by the '5-HT_{1D} agonist' (see below) RU 24969. Buspirone was less effective in the ferret against cisplatin indicating that there may be species differences in the degree of involvement of 5-HT_{1A} receptors in emesis.

Evidence is emerging implicating other types of 5-HT_1 receptor in emesis [6]. An extensive study in the cat showed that the 5-HT_1 receptor agonist RU 24969 (0.3-3.0 mg/kg sc) reliably induced emesis whereas 5-methoxytryptamine (0.3-3.0 mg/kg sc) was only moderately effective and trifluormethylphenylpiperizine (1-30mg/kg sc was ineffective. Using a battery of receptor antagonists with activity at 5-HT_{1A}, 5-HT_{1B}, 5-HT_{1C}, 5-HT_2, and 5-HT_3 receptors, the receptor involved in the emetic response to RU 24969 was, by default, proposed to be the 5-HT_{1D} subtype. One interesting observation was that neither RU 24969 nor 5-methoxytryptamine induced emesis in all animals studies and the author suggested that this may be because of an agonist antiemetic action at the 5-HT_{1A} site offsetting the 5-HT_{1D} emetic action in some animals. A similar antiemetic/emetic action has been invoked to account for the failure of 5-HT_3 receptor agonists to induce emesis when given into the brainstem or intravenously [7, see below].

A study of fictive emesis in the decerebrate cat demonstrated that buspirone (1-4 mg/kg IV) and 8-OH-DPAT (0.2-1.3 mg/kg IV) both reduced emesis induced by electrical stimulation of vagal afferents. In contrast to the motion studies, buspirone was considered to be more potent than 8-OH-DPAT [8]. The most probable way in which vagal afferent activation induces emesis is via the nucleus tractus solitarius and as 5-HT_{1A} binding sites and receptor mRNA are both abundant there [3], this is the most likely site of action of the 5-HT_{1A} agonists. In contrast to studies of 5-HT_3 receptors, the relative distribution of 5-HT_{1A} receptors on NTS neurones (postsynaptic) and vagal afferents (presynaptic) has not been studied.

In view of the broad spectrum antiemetic effects of the 5-HT_{1A} receptor agonists against emetic stimuli acting via different pathways (area postrema, vagus, and vestibular system), it is tempting to suggest that there is a single site of action in the emetic system. The location is unclear (nucleus tractus solitarius, reticular formation?) but its identification is of considerable importance as it may give important insights into the organization of the emetic reflex and identify a novel therapeutic target for antiemetic drugs.

In the dog, the nonselective 5-HT_1 receptor antagonist methysergide (1 mg/kg IV) was able to block the emetic response to apomorphine (a D2 receptor agonist acting via the area postrema) and intragastric copper sulphate (an irritant acting predominantly via vagal afferents) implicating activation of 5-HT$_1$ receptors in the brainstem emetic pathway, the convergence point for these two inputs [9]. Similar studies have not been published using more selective antagonists so these observations must be treated with caution.

3. 5-HT$_2$ Receptors

Evidence for the involvement of 5-HT$_2$ receptors is scant. In the dog 1-(1-naphthyl) piperazine (1 mg/kg IV) an antagonist at 5-HT$_2$ receptors (with some activity at 5-HT$_{1C}$ and alpha-adrenoreceptors) blocked emesis induced by both apomorphine and intragastric copper sulphate [9]. This indicates that 5-HT$_2$ receptor activation in the brainstem may be involved in generating the emetic response to these stimuli. Cinanserin was without effect but the authors argued that the dose used (5 mg/kg IV) may have been too low to produce an adequate blockade of the receptors. The involvement of 5-HT$_2$ receptors in emesis requires further study using more selective and potent antagonists.

4. 5-HT$_3$ Receptors

The 5-HT$_3$ receptor is perhaps the one for which there is the largest body of evidence for an involvement in emesis and, in addition, it is the only one for which there are clinical indications, i.e. emesis induced by anticancer chemo- and radiotherapy, postoperative nausea and vomiting. Studies of the spectrum of the antiemetic effects of the 5-HT$_3$ receptor antagonists (particularly granisetron, ondansetron, and tropisetron) and their site of action have considerably increased our knowledge of the basic neuropharmacology of emesis. The current working hypothesis is that both systemic cytotoxic drugs (e.g. cisplatin, cyclophosphamide) and total body radiation induce the release of 5-HT from the enterochromaffin cells (EC) in the mucosa of the upper gastrointestinal tract. This 5-HT acts locally on 5-HT$_3$ receptors located on vagal afferents terminating in close proximity to the EC cells or possibly enters the hepatic portal vein and acts on hepatic afferents [7]. This afferent activation induces emesis. There is no evidence to support a gross increase in systemic levels of 5-HT and hence an action on the area postrema ('chemoreceptor trigger zone for emesis') nor is there any indication for the antiemetic site of action being on the 5-HT$_3$ receptors in the area postrema. To date the weight of evidence suggests that the antiemetic site of action of the 5-HT$_3$ receptor antagonists is on the vagal afferent terminals in the gut wall with a possible additional site on 5-HT$_3$ receptors in the nucleus tractus solitarius in the brainstem, many of which are associated with vagal afferent terminal [7]. Although there is considerable literature on antiemetic effects of the 5-HT$_3$ receptor antagonists and several agents are in widespread clinical use, there are a number of issues that need to be addressed, a few of which are raised in brief below. While these questions relate specifically to anticancer therapy, they can be applied to other forms of emesis.

4.1 HOW IS THE EMETIC STIMULUS TRANSDUCED?

If, as appears likely, the emesis is caused by the release of 5-HT from the enterochromaffin cell then how is this brought about, particularly by a range of cytotoxic drugs with markedly different effects on the cellular chemistry, at least as indicated by their antitumor effects? Morphological studies of the effects of these stimuli have been undertaken but unfortunately the most detailed are in species lacking an emetic reflex (e.g. rat). However, they do reveal that there is little evidence of gross mucosal damage. It, therefore, appears that cytotoxic drugs and radiation (both 'cellular poisons' probably generating free radicals) can activate the normal calcium-dependent mechanism for degranulation. Although such a direct effect on the EC cell offers the simplest mechanism for 5-HT release, it is not the only possibility. In the guinea pig small intestine *in vitro* the cisplatin-induced release of 5-HT was blocked by TTX, hexamethonium, or scopolamine [10], suggesting that the primary effects of cisplatin are on enteric cholinergic neurones which activate the EC cells. The relevance of this effect to emesis is unclear: if such a mechanism predominated in animals with an emetic reflex, then scopolamine would be expected to be an excellent antiemetic against cisplatin-induced emesis, whereas it is only weakly effective. However, the study does indicate that cisplatin is neuroactive. The mechanism is not known; an investigation of this may shed light on the effects of cisplatin on the EC cell.

4.2 WHAT DETERMINES THE INTENSITY AND DURATION OF THE EMETIC RESPONSE?

The emetic potential of cytotoxic drugs differs considerably ranging from a few episodes in a small number of patients with 5-fluorouracil to intense emesis lasting as least 24 hours in > 90% patients with cisplatin. The intensity and duration of emesis is also related to the dose of radiation. Is the intensity of the emetic response a reflection of the release of 5-HT? This question has been addressed to a limited extent in the ferret. Analysis of the dose-response curve for granisetron using two doses of radiation [11] revealed that the higher dose of radiation (800 rads) required a greater dose of the antagonist (37.2 ± 6.8 μg/kg sc) to produce a 50% reduction in retching than that required (4.8 ± 1.6 μ/kg sc) for the lower dose (200 rads). A comparison of the effects of ondansetron [12] in the ferret against stimuli with differing emetic potential (emetine > cyclophosphamide > radiation 800 rads) revealed that the dose of ondansetron required for antiemesis was related to the intensity of the stimulus. Both of these animal studies support the contention that the magnitude of the emetic response is determined primarily by the ability of the stimulus to release 5-HT and hence activate the 5-HT$_3$ receptors.

One anomaly in this argument is that it implies that the release of 5-HT is continuous and yet the emetic response is phasic occurring in bursts of retches and vomits sometimes separated by many minutes or even hours in man. The mechanism by which this pattern of emesis is generated is unclear but presumably reflects the underlying release of 5-HT or other driving stimulus. Studies of the temporal pattern of 5-HT release into the interstitial space surrounding the vagal afferents are required to resolve this problem.

4.3 IS 5-HT A SUFFICIENT EMETIC STIMULUS?

Studies with $5\text{-}HT_3$ receptor agonists (2-me-5-HT, 1-PBG, chloro-PBG) in the ferret have revealed that while emesis can be induced by giving them intraperitoneally or intragastrically, the response is relatively short-lived ($<$ 30 minutes) in comparison to cytotoxic drugs and it is difficult to induce a response in all animals [13]. While it is possible that rapid desensitization of the $5\text{-}HT_3$ receptor may contribute to the variability of the response, it is also possible that it may be an indicator that some cofactor is required in addition to 5-HT. While activation of the $5\text{-}HT_3$ receptor is the critical step in the induction of emesis, as indicated by the antiemetic efficacy of the $5\text{-}HT_3$ antagonists, a cofactor may serve to modify the sensitivity of the emetic system to 5-HT or amplify the effects of $5\text{-}HT_3$ receptor activation. Another explanation could be that using 'bolus' administration of the agonists fails to mimic the pattern and magnitude of 5-HT release induced by cytotoxic drugs. In addition, systemic administration of the agonists while stimulating emetic pathways (e.g. abdominal vagal afferent) may simultaneously activate pathways suppressing emesis. Some pulmonary vagal afferents may be capable of inhibiting emesis by modulating the somatic outflow to the diaphragm and skeletal muscles [14].

4.4 WHAT DO THE $5\text{-}HT_3$ RECEPTORS IN THE AREA POSTREMA DO?

Autoradiography has identified $5\text{-}HT_3$ ligand-binding sites in the brainstem nucleus tractus solitarius and the area postrema with the density being considerably greater in the former [15]. To date there is little evidence to support the $5\text{-}HT_3$ receptors in the area postrema as the clinically relevant site of antiemetic action of the antagonists (see [7] for review). In addition to emesis, the area postrema is involved in the genesis of conditioned taste aversion, appetite, and salt intake. The effect of $5\text{-}HT_3$ receptor antagonists on these functions has not been extensively studied but it is possible that an action on the area postrema may contribute to the reduction of anorectic and related behavioral effects of cytotoxic drugs reported in animals and man.

4.5 WHY ARE THE ANTAGONISTS MORE EFFECTIVE IN ANIMALS THAN IN MAN?

The majority of preclinical studies with potent and selective 5-HT_3 receptor antagonists have been able to achieve complete control of emesis induced by cytotoxic drugs or radiation. However, in man such a high degree of complete control has not been achieved, although the vast majority (> 90%) of patients do gain some benefit, particularly when the 5-HT_3 receptor antagonist is combined with a steroid (e.g. dexamethasone). The reason for the difference is not known but it is not due to inadequate blockade of the 5-HT_3 receptors. It is most likely to involve psychological aspects of chemotherapy, the proemetic effects of a tumor burden, administration of concomitant medication (e.g. analgesics, fluid loading), emetic history (sensitivity to motion and pregnancy sickness being an index of the general sensitivity of the emetic pathways), and possibly the expression or induction of novel (non-5-HT_3) emetic pathways. Although the stimulus for this 'plasticity' is not known in man, in the ferret it can be induced by section of the abdominal vagi [11]. Studies of the pharmacology of the emetic pathways in patients insensitive to 5-HT_3 receptor antagonism are urgently required.

5. 5-HT_4 Receptors

Studies of the possible involvement of 5-HT_4 receptors are in their infancy because of the lack of potent and selective antagonists that can be used *in vivo*. Selectivity is a particular problem in this case as many of the antagonists (e.g. SDZ 205, 557) are also active at the 5-HT_3 receptor [6] making it difficult to assess the contribution of 5-HT_4 receptors to any antiemetic effects observed. In the ferret the benzamide zacopride induces emesis, particularly when given orally. The pharmacology of zacopride is complex with both antagonist and agonist effects at the 5-HT_3 receptor and agonist effects at the 5-HT_4 receptor which are probably responsible for its gastrointestinal motility stimulant properties. The emetic effects of zacopride were initially ascribed solely to activation of 5-HT_3 receptors but preliminary evidence has been presented for an involvement of 5-HT_4 receptors [17]. In the ferret the dose of granisetron (BRL 43694) and tropisetron (ICS-205-930) required to block emesis induced by total body X-radiation was identified from a detailed dose-response study and in both cases this was less than 100 μg/kg sc. This dose was taken as indicating the dose of both drugs at which full functional blockade of the 5-HT_3 receptors involved in emesis occurred. As tropisetron has been implicated from *in vitro* studies as a 5-HT_4 receptor antagonist when given at high concentrations (10-6 M), it was reasoned that it may have a similar effect *in vivo* at high doses. A comparison of the effect of 'low' (100 μg/kg sc) and 'high' (1 mg/kg sc) doses of granisetron had little additional effect whereas tropisetron blocked the emetic response in 3 out of 4 animals and reduced the response in the remaining animal. These results were taken to provide preliminary evidence for an involvement of both

5-HT$_3$ and 5-HT$_4$ receptors in zacopride-induced emesis. It is noteworthy that emesis is rarely induced with other orally administered benzamides such as renzapride and cisapride and as these have only been reported to have agonist effects at 5-HT$_4$ receptors, it raises the question whether some forms of emesis may involve activation of both receptor types: one receptor priming the system (5-HT$_4$?) and the other (5-HT$_3$?) driving the primed system. Studies in the ferret have also demonstrated that while granisetron and ondansetron at 'high' doses (1 mg/kg sc) have little effect on emesis induced by intragastric copper sulphate (40 mg%), tropisetron at the same dose blocked the response. These studies suggest that 5-HT$_4$ receptors may be involved in emesis most likely at a site in the gastrointestinal mucosa intimately associated with the vagal afferent mucosal chemoreceptor transduction mechanism. Further studies of the possible involvement of 5-HT$_4$ receptors are clearly warranted.

6. References

1. Miner, W.D. and Sanger, G.J. (1986) 'Inhibition of cisplatin-induced vomiting by selective 5-hydroxytryptamine M-receptor antagonism', Br. J. Pharmacol. 88, 497-499.
2. Lucot, J.B. and Crampton, G.H. (1986) 'Xylazine emesis, yohimbine and motion sickness susceptibility in the cat', J. Pharm. Exp. Ther. 237, 450-455.
3. Lucot, J.B. (1992) 'Prevention of motion sickness by 5-HT$_{1A}$ agonists in cats', in A. Bianchi, L. Grelot, A.D. Miller and G.L. King (eds.), Mechanisms and control of emesis, Colloque INSERM, John Libbey, Eurotext 223, 195-201.
4. Lucot, J.B. (1990) 'Effects of serotonin antagonists on motion sickness and its suppression by 8-OH-DPAT in cats', Pharmacol. Biochem. Behav. 37, 283-287.
5. Lucot, J.B. and Crampton, G.H. (1987) 'Buspirone blocks cisplatin-induced emesis in cats', J. Clin. Pharmacol. 27, 817-818.
6. Lucot, J.B. (1990) 'RU 24969-induced emesis in the cat: 5-HT$_1$ sites other than 5-HT$_{1A}$, 5-HT$_{1B}$, 5-HT$_{1C}$ implicated', Eur. J. Pharmacol. 180, 193-199.
7. Andrews, P.L.R. and Davis, C.J. (1993) 'The mechanism of emesis induced by anti-cancer therapies', in P.L.R. Andrews and G.J. Sanger (eds.), Emesis in anti-cancer therapy, mechanisms and treatment, Chapman and Hall Medical, London, p. 113-162.
8. Milano, S. and Grelot, L. (1992) 'Differential Blocking effects of buspirone and 8-OH-DPAT on vagal-induced emesis in decerebrate cats', in A. Bianchi, L. Grelot, A.D. Miller and G.L. King (eds.), Mechanisms and control of emesis, Colloque INSERM, John Libbey, Eurotext 223, 353-355.
9. Lang, I.M. and Marvig, J. (1989) 'Functional localization of specific receptors mediating gastrointestinal motor correlates of vomiting', Am. J. Physiol. 256, G92-G99.
10. Schwörer, H., Racke, K., and Kilbinger, H. (1991) 'Cisplatin increases the

release of 5-hydroxytryptamine (5-HT) from the isolated vascularly perfused small intestine of the guinea pig: Involvement of 5-HT_3 receptors', Naunyn Schmeidebergs Arch. Pharmacol. 344, 143-149.

11. Andrews, P.L.R., Bhandari, P., and Davis, C.J. (1992) 'Plasticity and modulation of the emetic reflex', in A.L. Bianchi, L. Grelot, A.D. Miller and G.L. King (eds.), Mechanisms and control of emesis, Colloque INSERM, John Libbey, Eurotext 223, p. 275-284.

12. Andrews, P.L.R. and Bhandari, P. (1993) 'The 5-hydroxytryptamine$_3$ receptor antagonists as antiemetics: preclinical evaluation and mechanism of action', Eur. J. Cancer 29A, S11-S16.

13. Ravenscroft, M., Wells, U., Bhandari, P.B., and Andrews, P.L.R. (1992) 'Agonist evidence for the involvement of 5-HT_3 receptors in emesis in the ferret', in A.L. Bianchi, L. Grelot, A.D. Miller and G.L. King (eds.), Mechanisms and control of emesis, Colloque INSERM, John Libbey, Eurotext 223, p. 251-252.

14. Andrews, P.L.R. and Lawes, I.N.C. (1992) 'A protective role for vagal afferents hypothesis', in S. Ritter, R.C. Ritter and C.D. Barnes (eds.), Neuroanatomy and Physiology of Abdominal vagal afferents, CRC Press, Boca Raton, Florida, p. 280-301.

15. Pratt, G.D., Bowery, N.G., Kilpatrick, G.J., Leslie, R.A., Barnes, N.M., Naylor, R.J., Jones, B.J., Nelson, D.R., Palacios, J.M., Slater, P., and Reynolds, D.J.M. (1990) 'Consensus meeting agrees distribution of 5-HT_3 receptors in mammalian hindbrain', Trends Pharmacol. Sci. 11, 135-137.

16. Eglen, R.M., Alvarez, R., Johnson, L.G., Leung, L.G., and Wong, E.H.F. (1993) 'The action of SDZ 205,557 at 5-hydroxytryptamine (5-HT_3 and 5-HT_4) receptors', Br. J. Pharmacol. 108, 376-382.

17. Bhandari, P. and Andrews, P.L.R. (1991) 'Preliminary evidence for the involvement of the putative 5-hydroxytryptamine$_4$ receptor in zacopride and copper sulphate-induced vomiting in the ferret', Eur. J. Pharmacol. 204, 273-280.

CLINICAL EFFECTS OF 5-HT$_3$ ANTAGONISTS IN TREATING NAUSEA AND VOMITING

M.E. BUTCHER
Glaxo Group Research Ltd.
Greenford, Middlesex UB6 0HE
UK

ABSTRACT. Chemotherapy, and to a lesser extent radiotherapy, are frequently associated with particularly severe and debilitating nausea and vomiting. The discovery of specific 5-HT$_3$ antagonists such as ondansetron has been a major breakthrough in the control of cytotoxic-induced emesis. In the control of acute cisplatin-induced emesis, ondansetron has been found to be significantly superior to placebo, high-dose metoclopramide, and alizapride plus methylprednisolone. Complete or major control (≤ 2 emetic episodes) has been achieved in 65-85% of patients. Similarly ondansetron is effective in preventing emesis resulting from chemotherapy other than cisplatin, e.g. cyclophosphamide/anthracyclin combinations, and high-dose radiotherapy to the abdomen. The addition of dexamethasone to ondansetron has been shown to significantly improve efficacy with complete control of emesis being achieved in $> 90\%$ of patients. Agents such as ondansetron have been shown to be safe and well tolerated and in particular they do not cause acute dystonic reactions.

1. Introduction

Nausea and vomiting are well recognized in a number of different situations with vomiting often acting as a protective mechanism against ingested poisons [1]. Vomiting can also occur when there would appear to be no physiological advantage, for example, in pregnancy and motion sickness. Although, emesis may be induced by a variety of stimuli, all involve the so-called 'vomiting center' with inputs from the gastrointestinal tract via the vagal afferent nerve, the chemoreceptor trigger zone (CTZ), the vestibular apparatus, and higher cortical centers.

The discovery of selective 5-HT$_3$ receptor antagonists [2] in the early 1980s has led to an increase in interest in the control of nausea and vomiting in clinical settings where this is desirable. Three agents, ondansetron, granisetron, and tropisetron are now commercially available. All are highly selective for the 5-HT$_3$ receptor with no clinically relevant activity at other receptor types [3].

2. Cytotoxic-Induced Emesis (CIE)

Chemotherapy and radiotherapy given for the treatment of cancer is frequently associated with particularly severe and debilitating emesis [4]; indeed, virtually all patients given high-dose cisplatin will experience emesis within 24 hours. For potentially curative chemotherapy to have its maximum opportunity to control or cure the disease the patient must be able to tolerate side-effects, particularly nausea and vomiting. Cancer patients have ranked nausea and vomiting as the two most distressing treatment-related side-effects [5]. In some cases these can be so severe that patients delay or even refuse further treatment [3]. Nausea and vomiting can also lead to other complications such as dehydration, electrolyte imbalance, anorexia, and psychological depression [6,7,8].

Many factors have to be considered when assessing the antiemetic schedule for patients [9]. These include patient factors such as gender (females tend to vomit more than males), age (young patients experience more emesis), and whether the patient has experienced emesis previously with chemotherapy. However, overriding this is the chemotherapy itself, as agents vary in their emetogenic potential. Cisplatin and dacarbazine are generally accepted as being the most emetogenic, causing severe emesis in more than 90% of patients [10,11].

Cytotoxic agents are usually given in combination and normally cause the most severe nausea and vomiting during the first 24 hours following chemotherapy (acute emesis). Nausea and vomiting may however continue for several days thereafter (delayed emesis) and it is vitally important to the patient's quality of life that effective antiemetics are given during this period. Failure to effectively prevent nausea and vomiting during the first course of cytotoxic treatment can lead to anticipatory emesis with later courses. This is a conditioned response and occurs before chemotherapy or radiotherapy is given.

2.1 ANTIEMETIC THERAPY

Many agents, with differing pharmacological profiles have some activity against emesis induced by mild to moderately emetogenic agents and are reviewed elsewhere [12]. These include dopamine antagonists (metoclopramide, droperidol, prochloroperazine), corticosteroids (dexamethasone, methylprednisolone), anxiolytic agents (lorazepam), and cannabinoids (nabilone, dronabinol). Of these, high-dose metoclopramide is probably the most effective against nausea and vomiting induced by highly emetogenic agents such as cisplatin although there are also unwanted side-effects. Particularly distressing are the extrapyramidal effects seen with agents such as metoclopramide.

The literature relating to the $5-HT_3$ antagonists is now extensive and the excellent efficacy of this class of compound is exemplified by ondansetron, the first such agent to be commercially available.

2.2 ONDANSETRON

For the control of acute cisplatin-induced emesis ondansetron has been found to be significantly superior to placebo [13] and high-dose metoclopramide [14,15,16]. Complete or major control of emesis (≤ 2 episodes of emesis [vomiting or retching]) has been achieved in approximately 65-85% of patients, depending on age and gender and the dose of cisplatin used. Examination of the pattern of emesis following cisplatin treatment (figure 1) revealed a peak of acute emesis over the first 10-12 hours [14]. Subsequent studies [17] have shown that a single IV dose of 8 mg is as effective as a 24-hour infusion or 3 divided doses (approximately 32 mg total dose) in controlling acute emesis and that a dose of 32 mg achieves superior efficacy with 73-88% of patients being well controlled (table 1).

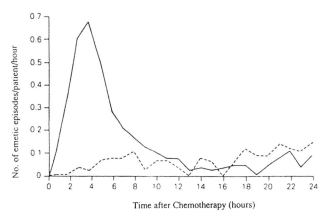

Figure 1. Episodes of emesis with metoclopramide (—) and with ondansetron (---) during the 24 hours after cisplatin administration (n=76).

TABLE 1. Efficacy of a Single Intravenous Dose of Ondansetron

	% Patients with ≤ 2 emetic episodes	
Ondansetron	Cisplatin 50-70 mg/m^2	Cisplatin ≥ 100 mg/m^2
0.15 mg/kg x 3	72%	60%
8 mg single dose	69%	57%
32 mg single dose	88%	73%

The addition of a single intravenous dose of dexamethasone significantly enhances the efficacy of ondansetron [18] with complete control of emesis being achieved in up to 90% of patients. Furthermore a recent study has shown this combination to be significantly superior to the traditional "optimum antiemetic cocktail" of metoclopramide, dexamethasone, and diphenhydramine [19]. Also in patients receiving a low dose of cisplatin (approximately 20 mg/m^2) daily for 5 days, the combination of ondansetron and dexamethasone has been shown to be superior to the standard therapy of metoclopramide plus dexamethasone [20].

Although many antiemetics have been shown to be effective with the first course of chemotherapy [21,22], the efficacy of oral ondansetron, 8 mg given three times a day for 5 days for acute and delayed emesis, has been shown to be excellent over 6 courses of intravenous cyclophosphamide-based chemotherapy in a randomized trial in breast cancer [23]. In this study patients' quality of life, measured by the Rotterdam Symptom Checklist, was also improved with ondansetron compared to metoclopramide.

Emesis induced by radiotherapy, although normally less severe and less predictable than that induced by chemotherapy, can also be troublesome for some patients. Oral ondansetron has been shown to be highly effective (97% complete control of emesis) and superior to metoclopramide following a single high-dose fraction of radiation (800-1000 cGY) to the upper abdomen [24]. When lower doses of radiation are given to the abdomen in individual daily fractions (up to 4 weeks of treatment) ondansetron has been shown to be well tolerated and superior to prochlorperazine when given orally over this prolonged period [25]. Ondansetron has also been shown to be effective in the control of emesis induced by total body irradiation in both adults and children [26,27].

In large open studies involving more than 500 children receiving a wide variety of chemotherapy [28], ondansetron has been shown to be effective and well tolerated. This is particularly important as the use of agents such as metoclopramide are often inappropriate due to the high rate of extrapyramidal reactions seen in young patients.

2.3   GRANISETRON

Large-dose ranging studies have demonstrated that a single intravenous dose of 40 μ g/kg of granisetron has equal efficacy to a larger dose of 160 μ g/kg in the control of acute cisplatin-induced emesis (57% and 60% complete control [including mild nausea] respectively) [29,30]. Patients were given a second or third dose for "breakthrough" vomiting.

Granisetron has been shown to have similar efficacy to combinations of metoclopramide and dexamethasone [31] and metoclopramide, dexamethasone, and diphenhydramide [32] in the control of acute cisplatin-induced emesis, and to be superior to chlorpromazine and dexamethasone [33] and to prochlorperazine and dexamethasone [34] in the control of noncisplatin chemotherapy-induced emesis. No data has been reported regarding its use in the pediatric population.

2.4 TROPISETRON

Tropisetron can also be given as a single daily intravenous dose for the control of acute emesis and as a single daily oral dose for delayed emesis. Dose ranging studies have indicated that a dose of 5 mg gives optimal antiemetic efficacy with no further benefit being derived from higher doses [35].

3. Postoperative Nausea and Vomiting (PONV)

PONV is a common problem after surgery with potentially serious medical complications such as aspiration pneumonia, dehydration, and wound rupture. It may also delay discharge from hospital, with obvious economic consequences, and at the very least is distressing to the patient [36].

Unlike CIE which is caused by one distinct stimulus and is predictable in its severity and duration, PONV is multifactorial and variable in both severity and duration. Some of the key differences are highlighted below:

	PONV	CIE
Cause	Many Factors	Chemotherapy/Radiotherapy
Nature	Unpredictable	Predictable
Duration	24 Hours	1-7 Days
Severity	Variable	Variable/Severe
Timing of Onset	1-8 Hours	1-24 Hours

Some of the factors that contribute to PONV are: specific patient characteristics, anesthetic agents, surgical procedures, and postoperative factors (e.g. postoperative opiod use). The incidence of PONV varies from 10-60% depending on these factors. However, an incidence as high as 80% has been observed after ophthalmic surgery in children [37] and 75% after gynecological surgery [38].

3.1 CURRENT ANTIEMETICS

The 4 main agents in current use are metoclopramide, droperidol, alizapride, and prochlorperazine. There is no clear drug of choice and all have variable efficacy and side-effects including sedation, dystonic reactions, and cardiovascular disturbances [39].

3.2 ONDANSETRON

Ondansetron is the first 5-HT$_3$ antagonist to be investigated for the prevention and treatment of PONV. Ondansetron given intravenously at a dose of 4 mg before surgery has been shown to be significantly superior to placebo in patients undergoing gynecological laparoscopy. Complete control of emesis was achieved in 76% of patients on ondansetron compared to 46% on placebo in the 24-hour post-surgical period [40].

Similarly a 4 mg intravenous dose is effective in treating established PONV in the post-surgical setting [41].

Oral ondansetron (8 mg tid) has also been shown to be effective in preventing PONV in gynecological surgery [42].

4. Other Emetic Challenges

4.1 MOTION SICKNESS

Ondansetron has been compared to hyoscine and placebo in a "close-coupled" stimulation motion sickness model in volunteers [43]. Hyoscine was shown to have a beneficial effect; however, ondansetron was not superior to placebo indicating that 5-HT$_3$ receptors may not be involved in the pathophysiology of motion sickness.

4.2 AIDS-RELATED TREATMENT

Patients with AIDS receive a variety of treatments that can cause nausea and vomiting, e.g. zydovudine treatment and co-trimoxazole treatment for pneumocystics carinii pneumonia (PCP). A recent pilot study has indicated that ondansetron may be effective in controlling emesis associated with high-dose co-trimoxazole treatment [44].

4.3 PROTRACTED EMESIS IN CANCER PATIENTS

Initial reports [45,46] have indicated that ondansetron may have a role in controlling the protracted nausea that sometimes occurs in terminally ill cancer patients.

5. Conclusions

The discovery of selective 5-HT$_3$ antagonists has led to an increased understanding of the pathophysiological pathways by which nausea and vomiting are caused. Drugs such as ondansetron, granisetron, and tropisetron are now proving to be a major breakthrough in adults for the control of the debilitating emesis associated with cytotoxic treatment. In addition ondansetron can also be used in children with cancer and for the prevention and treatment of postoperative nausea and vomiting.

These drugs have proved to be extremely well tolerated and in particular are not associated with extrapyramidal symptoms, a major advance clinically over agents such as metoclopramide.

6. References

1. Borison, H.L. and Wang, S.C. (1953) 'Physiology and Pharmacology of Vomiting', Pharmacol. Rev. 5, 193-230.
2. Fozard, J.R. (1984) 'MDL72222: A potent and highly selective antagonist at neuronal 5-hydroxytryptamine receptors', Naunyn Schmiedebergs Arch. Pharmacol. 340, 403-10.
3. Freeman, A.J., Cunningham, K.T., and Tyers, M.B. (1992) 'Selectivity of 5-HT_3 receptor antagonists and antiemetic mechanisms of action', Anti-cancer Drugs 3, 79-85.
4. Laszlo, J. (1982) 'Emesis as limiting toxicity in cancer chemotherapy', in J. Laszlo (ed.), Antiemetics and Cancer Chemotherapy, Williams and Wilkins, Baltimore, pp. 1-5.
5. Coates, A., Abraham, S., Waye, S.B. et al. (1983) 'On the receiving end - patient perception of the side effects of cancer chemotherapy', Eur. J. Cancer Clin, Oncol. 19, 203-308.
6. Morrow, G.R. (1987) 'Management of nausea in the cancer patient', in S. Rosenthal, J.R. Carignan and B.D.E. Smith (eds.), Medical Care of the Cancer Patient. W.B. Saunders Company, Philadelphia, pp. 381-388.
7. Harris, J.G. (1978) 'Nausea vomiting and cancer treatment', CA-A Cancer Journal for Physicians 28, 194-201.
8. Durant J.R. (1984) 'The problem of nausea and vomiting, in modern cancer chemotherapy', CA-A Cancer Journal for Clinicians 34, 2-6.
9. Morrow, G.R. (1989) 'Chemotherapy-related nausea and vomiting: Etiology and management', CA-A Cancer Journal for Clinicians 39, 89-104.
10. Gralla, R.J., Tyson, L.B., Kris, M.G., and Clarke, R.A. (1987) 'The management of chemotherapy-induced nausea and vomiting', Medical Clinics in North America 71, 289-301.
11. Olver, I.N., Simon, R.N., and Aisner, J. (1986) 'Antiemetic Studies: A methodological discussion', Cancer Treat. Rev. 70, 555-563.
12. Tortorice, P.V. and O'Connell, M. (1990) 'Management of chemotherapy-induced nausea and vomiting', Pharmacotherapy 10, 129-145.
13. Cubeddu, L.X., Hoffmann, I.S., Fuenmayor, N.T., and Finn, A.L. (1990) 'Efficacy of ondansetron (GR 38032F) and the role of serotonin in cisplatin-induced nausea and vomiting', N. Eng. J. Med. 322, 810-816.
14. Marty, M., Pouillart, P., School, S. et al. (1990) 'Comparison of the 5-hydroxytryptamine$_3$ (serotonin) antagonist odansetron (GR 38032F) with high-dose metoclopramide in the control of cisplatin-induces emesis', N. Engl. J. Med. 322, 816-821.

15. de Mulder, P., Seynaeve, C., Vermorken, J. et al. (1990) 'Odansetron compared with high-dose metoclopramide in the prophylaxis of acute and delayed cisplatin-induced nausea and vomiting - a multicentre randomised, double-blind crossover study', Ann. Intern. Med. 113, 832-843.
16. Hainsworth, J., Harvey, W., Pendergrass, K. et al. (1991) 'A single-blind comparison of intravenous ondansetron, a selective serotonin antagonist, with intravenous metoclopramide in the prevention of nausea and vomiting associated with high-dose cisplatin chemotherapy', J. Clin. Oncol. 9, 721-728.
17. Brown, G.W., Paes, D., Bryson, J., and Freeman, A.J. (1992) 'The effectiveness of a single intravenous close of ondansetron', Oncology 49, 273-278.
18. Roila, F., Tonato, M., Cognetti, F. et al. (1991) 'Prevention of cisplatin-induced emesis: a double-blind multi-centre randomised cross-over study comparing ondansetron and ondansetron plus dexamethasone', J. Clin. Oncol. 9, 675-678.
19. Italian Group for Antiemetic Research (1992) 'Ondansetron plus dexamethasone vs metoclopramide plus dexamethasone plus diphenhydramine in the prevention of cisplatin-induced emesis', Lancet 340, 96-99.
20. Weissbach, L. (1991) 'Odansetron', Lancet 338, 753.
21. Schmoll, H.-J. (1989) 'The role of ondansetron in the treatment of emesis induced by non-cisplatin-containing chemotherapy regimens', Eur. J. Cancer Clin. Oncol. 25 (Suppl.1), 535-539.
22. Harousseau, J.L., Gisselbrecht, C., Paillarse, J.M. et al. (1991) 'Comparison of ondansetron and Alizapride in the prophylaxis of lymphoma chemotherapy-induced emesis', in M. Hartig and M. Pappo (eds.), Odansetron and Chemotherapy Induced Emesis. 3rd International Congress on Neo-adjuvant Chemotherapy, Springer Verlag, France.
23. Soukop, M., McQuade, B., Hunter, E. et. al. (1990) 'Odansetron compared with metoclopramide in the control of emesis and quality of life during repeated chemotherapy for breast cancer', Oncology 49, 295-304.
24. Priestman, T.J., Roberts, J.T., Lucraft, H. et al. (1990) 'Results of a randomized, double-blind comparative study of ondansetron and metoclopramide in the prevention of nausea and vomiting following high-dose upper abdominal irradiation', Clin. Oncol. 2, 71-75.
25. Priestman, T.J., Roberts, J.T., and Upadhyaya, B.K. (1992) 'Randomized, double-blind trial of ondansetron (OND) and prochlorperazine (PCP) in the prevention of fractionated radiotherapy (RT) induced emesis', Proc. Am. Soc. Clin. Oncol. Abst. 1370, 393.
26. Crookewit, S. (1990) 'The efficacy of ondansetron in emesis induced by total body irradiation', Zofran (odansetron) Satellite Symposium - European Society for Medical Oncology (EMSO), Copenhagen, Dec. 1990, 15-17.
27. Hewitt, M., Cornish, J., Pamphilon, D., and Oakhill, A. (1991) 'Effective

emetic control during conditioning of children for bone marrow transplantation using ondansetron, a 5-HT$_3$ antagonist', Bone Marrow Transplantation 7, 431-438.
28. Jürgens, H. and McQuade, B. (1992) 'Odansetron as Prophylaxis for Chemotherapy and Radiotherapy induced Emesis in Children', Oncology 49, 279-285.
29. Smith, I.E., on behalf of the granisetron study group (1990) 'A dose-finding of granisetron in patients receiving moderately emetogenic chemotherapy', Eur. J. Cancer. 26 (Suppl. 1), 19-23.
30. Soukop, M. on behalf of granisetron study group (1990) 'A dose-finding study of granisetron in patients receiving high-dose cisplatin', Eur. J. Cancer 26 (Suppl. 1), 15-19.
31. Chevallier, B. on behalf of the granisetron study group (1990) 'Efficacy and safety of granisetron compared to high-dose metoclopramide plus dexamethasone in patients receiving high-dose cisplatin', Eur. J. Cancer 26 (Suppl. 1), 33-36.
32. Venner, P. for The Clinical Trials Group of the National Cancer Institute of Canada (1990) 'Granisetron for high dose cisplatin (HDCP) induced emesis: a randomized double-blind study (abst.)', Proc. Am. Soc. Clin. Oncol. 9, 320.
33. Marty, M. on behalf of the granisetron study group (1990) 'A comparative study of the use of granisetron, a selective 5-HT$_3$ antagonist, versus a standard antiemetic regimen of chlorpromazine plus dexamethasone in the treatment of cytostatic-induced emesis', Eur. J. Cancer 26 (Suppl. 1), 28-33.
34. Warr, D. (1989) 'A double-blind, randomized comparison of the novel antiemetic BRL43694A versus dexamethasone plus prochlorperaxine (abst.)', Proc. Eur. Congr. Clin. Oncol. 5.
35. Dogliotti, L., Fagginolo, R., Berrut, A., Antomacci, R.A., Ortega, C., and Lancranjan, J. (1990) 'Prevention of Nausea and Vomiting in cisplatin-treated patients by a selective (5-HT$_3$) receptor antagonist, ICS 205-930', Tumori 76, 595-598.
36. Van Wijk, M. and Smalhout, B. (1990) 'A postoperative analysis of the patient's view of anaesthesia in a Netherlands' teaching hospital', Anaesthesia 45, 679-682.
37. Abramowitz, M.D., Oh, T.H., Epstein, B.S., Ruttiman, U.E., and Friendly, D.S. (1983) 'The aniemetic effect of droperidol following outpatient stabisumus surgery in children', Anasthesiology 59(b), 579-583.
38. Mathia, W.J., Bell, S.K., and Leak, W.D. (1988) 'Effects of preoperative metoclopramide and droperidol on postoperative nausea and vomiting in ambulatory surgery patients', Am. Assoc. Nurse Anesth. J. 56(4), 325-333.
39. Palazzo, M.G.A. and Strunin, L. (1984). Anaesthesia and emesis.II: prevention and management', Can. Anaesth. Soc. J. 31(4), 407-415.
40. Kovac, A., McKenzie, R., and O'Connor, T. (1992) 'Single dose intravenous odansetron in the prevention of postoperative nausea and vomiting', Eur. J.

Anaesthesiology, in press.
41. DuPen, S., Scuderi, P., and Wetchler, B. (1992) 'Single dose intravenous ondansetron in the treatment of postoperative nausea and vomiting', Eur. J. Anaesthesiology, in press.
42. Kenny, G., Oates, J., Lesser, J. et al. (1992) 'Efficacy of orally administered ondansetron in the prevention of postoperative nausea and vomiting: A dose ranging study', Br. J. Anaesth. 68, 446-470.
43. Stott, J.R.R., Barnes, G.R., Wright, R.J., and Ruddock, C.J.S. (1989) 'The effect on motion sickness and occulomotor function of GR38932F, a 5-HT_3-receptor antagonist with anti-emetic properties', Br. J. Clin. Pharmac. 37, 147-157.
44. Main, J., Gompels, M., McWilliams, S. et. al. (1991) 'Odansetron as an anti-emetic in AIDS patients receiving high-dose co-trimoxazole: an open study', Abst. No. MB2186, VIIth International Operative Congress on AIDS, Florence, 16-21 June, 228.
45. Nicholson, S., Evans, C., and Mansi, J. (1992) 'Ondansetron in intractable nausea and vomiting', Lancet 339 (8791), 490.
46. Mulvenna, P.M., and Regnard, C.F.B. (1992) 'Subcutaneous ondansetron', Lancet 339 (8800), 1059.

INHIBITORS OF SEROTONIN UPTAKE

S.Z. LANGER and D. GRAHAM
Synthélabo Recherche
B.P. 110
92225 Bagneux Cedex
France

ABSTRACT. Over the past fifteen years selective serotonin uptake inhibitors such as fluoxetine, sertraline, citalopram, paroxetine, and litoxetine have been developed, some of which are currently widely used in the clinical management of affective disorders. These selective uptake inhibitors have provided powerful probes to undertake molecular pharmacological studies of the sodium-ion coupled serotonin transporter. Using affinity-chromatography purification the rat brain serotonin transporter was identified as a polypeptide of Mr = 73,000. In addition, PCR technology recently enabled the cloning of this transporter polypeptide which now opens up exciting perspectives for further fundamental and clinical research on this transport system.

The hypothesis that serotonin hypofunction might be implicated in certain forms of depressive illness provided the impetus to design selective inhibitors of sodium-ion coupled serotonin uptake. A number of these compounds have subsequently been developed and have now become clinically important drugs. In this report we discuss some of the current therapeutic applications of selective serotonin uptake inhibitors. In addition, we describe the utilization of these compounds as specific probes to characterize at a molecular level their primary target site, the sodium-ion coupled serotonin transporter.

1. General Properties of Sodium-Ion Coupled Serotonin Transport

Sodium-ion coupled serotonin transport systems or transporters play an important role in overall serotonin homeostasis where they function to reduce extracellular levels of this biogenic amine (for review see [1]). For example, the localization of these plasmalemma transporter molecules at presynaptic serotonergic nerve terminals provides an uptake system to reduce synaptic cleft concentrations of released serotonin, thereby serving as a neurotransmitter inactivating mechanism. Also, the presence of serotonin transporters in platelets and mast cells permits the concentration and storage of serotonin in these cells for subsequent secretion. In addition, a number of other cell-types including astrocytes and endothelial cells have been described to possess sodium-ion coupled serotonin transporters.

In fact, the sodium-ion coupled serotonin transporter is an example of a secondary transport process in that it depends upon the activity of a primary ion pump, the Na^+/K^+ ATPase. Electrochemical ion gradients for sodium and chloride generated across the plasma membrane and the coupling of these ions to the transporter provides the necessary energy for serotonin influx in a tightly-coupled symporter uptake system.

A number of compounds have been described which inhibit selectively at low nanomolar concentrations sodium-ion coupled serotonin transport. These selective serotonin uptake inhibitors derived from different chemical families include nontricyclic compounds such as fluoxetine, sertraline, citalopram, litoxetine, paroxetine, indalpine, and fluvoxamine.

2. Molecular Pharmacology of the Sodium-Ion Coupled Serotonin Transporter

The pronounced selectivity of the nontricyclic serotonin uptake inhibitors has led to the utilization of radiolabelled forms of these compounds as molecular probes to characterize the serotonin transporter. Studies with tritiated ligands including ^{3}H-paroxetine [2-3], ^{3}H-citalopram [4], and ^{3}H-6-nitroquipazine [5] *in vitro* have shown these compounds to be extremely selective serotonin transporter probes. As such, the selectivity of several of these tritiated inhibitors has been exploited for mapping serotonergic pathways in rat and human brain using *in vitro* and *in vivo* autoradio-graphic studies [6-7]. In addition, with the emergence of brain imaging techniques to measure quantitatively neurotransmitter systems in the intact human brain, the possibility of using carbon-11 radiolabelled serotonin uptake inhibitors in conjunction with positron emission tomography has been explored [8].

We have extensively used ^{3}H-paroxetine as a probe to investigate the site(s) of interaction of a variety of nontricyclic and tricyclic uptake inhibitors with the neuronal sodium-ion coupled serotonin transporter [9]. The properties of ^{3}H-paroxetine binding to membrane preparations of rat cerebral cortex in competitive inhibition, kinetic dissociation, and chemical modification studies indicate that the selective serotonin uptake inhibitors, as well as the tricyclic antidepressant, imipramine, bind to mutually-exclusive domains at the substrate recognition site of the serotonin transporter.

We have further characterized the sodium-ion coupled serotonin transporter using a protocol designed to purify this macromolecular entity [10]. The serotonin transporter of rat cerebral cortical membrane preparations was solubilized using the non-ion detergent, digitonin, in a conformational state which retained a ^{3}H-paroxetine pharmacological binding profile almost identical to that observed in membrane preparations. Subsequently, solubilized serotonin transporter was extensively purified by sequential affinity chromatography on a reduced citalopram-based agarose resin and wheat-germ aglutinin-sepharose. Starting from 159 pmol of transporter (corresponding to 8 g of rat cerebral cortex) a purification of approximately 2000-fold of ^{3}H-paroxetine binding activity was obtained with an overall yield of 9%. With this extensive purification procedure, the pharmacological properties of the transporter remained essentially similar to those displayed by the parent membrane-bound form. SDS-polyacrylamide gel electrophoresis of the affinity-purified material under reducing conditions revealed the presence of several major polypeptide bands in this preparation (figure 1). The serotonin transporter, however, was identified as a polypeptide of Mr = 73,000 by performing a parallel purification protocol in which the solubilized detergent extracts were first incubated with a saturating concentration of free citalopram before carrying out the affinity-chromatographic procedures.

Recently, two groups have also reported purification of the human platelet serotonin transporter. In each case significant overall purification was reported, although the molecular size estimates of the polypeptide identified as corresponding to the transporter were slightly different (Mr = 78,000 [11]; and Mr = 68,000 [12]).

3. Cloning of the Sodium-Ion Coupled Serotonin Transporter

Based upon the high degree of similarity (68% allowing for conservative amino acid substitutions) between the amino acid sequences of the human GABA and noradrenaline transporters, PCR technology was recently applied to clone the serotonin transporter of rat brain [13] and rat basophilic leukemic cells [14]. The isolated cDNA clones obtained predicted extremely highly homologous proteins of 617 and 653 amino acids for the rat brain and basophilic transporters, respectively, corresponding to proteins of Mr = 68,000 and 73,000. Transfection studies with these cloned cDNAs indicated that each gene product conferred the full pharmacological characteristics of the serotonin uptake system. The divergence in the primary amino acid sequences at the N and C terminals of these two transporters might reflect serotonin transporter heterogeneity, although a sequencing error should not be excluded.

The serotonin transporter is in fact a member of a larger neurotransmitter transporter supergene family whose gene products give rise to proteins of 12 putative transmembrane spanning regions. A common characteristic of these proteins whose members include transporters for GABA [15], L-proline [16], glycine [17], betaine [18], choline [19] as well as those for the biogenic amines, noradrenaline [20], dopamine [21], and serotonin [13,14] is that they each use sodium and chloride ions as co-substrates.

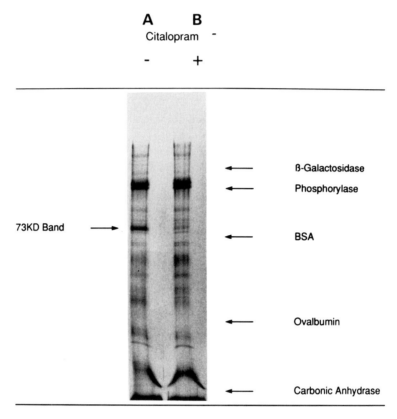

Figure 1. SDS-Polyacrylamide gel electrophoresis of affinity-purified serotonin transporter. Solubilized detergent extracts of rat cerebral cortical membranes were divided into two equal batches, and one batch (B) incubated with 1 mM citalopram. Both preparations then underwent parallel affinity-chromatography purifications on citalopram-based-agarose resin and WGA-sepharose. The final eluates were concentrated and electrophoresed on 10% SDS-polyacrylamide gels under reducing conditions. The gels were calibrated with known Mr markers. Reprinted by permission of Portland Press.

4. Serotonin Uptake Inhibitors - Therapeutic Uses

Many of the selective serotonin uptake inhibitors synthetized over the past fifteen years are now at various stages of evaluation as antidepressants (table 1). Clinical studies with these selective serotonin uptake inhibitors indicate that these drugs are as beneficial as the tricyclic antidepressants with the advantage that they produce a lower side effect spectrum. Thus, the highly selective properties of the nontricyclic serotonin uptake inhibitors limit interactions at α-adrenergic, histaminergic, and cholinergic receptors which greatly reduce side effects such as sedation, fatigue, orthostatic hypotension, dry mouth, constipation, blurred vision, and urinary retention. Moreover, the selective serotonin uptake inhibitors do not produce arrhythmias, tachycardia, or cardiac toxicity.

TABLE 1. Development status of selective serotonin uptake inhibitors as antidepressants.

Launched	Fluoxetine, Sertraline, Fluvoxamine, Citalopram, Paroxetine
Pre-registration	Femoxetine
Phase II Clinical Trials	Litoxetine

Selective serotonin uptake inhibitors are also proving to be effective in the clinical management of certain types of anxiety disorders. In particular, a number of clinical trials have revealed the efficacy of these agents in treating panic disorders, obsessive compulsive disorders, and anxiety disorders associated with depression [22-23]. In addition, selective serotonin uptake inhibitors are being evaluated in the clinic for their therapeutic potential in the treatment of obesity and alcoholism.

5. Future Directions

The recent cloning of the sodium-ion coupled serotonin transporter opens up a new dimension to study the physiological, pharmacological, and clinical importance of this biogenic amine transport system. The utilization of molecular techniques such as site-directed mutagenesis or the construction of chimeric transporters derived from different biogenic amine transporters will help to provide detailed molecular insights of how the serotonin transporter functions and is regulated. Thus by such means it is hoped to identify key amino amine residues and domains important for selective inhibitor blockade as well as for substrate and cosubstrate ion binding and translocation.

The existence of serotonin transporter cell-type diversity and the pharmacological diversity of other sodium-ion coupled transporters such as the GABA transporter [24], raises the question of serotonin transporter subtypes. Should such serotonin transporter heterogeneity be confirmed, this opens up the possibility to design novel subtype-selective serotonin uptake inhibitors.

Finally, given the suggestions that serotoninergic dysfunction might be associated with certain neuropsychiatric disorders, the utilization of *in vitro* diagnostic tools to detect possible serotonin transporter genetic anomalies, particularly point mutations and polymorphisms should be examined.

6. Acknowledgements

The authors thank Françoise Péchoux for expert secretarial assistance during the preparation of this manuscript.

7. References

1. Graham, D. and Langer, S.Z. (1988) '5-Hydroxytryptamine transport systems', in P.R. Saxena, D.I. Wallis, W. Wouters and P. Bevan (eds.), Cardiovascular Pharmacology of 5-Hydroxytryptamine, Kluwer, Dordrecht, pp. 15-30.
2. Habert, E., Graham, D., Tahraoui, L., Claustre, Y., and Langer, S.Z. (1985) 'Characterization of ^{3}H-paroxetine binding to rat cortical membranes', Eur. J. Pharmacol. 188, 107-114.
3. Segonzac, A., Schoemaker, H., and Langer, S.Z. (1987) 'Temperature-dependence of drug interaction with the platelet 5-HT transporter: a clue to the imipramine selectivity paradox', J. Neurochem. 48, 331-339.
4. D'Amato, R., Largent, B.L., Snowman, A.M., and Snyder, S.H. (1987) 'Selective labelling of serotonin uptake sites in rat brain by ^{3}H-citalopram contrasted to labelling of multiple sites by ^{3}H-imipramine', J. Pharmacol. Exp. Ther. 242, 364-371.
5. Hashimoto, K. and Goromaru, T. (1991) 'High-affinity binding of ^{3}H-6-nitroquipazine to cortical membranes in the rat: inhibition by 5-hydroxytryptamine uptake inhibitors', Neuropharmacology 30, 113-117.
6. De Souza, E.B. and Kuyatt, B.L. (1987) 'Autoradiographic localization of ^{3}H-paroxetine-labelled serotonin uptake sites in the rat brain', Synapse 1, 488-496.
7. Scheffel, U. and Harting, P.R. (1989) 'In vivo labelling of serotonin uptake sites with ^{3}H-paroxetine', J. Neurochem. 52, 1605-1612.
8. Hume, S.P., Pascali, C., Pike, V.W., Turton, D.R., Ahier, R.G., Myers, R., Bateman, D.M., Cremer, J.E., Manjil, L.G., and Dolan, R. (1991) 'Citalopram: labelling with carbon-11 and evaluation in rat as a potential radioligand for in vivo PET studies of 5-HT re-uptake sites', Int. J. Rad. Appl. Instrum. 18, 339-351.
9. Graham, D., Esnaud, H., Habert, E., and Langer, S.Z. (1989) 'A common binding site for tricyclic and nontricyclic 5-hydroxytryptamine uptake inhibitors at the substrate recognition site of the neuronal sodium-dependent 5-hydroxytryptamine transporter', Biochem. Pharmacol. 38, 3819-3826.

10. Graham, D., Esnaud, H., and Langer, S.Z. (1992) 'Partial purification and characterization of the sodium-ion coupled 5-hydroxytryptamine transporter of rat cerebral cortex', Biochem. J. 286, 801-805.
11. Biesson, E.A.L., Horn A.S., and Robillard G.T. (1990) 'Partial purification of the 5-hydroxytryptophan-reuptake system from human blood platelets using a citalopram-derived affinity resin', Biochemistry 29, 3349-3354.
12. Launay, J.-M., Geoffroy, C., Mutel, V., Buckle, M., Cesura, A., Alouf, J.E., and Da Prada, M., (1992) 'One-step purification of the serotonin transporter located at the human platelet plasma membrane', J. Biol. Chem. 267, 11344-11351.
13. Blakely, R.D., Berson, H.E., Fremeau Jr., R.T., Caron, M.G., Peek, M.M., Prince, H.K., and Bradley, C.C. (1991) 'Cloning and expression of a functional serotonin transporter from rat brain', Nature 354, 66-70.
14. Hoffman, B.J., Mezey, E., and Brownstein, M.J. (1991) 'Cloning of a serotonin transporter affected by antidepressants', Science 254, 579-580.
15. Guastella, J., Nelson, N., Nelson, H., Czyzyk, L., Kenana, S., Miedel, M.C., Davidson, N., Lester, H.A., and Kanner, B.I. (1990) 'Cloning and expression of a rat brain GABA transporter', Science 249, 1303-1306.
16. Fremeau Jr., R.T., Caron, M.G., and Blakely, R.D. (1992) 'Molecular cloning and expression of a high affinity L-proline transporter expressed in putative glutamatergic pathway of rat brain', Neuron 8, 915-926.
17. Smith, K.E., Borden, L.A., Hartig, P.R., Branchek, T., and Weinshank, R.L. (1992) 'Cloning and expression of a glycine transporter reveal colocalization with NMDA receptors', Neuron 8, 927-935.
18. Yamauchi, A., Uchida, S., Kwon, H.M., Preston, A.S., Robey, R.B., Garcia-Perez, A., Burg, M.B., and Handler, J.S. (1992) 'Cloning of a Na^+- and Cl^--dependent betaine transporter that is regulated by hypertonicity', J. Biol. Chem. 267, 649-652.
19. Mayser, W., Schloss, P., and Betz, H. (1992) 'Primary structure and functional expression of a choline transporter expressed in the rat nervous system', FEBS Lett. 305, 31-36.
20. Pacholczyk, T., Blakely, R.D., and Amara, S.G. (1991) 'Expression cloning of a cocaine- and antidepressant-sensitive human noradrenaline transporter', Nature 350, 350-354.
21. Shimida, S., Kitayama, S., Lin, C.-L., Patel, A., Nanthakumar, E., Gregor, P., Kuhar, M., and Uhl, G. (1991) 'Cloning and expression of a cocaine-sensitive dopamine transporter complementary DNA', Science 254, 576-578.
22. Coplan, J.D., Gorman, J.M., and Klein, D.F. (1992) 'Serotonin related functions in panic-anxiety: a critical overview', Neuropsychopharmacology 6, 189-200.
23. Barr, L.C., Goodman, W.K., Price, L.H., McDouglas, C.J., and Charney, D.S. (1992) 'The serotonin hypothesis of obsessive compulsive disorder: implications of pharmacologic challenge studies', J. Clin. Psychiat. 53, 17-28S.

24. Mabjeesh, N.J., Frese, M., Rouen, T., Jeserich, G., and Kanner, B.I. (1992) 'Neuronal and glial γ-aminobutyric acid transporters are distinct proteins', FEBS Lett. 299, 99-102.

PRECLINICAL EVIDENCE FOR THE ROLE OF SEROTONIN RECEPTORS IN ANXIETY

BEREND OLIVIER*, JAN MOS, ELLEN MOLEWIJK, THEO ZETHOF, and GUUS VAN DER POEL
Department of CNS-pharmacology, Solvay Duphar B.V.
P.O. Box 900
1380 DA Weesp
The Netherlands and
*Department of Psychopharmacology
Faculty of Pharmacy
State University Utrecht
The Netherlands

ABSTRACT. Animal paradigms are important to detect anxiolytic properties of psychoactive compounds. Two animal anxiety paradigms are presented, ultrasonic distress vocalizations (USV) in rat pups and stress-induced hyperthermia (SIH) in mice, to describe the putative anxiolytic properties of serotonergic ligands. Both paradigms indicate that 5-HT_{1A} receptor agonists are potent anxiolytics, whereas specific 5-HT reuptake blockers also exert some anxiolytic properties. $5\text{-HT}_{1C/2}$ antagonists may have some anxiogenic properties (in USV), whereas 5-HT_3 antagonists seem to be devoid of anxiolytic properties in the paradigms used.

1. Introduction

There are basically two types of animal behavior models used in the study of putative anxiolytic drugs. Models can be based on conditioned behavior and involve responses often controlled by operant conditioning procedures. On the other hand models based on unconditioned behavior make use of the natural behavioral repertoire of the animal spontaneous behavior and generally require no specific training. The latter models are often classified as "ethologically-based models."

Examples of conditioned behavior models are punishment or conflict procedures, potentiated startle, periaqueductal gray stimulation, conditioned taste aversion, and drug discrimination. Unconditioned models in use are exploratory behavior, elevated plus maze, defensive burying, ultrasonic vocalization, etc.

The present contribution focuses on the unconditioned models and specifically on two of them: ultrasonic distress vocalizations in rat pups and stress-induced hyperthermia (or anticipatory anxiety) in mice. We will attempt to delineate the effects of various serotonergic ligands (both agonists and antagonists) in both paradigms.

2. Ultrasonic Vocalization (USV) in Rat Pups

USV of infant rats is produced for example when pups are separated from the mother. Because these calls are associated with social isolation, they are often described as "distress vocalizations," as is also produced by a large variety of young mammals, including humans ("crying"). Young rat pups between the age of 1 to 16 days readily emit bouts of ultrasounds of 20 to 200 ms in the range of 35 to 50 kH$_2$ [1].

In particular, when pups are exposed to a cold environment, USV occurs at a high rate [2,3]. In our hands we used rat pups 1-2 weeks old and measured the USV under two conditions: on a warm (37°C) plate and on a cold (18°C) plate. Under these conditions drugs of various classes were tested and each condition had each own profile of "drug action" [4,5,6]. A test consisted of injecting a pup intraperitoneally with the compound 30 minutes before a 5-minute test on either a cold or a warm plate. During this 5-minute test the number of USVs was recorded. To measure the drug's specificity an animal was placed head downwards on an inclined screen and the time needed to move the head and body upwards was measured [5].

Table 1 shows our results with serotonergic ligands. 5-HT$_{1A}$ receptor agonists were able to reduce the USV under both conditions in a dose-dependent way, often at doses which did not interfere with normal sensomotoric functioning. Mixed 5-HT$_1$ receptor agonists in particular (such as TFMPP [1B, 1C, 1A], eltoprazine [1A, 1B, 1D], and RU 24969 [1B, 1A, 1C, 1D]) had effect on the cold plate and can be predicted on the basis of this model to have some mild anxiolytic activity. The effect of the 5-HT$_{1A}$ receptor agonists on the USV was robust; these results were also found by others [7,8,9].

The effects found using mixed receptor 5-HT agonists were less clear. We found that 5-HT$_1$ receptor agonists (RU 24969, eltoprazine, TFMPP) showed no effects on the warm plate and moderate decreases on the cold plate. However, Winslow and Insel [8] report that 5-HT$_{1B}$ receptor agonists (CGS12066B and TFMPP) increased the rate of calling, although at higher doses TFMPP decreased USV, probably due to its marked 5-HT$_{1C}$-agonistic character. Nastiti et al. [9] only found USV-decreasing effects of TFMPP in mice.

We found that the β-adrenergic blockers and 5-HT$_{1A}$ receptor antagonists (-) pindolol and (±)-propranolol decreased USV on the warm plate, but not on the cold plate, whereas the specific 5-HT$_{1A}$-antagonists WAY 100,139 had no effect on the warm plate and a small, bell-shaped increase on the cold plate (figure 1). WAY 100,139 was able to antagonize the decrease in USV induced by the 5-HT$_{1A}$ receptor agonist flesinoxan (figure 1). Winslow and Insel [8] also found, although at higher doses (10 mg/kg) a decrease of dl,l-propranolol, but this compound, at lower doses

(5.0) mg/kg) was able to antagonize the decrease in USV after both 8-OH-DPAT nd TFMPP, thereby confirming its 5-HT_{1A} and 5-HT_{1B} antagonistic activity.

TABLE 1. Effects of various serotonergic drugs on the number of ultrasonic vocalizations of rat pups of 9-11 days old. In the right column the effect on motoric capacity is shown (inclined screen test).

Drugs	Number of ultrasonic cries during 5 minutes		Effect on inclined screen
	warm plate (37°C)	cold plate (18°C)	specificity
5-HT_{1A} ligands			
8-OH-DPAT	↓↓↓	↓↓↓	++
Flesinoxan	↓↓↓	↓↓↓	++
Isapirone	↓↓	↓↓↓	++
Buspirone	↓↓↓	↓↓↓	+
(-)Pindolol	↓	0	+
(±)Propranolol	↓↓	0	++
WAY 100,139	-	0 ↑	+
Mixed 5-HT agonists			
RU 24969	0	↓↓	+
Eltoprazine	0	↓↓	++
TFMPP	0	↓↓	++
DOI	↓↓	↓↓	++
$5\text{-HT}_{1C/2}$ antagonists			
Ritanserin	↑	0	-
Ketanserin	↑	↑	0
5-HT_3 antagonists			
Ondansetron	0	0	0
5-HT reuptake blockers			
Fluvoxamine	↓	↓↓↓	++
Fluoxetine	↓↓	↓↓	+
Zimeldine	0	↓↓	+

↓↓↓ strong decrease; ↓↓ moderate decrease; ↓ weak decrease; 0 no effect; ↑ increase (anxiogenic); - not tested; ++ large distance between anxiolytic and side-effects; += reasonable distance; 0= no distance

DOI, a 5-$HT_{1C/2}$ receptor agonist reduced USV both under warm and cold conditions. Winslow and Insel [8] found similar effects after DOI and also after mCPP, which is primarily a 5-HT_{1C} receptor agonist, thereby suggesting that the USV-decrease is caused by an effect on the 5-HT_{1C} receptor. Curiously, DOI enhanced USV in mice [9]. Ritanserin and ketanserin, 5-$HT_{1C/2}$ receptor antagonists, enhance USV, particularly on the warm plate. Winslow and Insel also found an USV, particularly on the warm plate. Winslow and Insel also found an USV-increase after ritanserin, whereas in mice ritanserin decreased USV [9]. Clearly more work is needed to unravel the role of the 5-HT_{1C} and 5-HT_2 receptor in USV.

All studies reported sofar, including ours, are unanimous about 5-HT_3 antagonists: no effect on USV [4,7,9].

The specific 5-HT reuptake blockers all reduce USV, which is in line with the findings of Winslow and Insel [10] for CMI, paroxetine, and citalopram.

In summary, the data demonstrates an important role for serotonin in modulation of rat pup ultrasonic calling. Of course various other neurotransmitters are implicated in the mediation of maternal-infant attachment, including noradrenalin, the GABA-BDZ system, and the endorphin system.

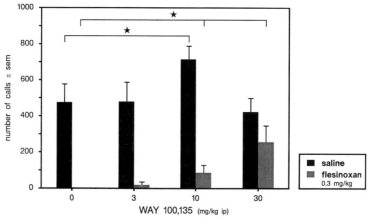

Figure 1. The effects of the 5-HT_{1A} receptor antagonists WAY 100,139 on saline- or flesinoxan (0.3 mg/kg, i.p.)-treated rat pup ultrasonic distress calls is shown. USV is measured under 18°C conditions (cold plate). Under this condition WAY 100,139 dose dependently antagonizes the flesinoxan-induced decrease in USV.*$p < 0.05$.

3. Anticipatory Anxiety in Mice (Stress-induced Hyperthermia)

When group-housed mice are removed one by one from their home cage, the last mouse removed always has a higher rectal temperature than those removed first. This phenomenon has been called stress-induced hyperthermia (SIH) or anticipatory anxiety [11]. The anticipatory increase in rectal temperature could be antagonized by various anxiolytic agents, including benzodiazepines, whereas several other centrally acting, nonanxiolytic drugs do not affect SIH [12,13].

We replicated and extended the original findings [14] and found SIH to be a robust phenomenon: the vehicle-treated test group displays increases of 1.1-1.3°C. We investigated the effects of various serotonergic ligands on SIH.

5-HT_{1A} receptor agonists (buspirone, ipsapirone, 8-OH-DPAT, flesinoxan) dose-dependently decreased SIH while having no or only slight effects on basal temperature [12,13,14].

5-HT_3 receptor antagonists have no effects in SIH [12,13,14] whereas 5-HT reuptake blockers have a very limited anxiolytic potential in this model [14].

This model seems to be a good supplement to the paradigms to describe putative anxiolytic properties of drugs, although the information about the involvement of the 5-HT system in SIH is only fragmentary.

It appears that the USV model in rat pups and the SIH model in mice can be used complementarily. Both paradigms seem to pick up different aspects of anxiety and therefore both paradigms are of good use in detecting putative serotonergic anxiolytics.

4. Acknowledgments

We thank Frieda Verberne for typing the manuscript and Ruud Oorschot for technical support and graphics.

5. References

1. Van der Poel, A.M., Molewijk, E., Mos, J., and Olivier, B. (1991) 'Is clonidine anxiogenic in rat pups?', in B. Olivier, J. Mos and J.L. Slangen (eds.), Animal models in Psychopharmacology, Birkhäuser Verlag, Basel, pp. 107-116.
2. Miczek, K.A., Tornatzky, W., and Vivian, J. (1991) 'Ethology and neuropharmacology: rodent ultrasounds', in B. Olivier, J. Mos and J.L. Slangen (eds.), Animal models in Psychopharmacology, Birkhäuser Verlag, Basel, pp. 409-427.
3. Insel, T.R. and Winslow, J.T. (1991) 'Rat pup ultrasonic vocalizations: an ethologically relevant behavior responsive to anxiolytics', in B. Olivier, J. Mos and J.L. Slangen (eds.) Animal models in Psychopharmacology, Birkhäuser

Verlag, Basel, pp. 15-36.
4. Mos, J. and Olivier, B. (1988) 'Ultrasonic vocalizations by rat pups as an animal model for anxiolytic activity: effects of antidepressants', in B. Olivier and J. Mos (eds.), Depression, Anxiety and Aggression, Preclinical and clinical interfaces, Medidact, Houten, pp. 85-93.
5. Mos, J. and Olivier, B. (1989) 'Ultrasonic vocalizations by rat pups as an animal model for anxiolytic activity: effects of serotonergic drugs', in P. Bevan, A.R. Cools and T. Archer (eds.), Behavioral Pharmacology of 5-HT, Lawrence Erlbaum, New Jersey, pp. 361-366.
6. Olivier, B., Tulp, M.Th.M., and Mos, J. (1991) 'Serotonergic receptors in anxiety and aggression; evidence from animal pharmacology', Human Psychopharmacology 6, S72-78.
7. Kehne, J.H., McCloskey, T.C., Baron, B.M., Chi, E.M., Harrison, B.L., Whitten, J.P., and Palfreyman, M.G. (1991) 'NMDA receptor complex antagonists have potential anxiolytic effects as measured with separation-induced ultrasonic vocalizations', Eur. J. Pharmacol. 193, 283-292.
8. Winslow, J.T. and Insel, T.R. (1991) 'Serotonergic modulation of the rat pup ultrasonic isolation call: studies with $5\text{-}HT_1$ and $5\text{-}HT_2$ subtype-selective agonists and antagonists', Psychopharmacology 105, 513-520.
9. Nastiti, K., Benton, D., Brain, P.F., and Haug, M. (1991) 'The effects of 5-HT receptor ligands on ultrasonic calling in mouse pups', Neurosci. Biobehav. Rev. 15, 483-487.
10. Winslow, J.T. and Insel, T.R. (1990) 'Serotonergic and catecholaminergic reuptake inhibitors have opposite effects on the ultrasonic isolation calls of rats pups', Neuropsychopharmacology 3, 51-59.
11. Borsini, F., Lecci, A., Volterra, G., and Meli, A. (1989) 'A model to measure anticipatory anxiety in mice?', Psychopharmacology 98, 207-211.
12. Lecci, A., Borsini, F., Mancinelli, A., D'Arranno, V., Stasi, M.A., Volterra, G., and Meli, A. (1990) 'Effects of serotoninergic drugs on stress-induced hyperthermia (SIH) in mice', J. Neural Transm. 82, 219-230.
13. Lecci, A., Borsini, F., Volterra, G., and Meli, A. (1990) 'Pharmacological validation of a novel animal model of anticipatory anxiety in mice', Psychopharmacology 101, 255-261.
14. Zethof, T.J.J., Van de Heyden, J.A.M., and Olivier, B. (1991) 'A new animal model for anticipatory anxiety?', in B. Olivier, J. Mos and J.L. Slangen (eds.) Animal models in Psychopharmacology, Birkhäuser Verlag, Basel, pp. 65-68.

THE EFFECTS OF TRYPTOPHAN DEPLETION IN DEPRESSION

RONALD M. SALOMON[1], HELEN L. MILLER[1], PEDRO L. DELGADO[2], DENNIS S. CHARNEY[1]
[1]*Department of Psychiatry*
Yale University School of Medicine
Department of Veterans Affairs Medical Center
Psychiatry Service 116-A1
West Haven, Connecticut 06516
and
[2]*Department of Psychiatry*
University of Arizona School of Medicine
1501 N. Campbell Avenue
Tucson, Arizona 85724

ABSTRACT. Hypotheses suggesting decreased serotonin function in patients with depression can be evaluated by using an amino acid drink to rapidly deplete levels of the serotonin precursor, tryptophan. Data from this experimental paradigm support an important role for serotonin in the mechanism of action of many antidepressant medications. In addition, there is preliminary data suggesting that this method may prove useful in predicting antidepressant drug response.

1. Introduction

Studies of serotonin (5-hydroxytryptamine, 5-HT) function over the past quarter century have brought to light clinical evidence which suggests that altered 5-HT function is involved in the pathophysiology of depression in at least some subgroups of patients. Many studies of depressed patients [1-8] and suicide victims [9-13] demonstrate evidence of decreased function of the 5-HT system in the brain. Recently, a method became available to acutely decrease brain 5-HT function using a dietary depletion of the 5-HT precursor, L-tryptophan (TRP). Effects of acute TRP depletion on mood have recently been evaluated in healthy subjects and depressed patients.

1.1 SEROTONIN METABOLISM IN DEPRESSION AND SUICIDE

The precursor for 5-HT synthesis is the essential dietary amino acid, TRP, which circulates in plasma in free and albumin-bound forms. Plasma TRP levels are lower and the ratio of TRP to other large neutral amino acids (LNAA) is smaller in depressed patients than healthy subjects in most studies [1,4,14-18]. A recent replication of this findings suggested, in addition, that plasma TRP levels normalize after remission of the depressive symptoms [19]. Plasma TRP levels may predict medication response. Low baseline TRP levels or TRP/LNAA ratios correlate with greater mood improvement after treatment with 5-HT-enhancing medications [20].

Levels of 5-HT metabolites appear to be decreased in depression. Ashcroft [2] reported that cerebrospinal fluid (CSF) concentrations of 5-hydroxyindoles were significantly lower in 9 depressed patients than in neurological patients. Subsequently, low CSF and brain tissue levels of another 5-HT metabolite, 5-hydroxyindoleacetic acid (5-HIAA), have been found in most [7,9], but not all [21,22], well-designed studies comparing depressed or suicidal patient groups to controls. CSF 5-HIAA levels are almost invariably low in suicide victims and other violent groups [9,23].

Additional evidence of altered 5-HT metabolism in depression is observed in circulating platelets, which accumulate and store a releasable pool of 5-HT. Less 5-HT is found in platelets of depressed patients than healthy subjects [6].

1.2 5-HT RECEPTORS IN DEPRESSION AND SUICIDE

The 5-HT_2 receptor has been quantified in autopsy studies of suicides and depressives. When compared to matched control tissues, frontal cortices from suicide victims and depressed patients display higher 5-HT_2 binding [11,24-27]. Mixed data from studies of cortical imipramine binding (marker of a 5-HT transporter) in depressed patients [28-30] may be explained by a finding of reversed asymmetry of imipramine binding among suicides (left > right hemisphere) compared to controls (right > left hemisphere) [13].

5-HT_2 receptors have also been measured on platelet membranes. Platelet 5-HT_2 binding is increased in depressed patients compared to healthy subjects [25,31-34]. Platelet imipramine binding may be decreased in depression [35-37], although several recent, larger studies have not found a difference from controls [38-40]. 5-HT uptake by platelets is decreased in depression, and, in most studies, normalizes in association with clinical remission [12,19,38].

1.3 NEUROENDOCRINE STUDIES IN DEPRESSION AND SUICIDE

Neurotransmitters modulate the secretion of most pituitary hormones, so that hormone release and circulating levels may be indicative of neurotransmitter function. In particular, 5-HT function has been monitored by measuring prolactin release. Acute administration of agents such as a) d-fenfluramine, which enhances 5-HT release, b) TRP, which increases 5-HT production and release, or c) clomipramine, which inhibits 5-HT reuptake, causes less prolactin release in depression [5,10,41-47] and also in patients with violent or suicide-attempt histories

[10,48] than in healthy subjects. The 5-HT post-synaptic receptor agonist meta-chlorophenylpiperazine (mCPP) did not mimic the other agents studied; it is active at several of the 5-HT receptor subtypes [49]. These data are consistent with an abnormality in presynaptic 5-HT function in some depressed patients. Most 5-HT challenge studies demonstrated normalization of prolactin responses after treatment. This suggests a 5-HT mechanism of action for many antidepressant treatments [50-55].

1.4 5-HT FUNCTION AND THE MECHANISM OF ACTION OF ANTIDEPRESSANT TREATMENTS

Many effective treatments for depression enhance 5-HT function. Fenfluramine- or TRP-induced prolactin release is increased by chronic treatment with certain antidepressant medications (desipramine, amitriptyline, clomipramine, carbamazepine, fluvoxamine, tranylcypromine, lithium) or electro-convulsive therapy. Consistent with these findings, a single night of sleep deprivation, which acutely improves mood in 60% of depressives, was found to significantly increase prolactin release following a TRP infusion (Salomon, in preparation). However, 5-HT-mediated prolactin release is not affected by chronic treatment with other antidepressants, including mianserin, bupropion, and trazodone [42,46,47,50,53-59].

2. Effect of Rapid Depletion of Plasma Tryptophan on Mood

Selective 5-HT depletion in humans results from inhibiting 5-HT synthesis by administering para-chlorophenylalanine (pCPA) or from restricting dietary TRP intake. In an early study using pCPA, Shopsin and Gershon described a rapid relapse of symptoms after the administration of pCPA to remitted depressives who had improved during chronic treatment with imipramine (5 patients [60]) or tranylcypromine (5 patients [61]). The symptoms abated quickly after pCPA was discontinued. However, prominent side effects of pCPA cause difficulties in assessing mood and also in preserving a double blind. Therefore, another newly available means of selective 5-HT depletion was tested. An acute restriction of dietary TRP depletes plasma TRP by limiting TRP intake and also forcing rapid utilization of existing pools of TRP.

The biologic mechanisms underlying the TRP depleting effects of amino acid mixtures are complex and have been recently reviewed [62]. Considerable evidence suggests that the ingestion of a mixture of 15 amino acids (without TRP) accelerates hepatic protein syntheses, tapping and depleting existing TRP pools. Competition among the large neutral amino acids for transport across the blood brain barrier also slows TRP distribution. Finally, the rate-limiting enzyme in the conversion of TRP to 5-HT (tryptophan hydroxylase), is sensitive to slight fluctuations in substrate availability, which further limits 5-HT synthesis.

Amino acid mixtures capable of reducing plasma TRP levels in humans were based on preclinical findings [63-68]. In primates, administration of seven of the ten essential amino acids (including cysteine, isoleucine, leucine, methionine, phenylalanine, threonine, and valine) produced a 60% decline in plasma TRP after 5 hours [69]. For human studies, the amino acid mixtures are prepared following the proportions found in human milk protein digests, without TRP (table 1). Carbohydrates are omitted since they induce insulin release, which increases peripheral uptake of all amino acids except TRP (because only TRP is protected by albumin carrier sites).

There is preclinical evidence that decreasing plasma TRP causes a decrease in brain TRP and 5-HT. Increases or decreases of plasma TRP in rats cause a corresponding increase or decrease in brain 5-HT synthesis [70]. In the rat, decreasing the levels of TRP in plasma caused decreased brain TRP levels [67,70,71], decreased 5-HT synthesis [67,70], and decreased 5-HT release [72,73]. CSF levels of TRP and 5-HIAA decreased in vervet monkeys during TRP depletion, while CSF levels of tyrosine (precursor of catecholamines), homovanillic acid (metabolite of dopamine), and 3-methoxy-4-hydroxyphenethyleneglycol (MHPG, metabolite of norepinephrine) remained unchanged [69].

Human studies correlating plasma and brain 5-HT levels are few. In one study, plasma-free 5-HT levels showed a significant correlation with simultaneously obtained CSF 5-HT levels from psychiatrically healthy patients undergoing minor surgical procedures [74]. In another study of drug-free volunteers and depressed patients treated with tricyclic antidepressants, TRP levels in blood and CSF were significantly correlated [75].

2.1   EFFECTS OF TRP DEPLETION IN HEALTHY SUBJECTS

Young et al. [76] adapted TRP-deficient amino acid mixtures for use in human studies, producing a 71% decline in plasma TRP in normal human males [77]. After ingesting amino acid mixtures that deplete TRP acutely, normal male subjects showed slightly elevated depression scores on the multiple affective adjectives checklist (MAACL). Mood changes were most apparent among in-dividuals who reported slightly dysphoric moods at baseline, while individuals with the fewest symptoms did not change [77-79].

2.2   EFFECTS OF TRP DEPLETION IN UNMEDICATED DEPRESSED PATIENTS

Delgado et al. [80], using a similar method to reduce plasma TRP evaluated the effects of TRP depletion in medication-free depressed patients. Patients consumed a low TRP (160 mg/day) diets for one day prior to ingesting the amino acid drink deficient in TRP at 8 a.m. [80,81]. Seven hours after the diet restriction and amino acid mixture there was a 90% decrease in plasma-free TRP levels from baseline. A balanced snack including milk was given after the afternoon mood ratings.

Sixty-nine depressed, medication-free patients were studied. Double-blind depletion tests or control tests were administered in random order about 7 days apart. For the control test, the low TRP diet (160 mg TRP/day) was supplemented with 500 mg TRP t.i.d. in unmarked capsules; placebo capsules were given with the depletion test to preserve the blind. The control, balanced amino acid meal included 2.3 gm TRP in addition to the 15 other amino acids (table 1). Behavioral ratings, using the 25-item Hamilton Depression Rating Scale (HDRS), were obtained prior to the beginning the diet, prior to the drink, 5 to 7 hours after the drink, and the day following the drink at 12:00 (noon).

TABLE 1. Amino acid mixture for tryptophan depletion drink

1.	l-alanine	5.5 gm
2.	l-arginine	4.9 gm
3.	l-cysteine	2.7 gm
4.	glycine	3.2 gm
5.	l-histidine	3.2 gm
6.	l-isoleucine	8.0 gm
7.	l-leucine	13.5 gm
8.	l-lysine (monohydrochloride)	11.0 gm
9.	l-methionine	3.0 gm
10.	l-phenylalanine	5.7 gm
11.	l-proline	12.2 gm
12.	l-serine	6.9 gm
13.	l-threonine	6.5 gm
14.	l-tyrosine	6.9 gm
15.	l-valine	8.9 gm

For the control drink add: L-TRP 2.3 gm

Unexpectedly, depressed patients did not become more depressed during TRP depletion, but showed changes in mood on the day following the depletion. Plasma levels of both total and free TRP dropped by 83% from baseline. There was no correlation between TRP levels and HDRS scores. However, on the day after testing, a clinically significant, bimodal mood change was observed. After the active test, mood scores improved (≥ 10 point decrease in the HDRS score) in 37% of the untreated patients, while mood scores declined in 23% (≥ 10 point increase in HDRS score). All 10 patients who were subsequently classified as antidepressant non-responders had shown a worsening of symptoms on the day after TRP depletion, while only 5 of 20 who subsequently responded to treatment showed this worsening. Further, 13 of 20 responders showed improved HDRS scores (p < 0.001). Prospective treatment studies are needed to determine if this test will have clinically significant predictive value [80-82].

2.3 TRP DEPLETION IN MEDICATED, REMITTED DEPRESSIVES

During the past several years, our research group has studied the effects of TRP depletion on depressive symptoms. In the first 46 antidepressant-remitted depressed patients studied [80], many experienced a precipitous return of depressive symptoms as the plasma TRP nadir was reached. Plasma-free and total tryptophan levels decreased by more than 85% during the acute TRP depletion. No patients relapsed during control testing. HDRS scores correlated significantly with free TRP levels. Relapse rate was related to antidepressant type. During depressive remissions induced by 5-HT enhancing agents (fluvoxamine, fluoxetine, or monoamine oxidase inhibitors) but not nonserotonergic agents (desipramine, bupropion), patients relapse transiently after ingesting a TRP depleting amino acid drink but show no changes after the control drink [80,81]. Patients treated with serotonergic treatments were significantly more likely to relapse during depletion testing. These findings suggest that 5-HT is involved in the mechanism of action of some, but not all, antidepressant medications [80].

3. Implications and Future Directions

The TRP depletion paradigm has proven to be an informative and potentially important tool for the exploration of 5-HT in the pathophysiology and treatment of psychiatric disorders. TRP depletion paradigms may aid in subtyping depressive illness and identifying relevant mechanisms of antidepressant action. Further, given the postulated role of 5-HT in other psychiatric disorders such as panic disorder schizophrenia, and disorders characterized by excessive aggression and impulsivity, this method may be of value in a variety of investigations.

4. References

1. Coppen, A. (1967) 'The biochemistry of affective disorders', Br. J. Psych. 113, 1237-1264.
2. Ashcroft, G.W. and Sharman, D.F. (1960) '5-Hydroxyindoles in human cerebrospinal fluids', Nature 186, 1050-1051.

3. Golden, R.N., Hsiao, J., and Lane, E. (1990) 'Abnormal neuroendocrine responsivity to acute IV clomipramine challenge in depressed patients', Psych. Res. 31, 39-47.
4. Coppen, A., Eccleston, E.G., and Peet, M. (1973) 'Total and free tryptophan concentration in the plasma of depressed patients', Lancet 2, 60-63.
5. Heninger, G.R., Charney D.S., and Sternberg, D.E. (1984) 'Serotonergic function in depression: prolactin response to intravenous tryptophan in depressed patients and healthy subjects', Arch. Gen. Psychiatry 41, 398-402.
6. Pare, C.M.B., Trenchard A., and Turner, P. (1974) '5-Hydroxytryptamine in depression', Adv. Biochem. Psychopharmacol. 11, 275-279.
7. Meltzer, H.Y. and Lowy, M.T. (1987) 'The serotonin hypothesis of depression', in H.Y. Meltzer (ed.), Psychopharmacology, The Third generation of Progress, Raven Press, New York.
8. Cowen, P.J., Parry-Billings, M., and Newsholme, E.A. (1989) 'Decreased plasma tryptophan levels in major depression', J. Affective Disorders 16, 27-31.
9. Åsberg, M., Schalling, D., Träskman-Bendz, L., Wägner, A. (1987) 'Psychobiology of suicide, impulsivity, and related phenomena', in H.Y. Meltzer (ed.), Psychopharmacology: The Third Generation of Progress, Raven Press, New York, pp. 655-668.
10. Coccaro, E.F., Siever, L.J., Klar, H.M., Maurer, G., Cochrane, K., Cooper, T.B., Mohs, R.C., and Davis, K.L. (1989) 'Serotonergic studies in patients with affective and personality disorders: Correlates with suicidal and impulsive aggressive behavior', Arch. Gen. Psychiatry 46 (7), 587-599 and erratum 47 (2), 124.
11. Mann, J.J., Stanley, M., MacBride, P.A., and McEwen B.S. (1986) 'Increased serotonin$_2$ and β-adrenergic binding in the frontal cortices of suicide victims', Arch. Gen. Psychiatry 43, 954-959.
12. Healy, D. and Leonard, B.E. (1987) 'Monoamine transport in depression: Kinetics and dynamics', J. Affective Disord. 12, 91-103.
13. Arató, M., Tekes, K., Tóthfalusi, L., Magyar, K., Palkovits, M., Frecska, E., Falus, A., and MacCrimmon, D.J. (1991) 'Reversed hemispheric asymmetry of imipramine binding in suicide victims', Biol. Psychiatry 29, 699-702.
14. Møller, S.D., Kirk, L., and Honore, P. (1979) 'Free and total plasma tryptophan in endogenous depression', J. Affective Disord. 1, 69-76.
15. Coppen, A. and Wood, K. (1978) 'Tryptophan and depressive illness', Psychol. Med. 8, 49-57.
16. Shaw, D.M., Tidmarsh, S.F., Johnson, A.L., Michalakeas, A.C., Riely, G.T., Blazek, T., and Francis, A.F. (1978) 'Multicompartmental analysis of amino acids: II. Tryptophan in affective disorders', Psychol. Med. 8, 487-494.
17. Cowen, P.J., McCance, S.L., and Cohen, P.R. (1989) 'Lithium increases 5-HT-mediated neuroendocrine responses in tricyclic resistant depression', Psychopharmacology 99, 230-232.

18. Curzon, G. (1979) 'Relationships between plasma, CSF, and brain tryptophan', J. Neural. Trans. 15 (suppl.), 93-105.
19. Quintana, J. (1992) 'Platelet serotonin and plasma tryptophan decreases in endogenous depression: Clinical therapeutic and biological correlations', J. Affective Disorders 24, 55-62.
20. Møller, S.D., Bech, P., Bjerrum, H., Bojholm, S., Butler, B., Rolker, H., Gram, L.F., Larsen, J.K., Lauritzen, L., Loldrup, D., Munk-Andersen, E., Odum, K., and Rafaelsen, O.J. (1990) 'Plasma ratio tryptophan/neutral amino acids in relation to clinical response to paroxetine and clomipramine in patients with major depression', J. Affective Disord. 18, 59-66.
21. Roy-Byrne, P., Post, R.M., Rubinow, D.R., Linnoila, M., Savard, R., and Davis, D. (1983) 'CSF 5HIAA and personal and family history of suicide in affectively ill patients: a negative study', Psychiatry Res. 10, 263-274.
22. Reddy, P.L., Khanna, S., Subhash, M.N., Channabasavanna, S.M., and Sridhara Rama Rao, B.S. (1992) 'CSF amine metabolites in depression', Biol. Psychiatry 31, 112-118.
23. Brown, G.L., Roy, A., Virkunen, M., and Linnoila, A. (1990) 'Serotonin in suicide, violence, and alcoholism', in E.F. Coccaro and D.L. Murphy (eds.), Serotonin in Major Psychiatric Disorders, American Psychiatric Press, Washington.
24. Stanley, M. and Mann, J.J. (1983) 'Increased serotonin-2 binding sites in frontal cortex of suicide victims', Lancet 1, 214-216.
25. Arango, V., Underwood, M.D., and Mann, J.J. (1992) 'Alterations in monoamine receptors in the brain of suicide victims', J. Clin. Psychopharmacol. 12 (suppl. 2) 8S-12S.
26. McKeith, I.G., Marshall, E.F., Ferrier, I.N., Armstrong, M.M., Kennedy, W.N., Perry, R.H., Perry, E.K., and Eccleston, D. (1987) '5-HT receptor binding in postmortem brains from patients with affective disorders', J. Affective Disord. 13, 67-74.
27. Arató, M., Tekes, K., Tothfalus, L., Frecska, E., Falus, A., Palkovits, M., and MacCrimnon, D.J. (1991) 'Serotonin dysregulation in suicide', in G.B. Cassano and H.S. Akiskal (eds.), Serotonin-related Psychiatric Syndromes: Clinical and Therapeutic Links, Royal Society of Medicine Services International Congress and Symposium Series No 165, Royal Society of Medicine Services Limited, pp. 41-46.
28. Stanley, M., Virgilio, J., and Gershon, S. (1982) 'Tritiated imipramine binding sites are decreased in the frontal coretex of suicide victims', Science 216, 1337-1339.
29. Gross-Isseroff, R., Israeli, M., and Biefon, A. (1989) 'Autoradiographic Analysis of Tritiated Imipramine Binding in the Human Brain Post Mortem: Effects of Suicide', Arch. Gen. Psychiatry 46, 237-241.
30. Arora, R.C. and Meltzer, H.Y. (1989) ^{3}H-Imipramine binding in the frontal cortex of suicides', Psychiatry Res. 30, 125-135.

31. Arora, R.C. and Meltzer, H.Y. (1989) 'Increased serotonin-2 (5-HT$_2$) receptor binding as measured by ^{3}H-lysergic acid diethylamide (^{3}H-LSD) in the blood platelets of depressed patients', Life Sci. 44, 725-734.
32. Biegon, A., Weizman, A., Karp, L., Ran, A., Tiano, S., and Wolff, M. (1987) 'Serotonin 5-HT$_2$ receptor binding on blood platelets: a peripheral marker for depression?', Life Sci. 41, 2485-2492.
33. Beigon, A., Grinspoon, A., Blumenfeld, B., Bleich, A., Apter, A., and Mester, R. (1990) 'Increased serotonin 5-HT$_2$ receptor binding on blood platelets of suicidal men', Psychopharmacology 100, 165-167.
34. Pandey, G.N., Pandey, S.C., Janicak, P.G., Marks, R.C., and Davis, J.M. (1990) 'Platelet serotonin-2 receptor binding sites in depression and suicide', Biol. Psychiatry 28, 215-222.
35. Briley, M.S., Langer, S.Z., Raisman, R., Sechter, D., and Zarifian, E. (1980) 'Tritiated imipramine binding sites are decreased in platelets of untreated depressed patients', Science 209, 303-395.
36. Paul, S.M., Rehavi, M., Skolnick, P., Ballenger, J.C., and Goodwin, F.K. (1981) 'Depressed Patients have decreased binding of tritiated imipramine to platelet serotonin "transporter"', Arch. Gen. Psychiatry 38, 1315-1317.
37. Langer, S.Z. and Briley, M. (1981) 'High affinity [^{3}H] imipramine binding: A new biological tool for studies in depression', Trends Neurosci. 4, 28-31.
38. Elliott, J.M. (1991) 'Peripheral markers in affective disorders' in R. Horton and C. Katona (eds.), Biological Aspects of Affective Disorders, Academic Press, San Diego, pp. 95-144.
39. Bech, P., Eplov, L., Gastpar, M., Gentsch, C., Mendlewicz, J., Plenge, P. Rielaert, C., and Mellerup, E.T. (1988) 'WHO pilot study on the validity of imipramine binding in afffective disorders: Trait versus state characteristics', Pharmacopsychiatry 21, 147-150.
40. Healy, D., Theodorou, A.E., Whitehouse, A.M., Lawrence, K.M., White, W., Wilton-Cox, H., Kerry, S.M., Horton, R.W., and Paykel, E.S. (1990) '^{3}H-Imipramine binding to previously frozen platelet membranes from depressed patients before and after treatment', Br. J. Psychiatry 157, 208-215.
41. Cowen, P.J. and Charig, E.M. (1987) 'Neuroendocrine responses to tryptophan in major depression', Arch. Gen. Psychiatry 44, 958-966.
42. Price, L.H., Charney, D.S., Delgado, P.L., and Heninger, G.R. (1991) 'Serotonin Function and Depression: Neuroendocrine and mood responses to intravenous L-tryptophan in depressed patients and healthy comparison subjects', Am. J. Psychiatry 148, 1518-1525.
43. O'Keane, V. and Dinan, T.G. (1991) 'Prolactin and cortisol responses to d-fenfluramine in major depression: Evidence for diminished responsivity of central serotonergic function', Am. J. Psychiatry 148, 1009-1015.
44. Upadhyaya, A.K., Pennell, I., Cowen, P.J., and Deakin, J.F. (1991) 'Blunted growth hormone and prolactin responses to L-tryptophan in depression; a state-dependent abnormality', J. Affective Disorders 21 (3), 213-218

45. Anderson, I.M., Ware, C.J., Da Roiza Davis, J.M., and Cowen, P.J. (1992) 'Decreased 5-HT-mediated prolactin release in major depression', Br. J. Psychiatry 160, 372-378.
46. Wetzler, S., Kahn, R.S., Asnis, G.M., Korn, M., and van Praag, H.M. (1991) 'Serotonin receptor sensitivity and aggresion', Psychiatry Res. 37, 271-279.
47. Shapira, B., Lerer, B., Kindler, S., Lichtenberg, P., Gropp, C., Cooper, T., and Calev, A. (1992) 'Enhanced serotonergic responsivity following electroconvulsive therapy in patients with major depression', Br. J. Psychiatry 160, 223-229.
48. O'Keane, V., Moloney, E., O'Neill, H., O'Connor, A., Smith, C., and Dinan, T.G. (1992) 'Blunted prolactin responses to d-fenfluramine in sociopathy: Evidence for subsensitivity of central serotonergic function', Br. J. Psychiatry 160, 643-646.
49. Olivier, B., Mos, J., Tulp, M.T.M., and van der Poel AM (1992) 'Animal models of anxiety and aggression in the study of serotonergic agents', in S.Z. Langer, N. Brunello, G. Racagni and J. Mendlewicz (eds.), Serotonin receptor subtypes: pharmacological significance and clinical implications, Int. Acad. Biomed. Drug Res., Karger, Basel, vol. 1, pp. 67-79.
50. Charney, D.S., Heninger G.R., and Sternberg, D.E. (1984) 'Serotonin function and the mechanism of action of antidepressant treatment', Arch. Gen. Psychiatry 41, 359-365.
51. Price, L.H., Charney, D.S., Delgado, P.L., and Heninger G.R. (1989) 'Effects of desipramine and fluvoxamine treatment on the prolactin response to L-tryptophan: A test of the serotonergic function enhancement hypothesis of depression', Arch. Gen. Psychiatry 46, 625-631.
52. Price, L.H., Charney, D.S., Delgado, P.L., Goodman, W.K., Krystal, J.H., Woods, S.W., and Heninger, G.R. (1990) 'Clinical data on the role of serotonin in the mechanism(s) of action of antidepressant drugs', J. Clin. Psychiatry 51 (suppl. 4), 44-50.
53. Price, L.H., Charney, D.S., Delgado, P.L., and Heninger, G.R. (1989) 'Lithium treatment and serotonergic function: neuroendocrine, behavioral, and physiologic repsonses to intravenous L-tryptophan in affective disorder patients', Arch. Gen. Psychiatry 46, 13-19.
54. Elphick, M., Yang, J., and Cowen, P.J. (1990) 'Effects of carbamazepine on dopamine- and serotonin-mediated neuroendocrine responses', Arch. Gen. Psychiatry 47, 135-140.
55. Cowen, P.J., McCance, S.L., Cohen, P.R., and Julier, D.L. (1989) 'Lithium increases 5-HT-mediated neuroendocrine responses in tricyclic resistant depression', Psychopharmacology 99, 230-232.
56. Anderson, I.M. and Cowen, P.J. (1986) 'Clomipramine enhances prolactin and growth hormone responses to L-tryptophan', Psychopharmacology 89, 131-133.
57. Price, L.H., Charney, D.S., and Heninger, G.R. (1985) 'Effects of

tranylcypromine treatment on neuroendocrine, behavioral, and autonomic responses to tryptophan in depressed patients', Life Sciences 37(9), 809-818.
58. Price, L.H., Charney, D.S., Delgado, P.L., and Heninger, G.R. (1990) 'Lithium and serotonin function: implications for the serotonin hypothesis of depression', Psychopharmacology 10, 3-12.
59. Golden, R.N., Ekstrom, D., Brown, T.M., Ruegg, R., Evans, D.L., Haggerty, J.J., Garbutt, J.C., Pedersen, C.A., Mason, G.A., Browne, J., and Carson, S.W. (1992) 'Neuroendocrine effects of intravenous clomipramine in depressed patients and healthy subjects', Am. J. Psychiatry 149, 1168-1175.
60. Shopsin, B., Gershon, S., Goldstein, M., Friedman, E., and Wilk, S. (1975) 'Use of synthesis inhibitors in defining a role for biogenic amines during imipramine treatment in depressed patients', Psychopharmacology Communications 1 (2), 239-249.
61. Shopsin, B., Friedman, E., and Gershon, S. (1976) 'Parachlorophenylalanine reversal of tranylcypromine effects in depressed patients', Arch. Gen. Psychiatry 33, 811-819.
62. Salomon, R.M., Delgado, P.L., Miller, H.L., and Charney D.S. (1992) 'Serotonin function in depression: Recent studies of the effect of reduced plasma tryptophan on depressive symptomatology', Psyche 1, 7-16.
63. Fernstrom, J.D. and Wurtman, R.J. (1971) 'Brain serotonin content: Physiological dependence on plasma tryptophan levels', Science 173, 149-152.
64. Gessa, G.L., Biggio, G., Fadda, A., Corsini, G.V., and Tagliamonte, A. (1974) 'Effect of oral administration of tryptophan-free amino acid mixtures on serum tryptophan, brain tryptophan, and serotonin metabolism', J. Neurochem. 22, 869-870.
65. Gessa, G.L., Biggio, G., Fadda, F., Corsini, G.U., and Tagliamonte, A. (1975) 'Tryptophan-Free Diet: A New Means for Rapidly Decreasing Brain Tryptophan Content and Serotonin Synthesis', Acta Vitamin. Enzymol. 29, 72-78.
66. Wurtman, R.J., Hefti, F., and Melamed, E. (1981) 'Precursor control of neurotransmitter systems', Pharmacol. Rev. 32, 315-335.
67. Moja, E.A., Cipolla, P., Castoldi, D., and Totanetti, O. (1989) 'Dose response decrease in plasma tryptophan and in brain tryptophan and serotonin after tryptophan-free amino acid mixtures in rats', Life Sci. 44, 971-976.
68. Biggio, G., Fadda, F., Fanni, P., Tagliamonte, A., and Gessa, G.L. (1974) 'Rapid depletion of serum tryptophan, brain tryptophan, serotonin and 5-hydroxyindoleacetic acid by a tryptophan-free diet', Life Sci. 14, 1321-1329.
69. Young, S.N., Ervin, F.R., Pihl, R.O., and Finn, P. (1989) 'Biochemical aspects of tryptophan depletion in primates', Psychopharmacology 98, 508-511.
70. Tagliamonte, A., Biggio, G., Vargiu, L., Gessa, G.L. (1973) 'Free Tryptophan in Serum Controls Brain Tryptophan Level and Serotonin Synthesis', Life Sci. 12 (2), 277-287.

71. Fernstrom, J.D., Larin F., and Wurtman R.J. (1973) 'Correlations between Brain Tryptophan and Plasma Neutral Amino Acid Levels Following Food Consumption in Rats', Life Sci. 13, 517-524.
72. Schaechter, J.D. and Wurtman, R.J. (1990) 'Serotonin release varies with brain tryptophan levels', Brain Res. 532 (1-2), 203-210.
73. Heslop, K., Portas, C.M., and Curzon, G. (1991) 'Effect of altered tryptophan availability on tissue and extracellular serotonin in the rat cortex', in H. Rollema and W. Ben Westerink (eds.), Proceedings Monitoring Molecules in Neuroscience, Krits Repro, Meppel, Netherlands, pp. 259-261.
74. Sarrias, M.J., Cabré, P., Martínez, E., and Artigas, F. (1990) 'Relationship between serotonergic measures in blood and cerebrospinal fluid simultaneously obtained in humans', J. Neurochem. 54, 783-786.
75. Larsson, M., Forsman, A., and Hallgren, J. (1988) 'HPLC assays of 5-HIAA and tryptophan in cerebrospinal fluid and 5-HT and tryptophan in blood: a methodological study with clinical applications', Methods & Findings in Experimental & Clinical Pharmacology 10 (7), 453-460.
76. Young, S.N., Smith, S.E., Pihl, R.O., and Ervin, F.R. (1985) 'Tryptophan depletion causes a rapid lowering of mood in normal males', Psychopharmacology 87, 173-177.
77. Young, S.N., Tourjman, S.V., Teff, K.L., Pihl, R.O., Ervin, F.R., and Anderson G.H. (1988) 'The effect of lowering plasma tryptophan on food selection in normal males', Pharmacol. Biochem. Behav. 31, 149-152.
78. Smith, S.E., Pihl, R.O., Young, S.N., and Ervin, F.R. (1987) 'A test of possible cognitive and environmental influences on the mood lowering effect of tryptophan depletion in normal males', Psychopharmacology 91, 451-457.
79. Abbott, F.V., Etienne, P., Franklin, K.B.J., Morgan, M.J., Sewitch, M.J., and Young, S.N. (1992) 'Acute tryptophan depletion blocks morphine analgesia in the cold-pressor test in humans', Psychopharmacology 108, 60-66.
80. Delgado, P.L., Charney, D.S., Price, L.H., Aghajanian, G.K., Landis, H., and Heninger, G.R. (1990) 'Serotonin function and the mechanism of antidepressant action: reversal of antidepressant induced remission by rapid depletion of plasma tryptophan', Arch. Gen. Psychiatry 47, 411-418.
81. Delgado, P.L., Price, L.H., Miller, H.L., Salomon, R.M., Licinio, J., Krystal, J.H., Heninger, G.R., Charney, D.S. (1991) 'Rapid serotonin depletion as a provocative challenge test for patients with major depression: relevance to antidepressant action and the neurobiology of depression', Psychopharmacology Bulletin 27, 321-330.
82. Delgado, P.L., Price, L.H., Miller, H.L., Salomon, R.M., Heninger, G.H., Charney, D.S. 'Serotonin and the neurobiology of depression: effects of tryptophan depletion in drug-free depressed patients', Arch. Gen. Psychiatry, in press.

ANXIOLYTIC EFFECTS OF DRUGS ACTING ON 5-HT RECEPTOR SUBTYPES

D.L. MURPHY, A. BROOCKS, C. AULAKH, and T.A. PIGOTT
Laboratory of Clinical Science,
National Institute of Mental Health
NIH Clinical Center, 10-3D41,
9000 Rockville Pike
Bethesda, Maryland   20892
USA

ABSTRACT. Anxiolytic effects have been well documented for the 5-HT_{1A} partial agonist azapirones (e.g. buspirone). Benzodiazepines (which have indirect serotonergic actions) continue to be more widely prescribed, however, despite side-effect profiles favoring the azapirones. Therapeutic effects of the azapirones have not been found in controlled sutdies of patients with panic disorder, are not yet clearly established for obsessive-compulsive disorder or social phobia, seem likely for anxiety associated with depression, and are validated for generalized anxiety disorder. Sparse clinical data have not yet confirmed anxiolytic actions suggested in animal model studies for $5\text{-HT}_{1C/2}$ antagonists (e.g., ritanserin) or 5-HT_3 antagonists (e.g. ondansetron). In human anxiety model studies based on anxiogenic effects of the mixed serotonin agonist/antagonist, m-chlorophenyl-piperazine, anxiolytic effects followed acute pretreatment with the $5\text{-HT}_{1/2}$ antagonists, metergoline, or ritanserin, and chronic treatment with clomipramine or fluoxetine, while only small, but statistically significant changes followed acute pretreatment with ondansetron. Among serotonergic agents, the selective serotonin reuptake inhibitors (SSRIs) appear more generally effective in most of the anxiety disorders than the serotonin receptor subtype-selective antagonists available at this time.

1.   Introduction

This paper provides an update on an area that has been extensively reviewed [1-8] but which has also been characterized by rapid, recent growth at the level of clinical studies as well as laboratory investigations of biochemical mechanisms and animal models. Because of space constraints, this review can only briefly summarize some of the major developments in the last several years and, in addition, provide some comments on a few areas of special interest.

2. Efficacy of 5-HT-Selective Agents in the Anxiety Disorders

As indicated in table 1, there are four major classes of putative anxiolytic agents whose major neurochemical mechanisms are thought to be initiated by effects at serotonin binding sites: (a) 5-HT_{1A} partial agonists, including buspirone, gepirone, ipsapirone, flesinoxan, and MDL 73005; (b) $5\text{-HT}_{1C/2}$ antagonists, including ritanserin and trazodone; (c) 5-HT_3 antagonists, including ondansetron, zacopride, granisetron, ICS 205-930, and BRL 46470A; and (d) the selective serotonin reuptake inhibitors (SSRIs), including fluoxetine, fluvoxamine, paroxetine, sertraline, citalopram, and clomipramine.

Table 1 also indicates several of the major anxiety disorders and anxiety-associated disorders that have been most extensively investigated for anxiolytic drug effects and that have clear, widely accepted diagnostic criteria (e.g. DSM-IIIR). Of note are that these symptomatically different anxiety disorders also seem to be distinguished by important psychotherapeutic drug treatment response differences.

Thus, anxiolytic effects have been most explicitly validated for the 5-HT_{1A} partial agonists in generalized anxiety disorder, with chronic administration leading to equivalent efficacy to that of the benzodiazepines such as diazepam [4,9]. Increasing evidence indicates that the 5-HT_{1A} partial agonists also have substantial antidepressant effects, with concomitant therapeutic benefit on anxiety symptoms associated with depression [14,17]. In contrast, several studies have failed to find therapeutic responses in patients with panic disorder following treatment with one 5-HT_{1A} partial agonist, buspirone [17-19]. The only controlled double-blind study using buspirone in obsessive compulsive disorder (OCD) found equivalent efficacy to that of the most established anti-OCD drug, clomipramine, but this small study was not placebo-controlled [20].

Ritanserin is a 5-HT_2 antagonist with nearly equal affinity for 5-HT_{1C} sites which was reported to be effective in several small studies of patients with generalized anxiety disorder some years ago [21,22]. Confirming data have not been forthcoming, and ritanserin was not found to be effective in patients with panic disorder [23]. The antidepressant, trazodone, has prominent 5-HT_2 antagonist properties with very weak uptake inhibiting properties; it was not found to be effective in panic disorder or obsessive compulsive disorder [22,24].

5-HT_3 antagonists have not yet been the subject of adequate clinical trials to evaluate whether the anxiolytic-type effects found in animal model studies validly predict therapeutic effects in any of the anxiety disorders [25]. One of two incompletely reported trials with ondansetron was recently interpreted as yielding significant antianxiety effects in generalized anxiety disorder patients [25].

It is particularly noteworthy that the SSRIs come the closest to being broad-spectrum agents across the different anxiety disorders, with beneficial effects demonstrated in a comprehensive fashion that are apparently greater than those of the 5-HT receptor subtype-selective agents (table 1).

TABLE 1. Efficacy of 5-HT-selective agents used in the treatment of anxiety disorders*

Anxiety Disorder	5-HT Selective Agents			
	5-HT$_{1A}$ Partial Agonists (e.g. buspirone)	5-HT$_{1C/2}$ Antagonists (e.g. ritanserin)	5-HT$_3$ Antagonists (e.g. ondansetron)	SSRIs (e.g. fluoxetine, fluvoxamine)
Panic Disorder, Agoraphobia	0	(0)		+++
Generalized Anxiety Disorder	++	+	(+)	(+)
Obsessive-Compulsive Disorder	(+)			+++
Social Phobia				(+)
Anxiety Associated with Depression	++			+++

*+++, Robust, clearly demonstrated clinical effects; ++, well-documented clinical effects; +, positive effects in a few controlled trials; (+) or (0), positive (+) or negative or inconclusive effects (0) in one controlled trial; 0, lack of clinical efficacy in several controlled trials [2-5,9-16].

3. Is There Evidence of Preferential Efficacy of 5-HT-Selective Agents in the Anxiety Disorders?

An interesting issue of some potential relevance to similarities and differences among the anxiety disorders and even to pathophysiological mechanisms involved in these disorders is the preferential efficacy of serotonin-selective versus non-serotonin selective agents, as summarized in table 2. Obsessive compulsive disorder occupies a position at one end of this continuum, as it is the only anxiety disorder responsive solely to serotonin-selective agents, the SSRI's [2,26]. Some partial evidence of preferential efficacy in panic disorder has been suggested by two studies demonstrating superior efficacy for medications with relatively greater serotonin selectivity [27,28]. However, efficacy in panic disorder has also been found with non-serotonin selective reuptake inhibitors, monoaine oxidase inhibitors and benzodiazepines (table 2).

TABLE 2. Preferential efficacy of serotonin-selective agents in the anxiety disorders*

Obsessive-Compulsive Disorder:	<u>Yes</u> SSRIs (clomipramine, fluvoxamine) are more effective than desipramine, clorgyline, etc., and are the only drugs with greater efficacy than placebo.
Panic Disorder:	<u>Possible, Partial</u> Fluvoxamine may be more effective than maprotiline, and clomipramine may be more effective than imipramine, but other antidepressants and benzodiazepines are also effective.
Generalized Anxiety Disorder:	<u>No</u> Azapirones are similar to benzodiazepines.
Anxiety Associated with Depression:	<u>No</u> SSRIs and azapirones are similar to other antidepressants which do not possess serotonin-selective properties.

*[2,4,9,14,17,23,26,27]

4. Anxiogenic Actions of Other 5-HT-Selective Agents

While, as summarized above, several classes of serotonin-selective agents are of therapeutic benefit in the anxiety disorders, other serotonergic agents have opposite, anxiogenic effects in these patients and in healthy controls. m-Chlorophenylpiperazine (m-CPP) is the most prominent of these agents (table 3). For the most part, other serotonin-selective agents including precursors, releasers and receptor agonists and antagonists are not anxiogenic, with the exception of possible panic attack precipitation by fenfluramine [29]. Studies of m-CPP's behavioral effects in animal models have attributed m-CPP's apparent anxiogenic actions to 5-HT_{1C}-mediated effects [30], with some qualifications [31, 32]. Partial validation of the m-CPP experimental model for human anxiety has come from several studies of serotonin-selective agents which reported attenuation of m-CPP's anxiogenic effects, by the following treatments: (a) the 5-$HT_{1C/2}$ antagonist ritanserin [33]; (b) a small but statistically significant effect of ondansetron [12]; (c) the 5-$HT_{1/2}$ antagonist, metergoline [34,35]; and (d) chronic treatment with clomipramine or fluoxetine [26,36,37]. This 5-HT-related cross-species m-CPP model for human anxiety has led to recent attempts to develop new potential anxiolytic agents with greater 5-HT_{1C} versus 5-HT_2 selectivity, and to long-term studies with m-CPP in attempts to down-regulate 5-HT_{1C}-related neurotransmitter processes.

TABLE 3. Anxiogenic responses to m-chlorophenylpiperazine (m-CPP), a mixed serotonin 5-HT$_{1C}$ agonist/5-HT$_{2,3}$ antagonist[*]

	Response to m-CPP
Panic Disorder (PD)	Significant increase in panic attacks at a m-CPP dose (0.25 mg/kg, oral) without behavioral effects in controls Oral m-CPP equivalent to caffeine in eliciting panic attacks in PD patients IV m-CPP elicits panic attacks in both PD patients and controls in one study, but not in controls in other studies
Obsessive-Compulsive Disorder (OCD)	Significant increases in OCD symptoms and/or anxiety after oral (2 of 2 studies) or IV (2 of 3 studies) m-CPP adiministration
Healthy Volunteers	Greater increases in anxiety after IV than oral m-CPP administration Anxiety responses to oral m-CPP are dose dependent

[*][7,26,38,39,40]

5. References

1. Romero, A.G. and McCall, R.B. (1989) 'Advances in central serotonergics', Ann. Rep. Med. Chem. 27, 41-50.
2. Murphy, D. L. and Pigott, T. A. (1990) 'A comparative examination of a role for serotonin in obsessive compulsive disorder, panic disorder, and anxiety', J. Clin. Psychiatry 51, 53-58.
3. Charney, D. S., Krystal, J. H., Delgado, P. L., and Heninger, G. R. (1990) 'Serotonin-specific drugs for anxiety and depressive disorders', Annu. Rev. Med. 41, 437-446.
4. Taylor, D. P. and Moon, S. L. (1991) 'Buspirone and related compounds as alternative anxiolytics', Neuropeptides 19, 15-19.
5. Cowen, P. J. (1991) 'Serotonin receptor subtypes: Implications for psychopharmacology', Br. J. Psychiatry 159, 7-14.
6. Eison, M. S. (1990) 'Serotonin: A common neurobiologic substrate in anxiety and depression', J. Clin. Psychopharmacology 10, 26S-30S.
7. Murphy, D. L. (1990) 'Neuropsychiatric disorders and the multiple human brain serotonin receptor subtypes and subsystems', Neuropsychopharmacology 3, 457-471.
8. Tunnicliff, G. (1991) 'Moleuclar basis of buspirone's anxiolytic action', Pharmacol. Toxicol. 69, 149-156.
9. Schweizer, E. and Rickels, K. (1991) 'Serotonergic anxiolytics: A review of their clinical efficacy', in R. J. Rodgers, and S. J. Cooper (eds.), 5-HT$_{1A}$ Agonists, 5-HT$_3$ Antagonists and Benzodiazepines, John Wiley and Sons, New York, pp. 365-376.

10. Dubovsky, S. L. (1990) 'Generalized anxiety disorder: New concepts and psychopharmacologic therapies', J. Clin. Psychiatry 51, 3-10.
11. Uhde, T. W., Tancer, M. E., Black, B., and Brown, T. M. (1991) 'Phenomenology and neurobiology of social phobia: Comparison with panic disorder', J. Clin. Psychiatry 52, 31-40.
12. Broocks, A., Pigott, T. A., Canter, S., Grady, T. A., L'Heureux, F., Hill, J. L., and Murphy, D. L. (1992) 'Acute administration of ondansetron and m-CPP in patients with obsessive-compulsive disorder (OCD) and controls: Behavioral and biological results', Biol. Psychiatry 31, 174A.
13. Laws, D., Ashford, J. J., and Anstee, J. A. (1990) 'A multicentre double-blind comparative trial of fluvoxamine versus lorazepam in mixed anxiety and depression treated in general practice', Acta Psychiatr. Scand. 81, 185-189.
14. Rickels, K., Amsterdam, J. D., Clary, C., Puzzuoli, G., and Schweizer, E. (1991) 'Buspirone in major depression: A controlled study', J. Clin. Psychiatry 52, 34-38.
15. Black, B., Uhde, T. W., and Tancer, M. E. (1992) 'Fluoxetine for the treatment of social phobia', J. Clin. Psychopharmacol. 12, 293-295.
16. Cox, B. J., Swinson, R. P., and Lee, P. S. (1992) 'Meta-analysis of anxiety disorder treatment studies', J. Clin. Psychopharmacol. 12, 300-301.
17. Robinson, D. S. (1991) 'Buspirone in the treatment of anxiety', in G. Tunnicliff, A. A. Eison and D. P. Taylor (eds.), Mechanism and Clinical Aspects, Academic Press, San Diego, pp. 3-17.
18. Sheehan, D. V., Raj, A. B., Sheehan, K. H., and Soto, S. (1990) 'Is buspirone effective for panic disorder?', J. Clin. Psychopharmacol. 10, 3-11.
19. Pohl, R., Balon, R., Yeragani, V. K., and Gershon, S. (1989) 'Serotonergic anxiolytics in the treatment of panic disorder: A controlled study with buspirone', Psychopathology 22, 60-67.
20. Pato, M. T., Pigott, T. A., Hill, J. L., Grover, G. N., Bernstein, S., and Murphy, D. L. (1991) 'Controlled comparison of buspirone and clomipramine in obsessive-compulsive disorder', Am. J. Psychiatry 148, 127-129.
21. Westenberg, H. G. M. and denBoer, J. A. (1989) 'Serotonin-influencing drugs in the treatment of panic disorder', Psychopathology 22, 68-77.
22. 'Ritanserin', (1992) Drugs Fut. 17, 431-433.
23. denBoer, J. A. and Westenberg, H. G. M. (1990) 'Serotonin function in panic disorder: A double blind placebo controlled study with fluvoxamine and ritanserin', Psychopharmacology 102, 89-94.
24. Pigott, T. A., L'Heureux, F., Rubenstein, C. S., Bernstein, S. E., Hill, J. L., and Murphy, D. L. (1992) 'A double-blind, placebo controlled study of trazodone in patients with obsessive-compulsive disorder', J. Clin. Psychopharmacol. 12, 156-162.
25. Costall, B. and Naylor, R. J. (1992) 'Anxiolytic potential of $5-HT_3$ receptor antagonists', Pharmacol. Toxicol. 70, 157-162.

26. Murphy, D. L., Pigott, T. A., Grady, T. A., Broocks, A., and Altemus, M. (1993), 'Neuropharmacological investigations of brain serotonin subsystem functions in obsessive-compulsive disorder', in P.B. Bradley, et al. (eds.), Serotonin, CNS Receptors and Brain Function, Pergamon Press, Oxford, U.K., in press.
27. Modigh, K., Westberg, P., and Eriksson, E. (1992) 'Superiority of clomipramine over imipramine in the treatment of panic disorder: A placebo-controlled trial', J. Clin. Psychopharmacol. 12, 251-261.
28. denBoer, J. A. and Westernberg, H. G. (1988) 'Effect of a serotonin and noradrenaline uptake inhibitor in panic disorder', Int. Clin. Psychopharmacol. 3, 59-74.
29. Targum, S. D. (1990) 'Differential responses to anxiogenic challenge studies in patients with major depressive disorder and panic disorder', Biol. Psychiatry 28, 21-34.
30. Kennett, G. A. (1992) '5-HT_{1C} receptor antagonists have anxiolytic-like actions in the rat social interaction model', Psychopharmacology 107, 379-384.
31. Murphy, D. L., Lesch, K. P., Aulakh, C. S., and Pigott, T. A. (1991) 'Serotonin-selective arylpiperazines with neuroendocrine, behavioral, temperature, and cardiovascular effects in humans', Pharmacol. Rev. 43, 527-552.
32. Blackburn, T. P. and Kennett, G. A. (1990) 'Anxiolytic-like actions of BRL 46470 - a novel 5-HT_3 antagonist', J. Psychopharmacol. 4, 4.
33. Seibyl, J. P., Krystal, J. H., Price, L. H., Woods, S. W., D'Amico, C., Heninger, G. R., and Charney, D. S. (1991) 'Effects of ritanserin on the behavioral, neuroendocrine, and cardiovascular responses to meta-chlorophenylpiperazine in healthy human subjects', Psychiatry Res. 38, 227-236.
34. Pigott, T. A., Zohar, J., Hill, J. L., Bernstein, S. E., Grover, G. N., Zohar-Kadouch, R. C., and Murphy, D. L. (1991) 'Metergoline blocks the behavioral and neuroendocrine effects of orally administered m-chlorophenylpiperazine in patients with obsessive-compulsive disorder', Biological Psychiatry 29, 418-426.
35. Pigott, T. A., Hill, J. L., Grady, T. A., L'Heureux, F., Bernstein, S. E., Rubenstein, C. S., and Murphy, D. L. (1993) 'A comparison of the behavioral effects of oral versus intravenous m-CPP administration in OCD patients and a study of the effect of metergoline prior to IV m-CPP', Biol. Psychiatry, in press.
36. Zohar, J., Insel, T. R., Zohar-Kadouch, R. C., Hill, J. L., and Murphy, D. L. (1988) 'Serotonergic responsivity in obsessive-compulsive disorder: Effects of chronic clomipramine treatment', Arch. Gen. Psychiatry 45, 167-172.
37. Hollander, E., DeCaria, C., Gully, R., Nitescu, A., Suckow, R. F., Gorman, J. M., Klein, D. F., and Liebowitz, M. R. (1991) 'Effects of chronic

fluoxetine treatment on behavioral and neuroendocrine responses to meta-chloro-phenylpiperazine in obsessive-compulsive disorder', Psychiatry Res. 36, 1-17.
38. Kahn, R. S., Wetzler, S., Asnis, G. M., Kling, M. A., Suckow, R. F., and vanPraag, H. M. (1990) 'Effects of m-chlorophenylpiperazine in normal subjects: A dose-response study', Psychopharmacology 100, 339-344.
39. Kahn, R. S. and Wetzler, S. (1991) 'm-Chlorophenylpiperazine as a probe of serotonin function', Biol. Psychiatry 30, 1139-1166.
40. Murphy, D. L., Mueller, E. A., Hill, J. L., Tolliver, T. J., and Jacobsen, F. M. (1989) 'Comparative anxiogenic, neuroendocrine, and other physiologic effects of m-chlorophenylpiperazine given intravenously or orally to healthy volunteers', Psychopharmacology 98, 275-282.

OBSESSIVE COMPULSIVE DISORDERS: A NEUROBIOLOGICAL HYPOTHESIS

BARRY L. JACOBS
Program in Neuroscience
Princeton University
Princeton, New Jersey 08544
USA

ABSTRACT. This review covers basic research on brain serotonin, especially the neurophysiology of serotonergic neurons in behaving animals. An attempt is made to relate these data to obsessive compulsive disorders in humans.

1. Introduction

The pathophysiology of obsessive compulsive disorders (OCDs) is poorly understood, as is the case for depression and anxiety which frequently display comorbidity with OCDs. Nonetheless, it is clear that brain serotonin (5-HT) plays an important role in all three of these disorders. The strongest support for this comes from studies demonstrating that a variety of serotonergic drugs, but especially those that block the reuptake of synaptically released 5-HT into the presynaptic terminal, are frequently effective in treating OCDs, depression, and anxiety. However, even with this evidence regarding a specific neurochemical dysfunction in OCDs, an underlying pathophysiology in 5-HT neurotransmission in these patients has not been established.

We have approached this issue from a different perspective. For the past 15 years we have examined the basic operating characteristics of the brain 5-HT system in behaving animals. This work has focussed on the relationship between behavioral, environmental, and physiological variables and the electrophysiological activity of individual brain 5-HT-containing neurons (extracellularly recorded action potentials, or single unit activity).

This brief review will focus on these single unit studies of brain serotonergic neurons in behaving animals, emphasizing their relationship to tonic and phasic motor activity. At the conclusion of this paper, I will discuss these data within the context of an hypothesis that attempts to account for the role of 5-HT in OCDs in humans.

2. Background

2.1 NEUROANATOMY

The anatomy of the CNS 5-HT system has recently been reviewed in detail [1]. The cell bodies of virtually all of the serotonergic neurons in the CNS are located in the brainstem, primarily on the midline, and often in association with the raphe nuclei, e.g. dorsalis (DRN), magnus (NRM), pallidus (NRP), etc. From these brainstem sites, a widespread axonal arborization innervates virtually the entire neuraxis.

The involvement of 5-HT in motor function in vertebrates is indicated initially by the dense axon terminal innervation of α-motoneurons in both the brainstem and spinal cord. Secondary motor structures such as the substantia nigra, basal ganglia, and habenula also receive dense serotonergic inputs. However, many other structures that have at best a distant relation to motor function also receive strong serotonergic inputs, e.g. lateral geniculate nucleus, amygdala, etc. Two things probably account for this anatomical pattern. First, a number of CNS structures are normally activated in conjunction with motor activity, for example, portions of the autonomic nervous system. Thus, in a coordinating fashion, the 5-HT system modulates these outputs simultaneously with its motor effect. Second, when the facilitatory influence upon motor neurons is being exerted, the 5-HT system is concurrently inhibiting information processing in sensory systems.

2.2. MODEL SYSTEMS

Administration of a variety of serotonergic drugs produces a motor syndrome in rats [2]. Its most conspicuous signs are head shakes or "wet dog" shakes, hyperreactivity, tremor, hindlimb abduction, lateral head weaving, and reciprocal forepaw treading. A similar syndrome is seen in a number of vertebrate species [2], as well as in human patient populations administered any of a variety of serotonergic drugs [3]. One of the most interesting aspects of these clinical cases is the fact that these drug effects are restricted almost exclusively to motor signs (myoclonus, tremor, shivering, etc.), with few, if any, indications of significant sensory alterations. The phenomenological similarity of the syndrome across a variety of species supports the view that this system subserves a common functional role in vertebrates, at least one important aspect of which is motoric.

Some investigators have examined these same drug effects upon specific functional systems within the CNS. In a series of papers (see [4] for an overview), the influence of serotonergic drugs upon treadmill-induced hindlimb locomotion was studied in the spinal cat. The results of these experiments indicate that the action of 5-HT is to increase the amplitude of the motor response (by contrast, noradrenergic drugs are capable of influencing the timing of the responses as well as actually initiating locomotion in the acute stage of spinalization). In my laboratory we have examined the effects of injecting serotonergic drugs directly into

the motor nucleus of V in behaving cats [5]. Consistent with the results just described, we observed an increase in the amplitude of both the electromyography (EMG) of the masseter muscle, and of an externally elicited jaw-closure (masseteric) reflex. The influence of 5-HT upon
the neuronal mechanisms mediating fictive locomotion (swimming) in the isolated spinal cord of the lamprey has also been examined [6]. When 5-HT is applied to the spinal cord or to reticulospinal neurons it elicits a depression of the afterhyperpolarization which normally follows the action potential. (By itself 5-HT elicits no effect on the membrane resting potential.) This depression produces an increase in the motoneuron discharge frequency. If 5-HT is applied to the solution bathing the isolated spinal cord during fictive locomotion, motoneuronal bursts become more intense and longer, and the burst rate increases.

In sum, these data, derived from several model systems, indicate that 5-HT exerts an important facilitatory modulating influence upon various aspects of motor function.

2.3. CELLULAR ANALYSES

Several studies have directly examined 5-HT's effects upon motoneurons. These data derive primarily from studies employing extracellular recordings in conjunction with microiontophoresis of 5-HT onto motoneurons in the rat motor nucleus VII or in the ventral horn [7,8]. Consistent with the results described in the preceding section, 5-HT alone produced little or no change in neuronal activity. However, when 5-HT is interacted with excitatory influences on motoneurons it produces a strong facilitation of neuronal activity. Pharmacologic analyses indicate that an action at 5-HT$_2$ receptors appears to mediate this effect [9]. Thus, 5-HT appears to shift motoneurons from a stable hyperpolarized state, with little or no neuronal activity, to a stable depolarized "plateau" state, with tonic neuronal activity [10].

3. Single Unit Studies in Behaving Animals

3.1 BASIC PROPERTIES

The activity of serotonergic neurons displays a strong state-dependency in the domestic cat, the only species in which these neurons have been extensively investigated [11]. From a rate of approximately 3 spikes/second during quiet waking, the activity of these neurons shows a gradual decline as the cat becomes drowsy and enters slow wave sleep. The culmination of this state-dependent decrease in activity occurs as the cat enters REM sleep, when the activity of these neurons typically falls silent. On the other hand, during active waking, the discharge rate is elevated above that in quiet waking. This general pattern of activity across the sleep-wake-arousal cycle displayed by neurons in the DRN is similar to that seen in the other groups of serotonergic neurons, such as those in the NRM and NRP

[12].

3.2 STRESSORS/CHALLENGES

Brain 5-HT is thought to be important in a variety of behavioral (aggression, feeding, sleep, etc.) and physiological (thermoregulation, cardiovascular control, glucoregulation, etc.) processes, therefore, it
seemed imperative to us to examine serotonergic neuronal activity under a diversity of conditions.

In the first series of studies, cats were: exposed to a loud white noise (100 dB) for 15 minutes; physically restrained for 15 minutes; or, had a dog brought into their proximity for 5 minutes. All three of these manipulations evoked strong sympathetic activation as evidenced by significant increases in tonic heart rate and in plasma catecholamines levels. Despite this organismic activation, the activity of DRN serotonergic neurons was not significantly increased above that observed during an undisturbed active waking baseline [13].

Paralleling these studies of behavioral/environmental challenges, we also examined the response of DRN serotonergic neurons during perturbation of three physiologic systems-thermoregulatory [14], cardiovascular [15], and glucoregulatory [16]. As we saw with environmental stressors, physiologic challenges also failed to activate brain serotonergic neurons above the level seen during an undisturbed active waking state. Once again, this is in spite of the fact that these manipulations produce behavioral arousal as well as activate the organism's sympathetic nervous system. Additionally, as with environmental stressors, these same physiologic challenges did significantly activate noradrenergic neurons in the cat locus coeruleus (LC) (reviewed in [17]). Overall, these data indicate that the electrophysiologic activity of brain serotonergic neurons is not easily perturbed by any of a variety of challenges to the organism.

3.3 TONIC MOTOR CONTROL

As mentioned above, the most impressive change in serotonergic neuronal activity is the complete suppression that is observed during REM sleep. A unique aspect of this state is the profound atonia of the antigravity muscles. In order to examine this relationship more carefully, we employed a method for dissociating REM sleep from the atonia. When lesions are placed bilaterally in the pontine tegmentum of cats, the animals display REM sleep without atonia [18]. By all criteria, these animals appear to be in REM sleep, except that they display overt behavior because the mechanism responsible for producing atonia has been disrupted. When these cats entered REM sleep, rather than showing the complete suppression of serotonergic neuronal activity typical of normal cats, the activity of their DRN neurons increased above the level seen in slow wave sleep, and occasionally reached the level achieved during waking [19].

We also approached this issue in a reciprocal manner, i.e. by producing atonia without REM sleep. This is accomplished by injecting a small amount of the cholinomimetic agent carbachol into a particular region of the cat pontine tegmentum. Important from our perspective is the fact that these animals often manifest unambiguous periods of waking during this atonia, as reflected by the animal's ability to track a visual stimulus or respond to a looming object. Consistent with the results just described, during these periods of carbachol-induced atonia during waking, DRN serotonergic neuronal activity was nearly completely suppressed [20].

These, and other data from our laboratory, demonstrated that an important relationship exists between tonic level of motor activity (or muscle tone) and the activity of brain serotonergic neurons.

3.4 PHASIC MOTOR CONTROL

We have recently discovered two conditions in which the activity of serotonergic neurons is dramatically increased. As many as 30-50% of serotonergic neurons in the DRN of cats are tonically activated (2-5-fold increase) during repetitive motor activities involving the oral-buccal region: chewing/biting, lapping, and grooming the body surface with the tongue [21]. These neurons do not appear to be activated during any other types of behavior. In fact, their activity typically decreases during many active behaviors, such as locomotion and orientation.

One hypothesis to account for these data is that these neurons are co-activated in association with rhythmic, central pattern generator mediated behaviors, and by the afferent inputs that normally facilitate such behaviors. Recall also that the motor neurons that mediate at least some of these oral-buccal behaviors in mammals receive very dense serotonergic inputs (e.g. motor nuclei V and VII).

We believe that when repetitive motor activity is ongoing, brain serotonergic neurons are activated in order to facilitate this output, while simultaneously suppressing transmission in sensory pathways. Reciprocally, during orientation, motor activity is suppressed, the discharge of brain serotonergic neurons is correspondingly suppressed resulting in dysfacilitation of motor output, while transmission of afferent information through sensory pathways is enhanced due to disinhibition.

4. Discussion

Do these results from basic research on brain 5-HT provide any insights into the etiology or treatment of clinical disorders? Recall that the activity of brain serotonergic neurons is at its maximal level during increased tonic motor output, and, for a subgroup of neurons, achieves an even higher level of activity during repetitive motor acts. These results suggest a novel way of viewing OCD. Since repetitive or compulsive motor acts increase serotonergic neuronal activity, I believe that patients with OCD may be engaging in such behaviors as an act of self-medication. In other words, they are activating their brain serotonergic system in a nonpharmacologic manner in order to derive some benefit or rewarding effect. Treating them with drugs accomplishes the same neurochemical endpoint and allows them to disengage from time consuming, socially unacceptable, and often harmful behaviors. A similar case could also be made for repetitive obsessional thoughts.

5. Acknowledgements

Supported by the AFOSR (90-0294) and the NIMH (MH-23433).

6. References

1. Jacobs, B.L. and Azmitia, E.C. (1992) 'Structure and function of the brain serotonin system', Physiol. Rev. 72, 165-229.
2. Jacobs, B.L. (1976) 'An animal behavior model for studying central serotonergic synapses', Life Sci. 19, 777-785.
3. Sternbach, H. (1991) 'The serotonin syndrome', Am. J. Psychiatry 148, 705-713.
4. Barbeau, H. and Rossignol, S. (1991) 'Initiation and modulation of the locomotor pattern in the adult chronic spinal cat by noradrenergic, serotonergic and dopaminergic drugs', Brain Res. 546, 250-260.
5. Ribeiro-do-Valle, L.E., Metzler, C.W., and Jacobs, B.L. (1991) 'Facilitation of masseter EMG and masseteric (jaw closure) reflex by serotonin in behaving cats', Brain Res. 550, 197-204.
6. Grillner, S., Wallen, P., Brodin, L., and Lansner, A. (1991) 'Neuronal network generating locomotor behavior in lamprey: circuitry, transmitters, membrane properties', Ann. Rev. Neurosci. 14, 169-199.
7. McCall, R.B. and Aghajanian, G.K. (1979) 'Serotonergic facilitation of facial motoneuron excitation', Brain Res. 169, 11-27.
8. White, S.R. and Neuman, R.S. (1980) 'Facilitation of spinal motoneurone excitability by 5-hydroxytryptamine and noradrenaline', Brain Res. 188, 119-127.
9. White, S.R. and Neuman, R.S. (1983) 'Pharmacological antagonism of facilitatory but not inhibitory effects of serotonin and norepinephrine on excitability of spinal motoneurons', Neuropharmacology 22, 489-494.
10. Hounsgaard, J., Hultborn, H., Jesperson, B., and Kiehn, O. (1988) 'Bistability of α-motoneurons in the decerebrate cat and in the acute spinal cat after intravenous 5-hydroxytryptophan', J. Physiol. 405, 345-367.
11. Trulson, M.E. and Jacobs, B.L. (1979) 'Raphe unit activity in freely moving cats: correlation with level of behavioral arousal', Brain Res. 163, 135-150.
12. Fornal, C.A. and Jacobs, B.L. (1988) 'Physiological and behavioral correlates of serotonergic single unit activity', in N.N. Osborne and M. Hamon (eds.) Neuronal Serotonin, John Wiley and Sons, New York, pp. 305-345.

13. Wilkinson, L.O. and Jacobs, B.L. (1988) 'Lack of response of serotonergic neurons in the dorsal raphe nucleus of freely moving cats to stressful stimuli', Exper. Neurol. 101, 445-457.
14. Fornal, C.A., Litto, W.J., Morilak, D.A., and Jacobs, B.L. (1987) 'Single-unit responses of serotonergic dorsal raphe nucleus neurons to environmental heating and pyrogen administration in freely moving cats', Exp. Neurol. 98, 388-403.
15. Fornal, C.A., Litto, W.J., Morilak, D.A., and Jacobs, B.L. (1990) 'Single-unit responses of serotonergic dorsal raphe neurons to vasoactive drug administration in freely moving cats', Am. J. Physiol. 259, R963-R972.
16. Fornal, C.A., Litto, W.J., Morilak, D.A., and Jacobs, B.L. (1989) 'Single-unit responses of serotonergic neurons to glucose and insulin administration in behaving cats', Am. J. Physiol. 257, R1345-R1353.
17. Jacobs, B.L. (1990) 'Locus coeruleus neuronal activity in behaving animals', in D.J. Heal and C.A. Marsden (eds.) The Pharmacology of Noradrenaline in the Central Nervous System, Oxford University Press, New York, pp. 248-265.
18. Jouvet, M. and Delorme, F. (1965) 'Locus coeruleus et sommeil paradoxal', C.R. Soc. Biol. 159, 895-899.
19. Trulson, M.E., Jacobs, B.L., and Morrison, A.R. (1981) 'Raphe unit activity during REM sleep in normal cats and in pontine lesioned cats displaying REM sleep without atonia', Brain Res. 226, 75-91.
20. Steinfels, G.F., Heym, J., Strecker, R.E., and Jacobs, B.L. (1983) 'Raphe unit activity in freely moving cats is altered by manipulations of central but not peripheral motor systems', Brain Res. 279, 77-84.
21. Ribeiro-do-Valle, L.E., Fornal, C.A., Litto, W.J., and Jacobs, B.L. (1989) 'Serotonergic dorsal raphe unit activity related to feeding/grooming behaviors in cats', Soc. Neurosci. Abstr. 15, 1283.

ROLE OF THE SEROTONERGIC SYSTEM IN PITUITARY HORMONE SECRETION: THE PHARMACOLOGIC CHALLENGE PARADIGM IN MAN

H.Y. MELTZER
Case Western Reserve University
School of Medicine
Department of Psychiatry
Cleveland, Ohio 44106
USA

ABSTRACT. Multiple agents are available to study the serotonergic system in man using a neuroendocrine challenge paradigm, including precursors such as TRP and 5-HTP, a group of direct-acting agonists such as mCPP, MK-212, ipsapirone, and buspirone, and indirect agents such as fenfluramine. The interpretation of the hormone (and behavioral) responses to these agents in terms of specific receptor subtypes is by no means straight-forward. Specific antagonists are essential to define the receptors involved. Multiple receptor subtypes may contribute to a specific response if the agent is not specific or if a synergistic effect of receptor stimulation is involved. Inclusion of a placebo study condition may be needed to rule out stress and other nonspecific effects. Despite these concerns, neuroendocrine challenge studies may reveal serotonergic dysfunction in specific individuals, its relation to psychopathology and treatment-response, and help to clarify the specificity of drugs for particular receptor subtypes in man.

1. Introduction

The role of serotonin (5-HT) in the etiology and pathophysiology of a variety of psychiatric disorders, including major depression, bipolar disorder, obsessive-compulsive disorder, anxiety disorders, alcoholism, impulse disorders, and schizophrenia is of considerable interest [1]. Similarly, there is the possibility that effects on the serotonergic system are, in part, the basis for the therapeutic effects of tricyclic antidepressants, specific 5-HT reuptake inhibitors, monoamine oxidase inhibitors, lithium, electroconvulsive treatment, azapirone anxiolytic and antidepressant drugs, and atypical antipsychotic drugs, e.g. clozapine, ripseridone, and amperozide [1]. The purpose of this article is to consider some aspects of neuroendocrine studies used to help elucidate the role of 5-HT in psychiatric disorders and the mechanism of action of psychotropic drugs. The focus will be on receptor subtypes involved in secretion of prolactin (PRL) and cortisol in man, the receptors stimulated by drugs of specific interest, and problems in interpreting neuroendocrine challenge studies.

2. Serotonin and Hormone Secretion

Secretion of a large number of pituitary hormones can be influenced by manipulation of the serotonergic system. Serotonergic neurons from the dorsal and median raphe as well as other nuclei stimulate the secretion of many pituitary hormones, including PRL, adrenocorticotrophin (ACTH), growth hormone (GH), thyroid-stimulating hormone, gonadotropic-releasing hormones (i.e. LH-RH) and beta-endorphin from the anterior pituitary, as well as oxytocin and vasopressin from the posterior pituitary, and renin from the kidney [2,3]. This article will focus on secretion of PRL and ACTH/corticosteroids. Multiple serotonergic receptors have been suggested to contribute to the secretion of these hormones [2,3]. For this reason, the effect of specific serotonergic agonists and antagonists as well as pre- and postsynaptic mechanisms can be studied using hormone secretion as an indicator of specific aspects of serotonergic function [2,4].

3. Problems in Interpretation of Neuroendocrine Challenge Studies in Man

The strategy of measuring changes in basal output of pituitary and adrenal hormones in man following serotonergic agents has been intensively utilized in the last decade [2] because it provides a convenient and generally safe way to assess responses to serotonergic agents in patients (before as well as after treatment) and controls. As has been discussed elsewhere [2], the challenge agent should stimulate or inhibit hormone secretion through a direct, highly selective mechanism. The precursors of 5-HT, i.e. tryptophan (TRP) and 5-hydroxytryptophan (5-HTP), may be used to assess the ability to synthesize 5-HT and the effect of putative 5-HT reuptake inhibitors [4], respectively. Secondly, specific 5-HT antagonists are needed to establish the receptor specificity of putative agonists. Thirdly, it is necessary to show, at least in control groups, that a dose-dependent relationship does exist for a given agonist and the dose of agonist chosen for subsequent study is able to identify an increase or decrease in responsivity in the experimental group. It is also necessary to establish that the putative antagonist has been administered at a high enough dose [2]. It should also be shown that plasma levels of the agonists or precursors are equivalent between groups or across conditions before differences in hormone response can be attributed to differences in serotonergic mechanisms. Thus, we have demonstrated that the plasma levels of TRP following an IV TRP infusion are significantly lower in depressed patients than normal controls and that this difference can account for the diminished PRL responses in the depressed patients [5].

Another potential pitfall in such studies is failure to account for the effects of differences in the effect of stress, circadian rhythms, and basal levels of hormones by inclusion of a placebo control study. In many published neuroendocrine challenge studies, the only condition studied is administration of the active agent. A challenge study with placebo is often not included because of the difficulty of getting patient agreement or the expense.

We have now compared the outcomes of placebo-controlled versus nonplacebo-controlled experiments, using the results of 28 studies of hormone responses of normal controls compared to unmedicated schizophrenic or depressed subjects or following treatment with various drugs [6]. Using the area under the curve as the outcome measure, and an analysis of covariance with baseline hormone levels covaried, with both placebo and active drugs included, significant group x protocol interactions were found in 10/48 (20.8%) cases. This indicates that there was a group difference after accounting for variance due to nonspecific effects, including stress. Of these 10 positive studies, a significant group effect was found when only the active challenge study was included in the analysis in seven of the 10 (70%) cases, giving a false negative rate of 3/48 (6.25%). Furthermore, four of the 48 analyses (8.3%) of only the active agent showed a significant group effect, whereas inclusion of the placebo study data showed no significant group x protocol effect. Thus, a total of 7/48 (14.6%) of the cases examined with active challenge only would have produced errors of inference.

These results strongly support the importance of inclusion of a placebo challenge in all neuroendocrine studies until there is evidence that this is unnecessary under the conditions of a given experiment.

4. Effect of Direct Agonists and Serotonin Antagonists on Corticosterone in the Rat

4.1 5-HT_{1A} MECHANISM

The direct-acting 5-HT_{1A} agonists, 8-OH-DPAT, ipsapirone, buspirone, gepirone, and LY 165163 have been found to produce dose-dependent increases in corticosterone and/or ACTH secretion in rodents [7-12]. The increases usually peak at 60 minutes and last a few hours. There is evidence that both a corticotrophin releasing factor (CRF)-mediated and a direct effect on the adrenal gland may be involved [13]. Thus, inference about central 5-HT_{1A} mechanisms following administration of a 5-HT_{1A} agonist may also require a separate analysis of the peripheral effect of these drugs. There is some evidence that pre- and postsynaptic effects of 5-HT_{1A} agonists may mediate the central effects of these agents since the corticosterone response to a high dose of 8-OH-DPAT was partially blocked by pretreatment with para-chlorophenylalanine (PCPA), an inhibitor of 5-HT synthesis. Low-dose 8-OH-DPAT-induced increases in ACTH/corticosterone secretion is not blocked by PCPA pretreatment [13,14]. These data indicate that neuroendocrine responses even to direct acting agonists reflect both pre- and postsynaptic serotonergic mechanisms and that this mix might vary as a function of the dose of the 5-HT agonist.

In the rat, 5-HT$_{1A}$ antagonists can inhibit the agonist-induced increase in corticosterone secretion [9-11,15]. This has been frequently demonstrated with (-)-pindolol which is also a nonselective beta-adrenergic blocker. However, betaxolol, another beta-adrenergic blocker which does not block 5-HT$_{1A}$ receptors, was without effect [11].

Other 5-HT antagonists which are not specific 5-HT$_{1A}$ receptor antagonists (e.g. cyproheptadine, pizotifen, or metergoline) or which are not active as 5-HT$_{1A}$ antagonists (altanserin, mianserin, or ritanserin) did not block the effect of the 5-HT$_{1A}$ agonists on corticosterone [9]. This suggests that blockade of the 5-HT$_2$/5-HT$_{1C}$ receptors in the rat does not interfere with the effect of 5-HT$_{1A}$ receptor stimulation on corticosterone secretion.

4.2   5-HT$_{2/1C}$ RECEPTOR MECHANISMS

The 5-HT$_2$/5-HT$_{1C}$ agonists DOI, MK-212 as well as quipazine, have been reported to stimulate rat corticosterone secretion by a 5-HT$_2$/5-HT$_{1C}$-dependent mechanism, as indicated by the ability of a variety of 5-HT$_2$/5-HT$_{1C}$ receptor antagonists, e.g. clozapine, LY 53857, ketanserin, MDL 11939, ritanserin, setoperone, and spiperone, to block their effects [9,15]. 5-HT$_2$/5-HT$_{1C}$ receptor blockers such as altanserin, ketanserin, and ritanserin completely block the 5-HTP-induced increase in rat corticosterone levels (Meltzer, H.Y., unpublished data).

5.   Effect of 5-HT$_{1A}$ and 5-HT$_{s/1C}$ Agonists on Rat PRL Secretion

5-HT$_2$/5-HT$_{1C}$ agonists, e.g. DOI and MK-212, stimulate rat PRL secretion in a dose-dependent manner [9,10]. These effects are blocked by 5-HT$_{2/1C}$ antagonists such as ketanserin, ritanserin, and clozapine [9,10].

6.   Role of 5-HT$_{1A}$ and 5-HT$_{2/1C}$ Receptor Stimulation on the HPA Axis and Prolactin Secretion in Man

The 5-HT$_{1A}$ agonist ipsapirone has been reported to produce a dose-dependent increase in plasma ACTH and cortisol in normal volunteers with the threshold being 0.2 mg/kg p.o. [16]. These responses were partially blocked by the nonselective 5-HT$_1$/5-HT$_2$ blocker, metergoline. However, (±)pindolol (dose 30 mg), which was reported to be without effect on basal HPA activity, completely antagonized the ipsapirone-induced plasma ACTH and cortisol responses. Betaxolol had no effect on the ACTH/cortisol response to ipsapirone [16]. Ipsapirone did not affect plasma PRL levels in man [17]. However, we have previously presented evidence that ipsapirone decreased plasma PRL levels in rats, possibly due to a dopamine agonist action [18].

Buspirone, another azapirone with 5-HT$_{1A}$ partial agonist properties, also dose-dependently increased cortisol secretion in man, but it has a more potent effect to stimulate PRL secretion [19]. We attributed this to the weak D_2 dopamine receptor blocking properties of buspirone but others have suggested this may be due to its 5-HT_{1A} agonist properties [20]. We have reported no significant effect of pindolol pretreatment (30 mg) on the PRL and cortisol responses to buspirone, 30 mg in 8 normal volunteers [21] and extended this to an additional four subjects. Pindolol pretreatment did not block the buspirone-induced increase in cortisol or PRL secretion, suggesting neither response is 5-HT_{1A}-mediated (Meltzer, H.Y., in preparation).

We have now studied the ability of pindolol, 30 mg p.o., to block the L-5-HTP (200 mg, p.o.)-induced increase in cortisol and PRL secretion in 12 normal volunteers. Pindolol inhibited the PRL but not the cortisol response, consistent with a 5-HT_{1A} basis for the PRL but not the cortisol response (Meltzer et al., in preparation).

We have previously reported that ritanserin, a 5-$HT_{2/1C}$ antagonist, blocked the 5-HTP-induced increase in cortisol [22]. In that study, the effect of 5-HTP on plasma PRL was not significant. Pindolol had no effect (Meltzer, H.Y., in preparation).

We have also studied the ability of pindolol to block the effect of MK-212, a drug usually considered to be a 5-HT_2/5-HT_{1C} agonist [23-25]. Pindolol, 30 mg, blocked the effect of MK-212 to stimulate PRL but not cortisol secretion (Meltzer et al., in preparation) suggesting that there is a 5-HT_{1A} component to the MK-212 effect on PRL but not cortisol secretion. Clozapine, a 5-$HT_{2/1C}$ antagonist, also blocked the effect of MK-212 to stimulate PRL secretion in patients. This suggests a synergistic effect between 5-HT_{1A} and 5-HT_2/5-HT_{1C} receptors is needed for PRL secretion to occur with MK-212.

In summary, there is evidence from studies with pindolol for a 5-HT_{1A}-mediated PRL response in man that can be elicited by 5-HTP or MK-212 but not buspirone. Similarly, there is evidence for a 5-HT_{1A}-mediated ACTH cortisol response elicited by ipsapirone but for reasons that need to be further elucidated, the cortisol responses to buspirone, 5-HTP, and MK-212 cannot be blocked by pindolol alone. Since ritanserin but not pindolol can block the cortisol response to 5-HTP, it is reasonable to conclude that there is a 5-$HT_{2/1C}$-mediated cortisol response in man.

7. Comparison of PRL and Cortisol Responses to Serotonergic Agonists

We have compared the magnitude of the plasma cortisol and PRL responses to five serotonergic agents: buspirone, 5-HTP, ipsapirone, mCPP (a 5-$HT_{1B/1C}$ agonist) [23], MK-212, and buspirone in male normal volunteers. The doses were not chosen to be equally effective so what is most informative are the relative magnitude of the responses of the two hormones to each agonist. As can be seen in figure 1, ipsapirone, 5-HTP, and mCPP produced the largest cortisol responses followed by MK-212 and buspirone. mCPP and buspirone produced the largest prolactin responses followed by 5-HTP and MK-212. There was no PRL response to ipsapirone. Thus, 5-HTP and MK-212 were relatively more potent with regard to the HPA axis; mCPP had an equal effect on both hormones, whereas buspirone and ipsapirone were mainly active to stimulate prolactin and cortisol secretion, respectively. These results suggest that the ipsapirone-induced cortisol response may be the best choice to evaluate 5-HT$_{1A}$-mediated cortisol responses, the MK-212 PRL response may be the best choice for a measure of 5-HT$_{1A}$-mediated PRL secretion while the 5-HTP-induced cortisol response may be the choice for a 5-HT$_{2/1C}$-mediated response.

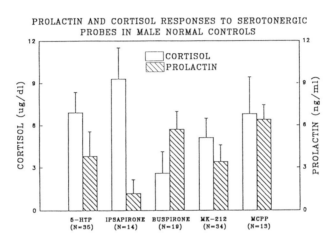

Figure 1. Effect of 5-HTP (200 mg p.o.), ipsapirone (0.3 mg/kg), buspirone 30mg, MK-212 (20 mg), and mCPP (0.5 mg/kg)

We examined the correlations between these five hormone responses for cortisol or PRL. The cortisol response to mCPP was significantly correlated with that due to ipsapirone, buspirone, MK-212, and 5-HTP (table 1). The cortisol response to ipsapirone was also significantly correlated with that due to buspirone and MK-212 but not 5-HTP. None of the other correlations were significant. This pattern of responses suggests that the magnitude of the cortisol response to a variety of serotonergic agents may be related to each other because of pituitary-adrenal stores of ACTH and cortisol as well as some overlap in the nature of the receptors that are stimulated. Only buspirone- and ipsapirone-induced PRL responses (rho=0.74, N=8) were significantly correlated at the 0.05 level of significance. This suggests much more diversity in the factors which influence the PRL responses, and, hence, relatively less importance of the serotonergic component.

TABLE 1. Correlations Between Cortisol Response to Serotonergic Agents in Male Normal Controls

	Ipsirone	Buspirone	MK-212	5-HTP
mCPP	0.56†	0.89*	0.65*	0.79*
Isapirone		0.89**	0.67†	NS
Buspirone			NS	NS
MK-212				NS

†p=0.06; *p < 0.05; **p < 0.01

We also examined the correlations between cortisol and PRL responses to the five serotonergic agents in normal control subjects. The cortisol and PRL responses were significantly correlated for MK-212 (N=34, rho=0.57, p=0.0004), mCPP (N=13, rho=0.61, p=0.03), 5-HTP (N=35, rho=0.34, p=0.035). There was a trend for a correlation with buspirone (N=18, rho=0.41, p=0.09) but no correlation was noted for ipsapirone (N=12, rho=0.03, p=0.91). The variance in these relationships suggest that factors other than plasma concentrations of these agents or 5-HT influence the PRL and cortisol responses to these agents. This may include variations in the type, density, and affinity of receptors which mediate the two hormone responses within a given individual. It is, therefore, difficult to utilize either response as a definitive measure of responsivity of a specific receptor.

8. Conclusions

Currently available agents for studying the serotonergic system in man using a neuroendocrine response paradigm include precursors such as TRP and 5-HTP, a group of direct-acting agonists such as mCPP, MK-212, ipsapirone, and buspirone, and indirect agents such as fenfluramine. The interpretation of the hormone responses to these agents in terms of specific receptor subtypes is by no means straight-forward. Specific antagonists are needed to define the nature of the response. Multiple receptor subtypes may contribute to a specific response if the agent is not specific or synergistic effects of receptor stimulation are involved. Inclusion of a placebo study condition may be needed to rule out stress and other nonspecific effects. Despite these concerns, neuroendocrine challenge studies may reveal serotonergic dysfunction in specific individuals, its relation to psychopathology and treatment-response, and help to clarify the specificity of drugs for particular receptor subtypes in man.

9. Acknowledgment

The research reported was supported in part by USPHS MH 41684, GCRC MO1RR00080, the Department of Veterans Affairs and grants from the Elisabeth Severance Prentiss and John Pascal Sawyer Foundations. H.Y.M. is the recipient of a USPHS Research Career Scientist Award MH 47808. The secretarial assistance of Ms. Lee Mason is greatly appreciated.

10. References

1. Brown, S.L. and van Praag, H.M. (eds.) (1991) 'The Role of Serotonin in Psychiatric Disorders', Brunner/Mazel, Inc., New York, NY.
2. Nash, J.F. and Meltzer, H.Y. (1991) 'Chapter 4: Neuroendocrine studies in psychiatric disorders: The role of serotonin', in S.L. Brown and H.M. van Praag (eds.), The Role of Serotonin in the Psychiatric Disorders, Brunner-Mazel, New York, pp. 57-90.
3. Levy, A.D. and Van de Kar, L.D. (1992) 'Endocrine and receptor pharmacology of serotonergic anxiolytics, antipsychotics and antidepressants', Life Sci. 51, 83-94.
4. Fuller, R.W. (1992) 'The involvement of serotonin in regulation of pituitary-adrenocortical function', Frontiers in Neuroendocrinol. 13, 250-270.
5. Koyama, T. and Meltzer, H.Y. (1986) 'A biochemical and neuroendocrine study of the serotonergic system in depression', in H. Hippius et al. (eds.), New Results in Depression, Springer-Verlag, Berlin, pp. 169-188.
6. Thompson, P.A., Maes, M., and Meltzer, H.Y. (1992) 'Methodology of neuroendocrine studies: use of a placebo-controlled design', submitted for publication.
7. Fuller, R.W., Snoddy, H.D., and Molloy, B.B. (1986) 'Central serotonin agonist actions of LY 165163, 1-(m-trifluoromethylphenyl)-4(p-aminophenylethyl) (piperazine), in rats', J. Pharmacol. Exp. Ther. 239, 454-459.
8. Urban, J.H., Van de Kar, L.D., Lorens, S.A., and Bethea, C.L. (1986) 'Effect of the anxiolytic drug buspirone on prolactin and corticosterone secretion in stressed and unstressed rats', Pharmacol. Biochem. Behav. 25, 457-462.
9. Koenig, J.I., Gudelsky, G.A., and Meltzer, H.Y. (1987) 'Stimulation of corticosterone and β-endorphin secretion by selective 5-HT receptor subtype activation', Eur. J. Pharmacol. 137, 1-8.
10. Koenig, J.I., Meltzer, H.Y., and Gudelsky, G.A. (1988) 'Hormone responses to selective serotonin receptor stimulation in the rat', in R.H. Reich and G.A. Gudelsky (eds.), 5-HT Agonists As Psychoactive Drugs, NPP Books, Ann Arbor, Michigan, pp. 283-298.
11. Lorens, S.A. and Van de Kar, L.D. (1987) 'Differential effects of serotonin (5-HT$_{1A}$ and 5-HT$_2$) agonists and antagonists on renin and corticosterone

secretion', Neuroendocrinology 45, 305-310.
12. Gilbert, F., Brazell, C., Tricklebank, M.D., and Stahl, S.M. (1988) 'Activation of the 5-HT_{1A} receptor subtype increases rat plasma ACTH concentrations', Eur. J. Pharmacol. 147, 431-437.
13. Matsuda, T., Kanda, T., Seong, Y.H., Baba, A. and Iwata, H. (19..) 'p-Chlorophenylalanine attenuates the pituitary-adrenocortical response to 5-HT_{1A} receptor agonists in mice,' Eur. J. Pharmacol 181, 295-297.
14. Gilbert, F., Dourish, C.T., Brazell, C., McClue, S., and Stahl, S.M. (1988) 'Relationship of increased food intake and plasma ACTH levels to 5-HT_{1A} receptor activation in rats', Psychoneuroendocrinology 13, 471-478.
15. Fuller, R.A. and Snoddy, H.D. (1979) 'The effects of metergoline and other serotonin receptor antagonists on serum corticosterone in rats', Endocrinology 105, 923-928.
16. Lesch, K.-P., Söhnle, K., Poten, B., Schoellnhammer, G., Rupprecht, R., and Schulte, H.M. (1990) 'Corticotropin and cortisol secretion after central 5-hydroxytryptamine-1A (5-HT_{1A}) receptor activation: effects of 5-HT receptor and ß-adrenoceptor antagonists', J. Clin. Endocrinol. Metab. 70, 670-674.
17. Lesch, K.-P., Rupprecht, R., Poten, B., Muller, U., Söhule, K., Fritze, J., and Schulte, H.M. (1989) 'Endocrine responses to 5-HT_{1A} receptor activation by ipsapirone in humans', Biol. Psychiatry 26, 203-205.
18. Nash, Jr., J.F. and Meltzer, H.Y. (1989) Effect of gepirone and ipsapirone on the stimulated and unstimulated secretion of prolactin in the rat', J. Pharmacol. Exp. Ther. 249, 236-241.
19. Meltzer, H.Y., Fleming, R., and Robertson, A. (1983) 'The effect of buspirone on prolactin and growth hormone secretion in man', Arch. Gen. Psychiatry 40, 1099-1104.
20. Cocarro, E.F., Gabriel, S., Mahon, T., Macauso, J., and Siever, L.J. (1990) 'Preliminary evidence of a serotonin (5-HT_1-like) component to the prolactin response to buspirone challenge in humans', Arch. Gen. Psychiatry 47, 594-595.
21. Meltzer, H.Y., Lee, H.S., and Nash, J.F. (1991) 'Letter to the Editor. Effects of buspirone on prolactin secretion is not mediated by 5-HT_{1A} receptor stimulation', Arch. Gen. Psychiatry 49, 163.
22. Lee, M.A., Nash, J.F., Barnes, M., and Meltzer H.Y. (1991), 'Inhibitory effect of ritanserin on the 5-hydroxytryptophan-mediated cortisol, ACTH and prolactin secretion in humans', Psychopharmacology 103, 258-264.
23. Hoyer, D. (1988) 'Functional correlates of serotonin 5-HT_1 recognition sites', J. Recept. Res. 8, 59-81.'
24. King, B.H., Brazell, C., Dourish, C.T. and Curzon, G. (1989) 'MK-212 increases rat plasma ACTH concentration by activation of the 5-HT_{1C} receptor subtype', Neurosci. Lett. 105, 174-176.
25. Lowy, M.T., Koenig, J.I., and Meltzer, H.Y. (1988) 'Stimulation of serum cortisol and prolactin in humans by MK-212, a centrally active serotonin agonist', Biol. Psychiatry 23, 818-828.

SEROTONIN IN ANXIETY AND RELATED DISORDERS

H.G.M. WESTENBERG and J.A. DEN BOER
Department of Biological Psychiatry
P.O. Box 85500
3508 GA Utrecht
The Netherlands

ABSTRACT. Anxiety disorders have recently become a major focus of research in psychiatry. The delineation of distinct diagnostic entities and the advent of selective serotonin (5-HT) drugs has resulted in a burgeoning of scientific interest. Clinical evidence on the role of 5-HT in anxiety disorders is mainly supported by the therapeutic effects of selective serotonin reuptake inhibitors (SSRIs) in panic disorder and obsessive compulsive disorder. SSRIs appear to have potent antipanic and antiobsessional effects in the absence of depressive symptoms. Clinical evidence that 5-HT is involved in generalized anxiety disorder is derived from studies with selective 5-HT agonists. The role of 5-HT in various anxiety states is further highlighted by the results of challenge tests with 5-HT selective drugs. This paper reviews the role of 5-HT in anxiety disorders with special emphasis on its differential role in the various symptom domains.

1. Introduction

A substantial body of animal studies suggests that the activity of the ascending serotonin (5-HT) pathways in the brain might play a role in regulating certain forms of anxiety in laboratory animals [1]. These data converge to suggest that decreasing 5-HT function may produce behavioral responses reminiscent of anxiety. Animal research is confined, however, by the fact that extrapolations from animal behavior to human emotions has limited validity. Human anxiety can range from normal emotions to clinical syndromes. It can be a component of different physical and mental disorders, but also the central feature of a syndrome.

2. Diagnostic Aspects

In the current nomenclature (DSM-IIIR), anxiety disorders encompass, among others, the following major categories: panic disorder (PD), generalized anxiety disorder (GAD), social phobia, and obsessive compulsive disorder (OCD). When anxiety occurs in a paroxysmal fashion it is called a panic attack, the hallmark of PD. Patients with chronic high levels of anxiety that do not have panic attacks are referred to as GAD. Social phobics also present with a high degree of anxiety, particularly in situations in which these subjects are exposed to the scrutiny of others. Anxiety symptoms also undoubtedly feature prominently in OCD, but there is little evidence to support its categorization as an anxiety disorder. Most data support its perception as a separate diagnostic entity. Of particular interest is the distinction between GAD and PD. This distinction is based on the observation that antidepressants, such as imipramine, can block spontaneous panic attacks, whereas GAD patients seem to respond best to anxiolytics, such as benzodiazepines. Although later studies do not fully confirm this differential therapeutic response, the separation is supported by a number of epidemiological, family, and twin studies [2].

3. Selective 5-HT Reuptake Inhibitors in Anxiety Disorders

Selective serotonin reuptake inhibitors (SSRIs) were originally developed as antidepressants, but many studies have revealed that this class of drugs present therapeutic potentials beyond the treatment of depression [3-5]. SSRIs provide an effective treatment for OCD and PD, whereas recent studies also suggest their efficacy in social phobia. The therapeutic potential of SSRIs in PD was inspired by the notion that the efficacy of antidepressants to block panic attacks may be related to their effect on 5-HT systems in the brain. To study this relationship we [4] compared the efficacy of maprotiline, an antidepressant which selectively inhibits the reuptake of noradrenaline, and fluvoxamine, which has selective effects for 5-HT. The number of panic attacks and the level of anxiety decreased significantly in patients on fluvoxamine. In the maprotiline group there was no significant change in the number of panic attacks and only a slight reduction in anxiety. Subjects were selected carefully, excluding those with concomitant depressive symptoms, to find a pure PD sample. Therefore, the results indicate a specific antipanic effect for fluvoxamine, which may be related to effects on 5-HT. Generally, SSRIs show a latency of 2-4 weeks before the therapeutic effect occurs. In addition, a transient increase in anxiety is seen in the majority of patients at the beginning of treatment. This somewhat puzzling phenomenon, which appears to be unique to PD patients, suggests an acute anxiogenic effect of SSRIs followed by an anxiolytic effect after chronic use. This concurs with experimental data, indicating that acute administration of SSRIs reduces the firing of 5-HT neurons, while due to desensitization of the somatodendritic autoreceptors, an enhanced 5-HT function is seen after repeated administration [6]. These data suggest that stimulation of an as yet unknown subset of 5-HT receptors may underlie the mechanism of action of SSRIs in PD. A selective therapeutic response to SSRIs has also been demonstrated in OCD. Several controlled studies with fluvoxamine and clomipramine have all shown clear evidence of efficacy for these drugs. In contrast, other antidepressants such as desimipramine and amitriptyline have been found ineffective [3]. Studies with SSRIs in GAD patients are lacking, but there are indications that patients suffering from social phobia may respond favorably as well. Recently, we have demonstrated the efficacy of fluvoxamine on social anxiety (figure 1). A clinical relevant improvement was also seen on measures of phobic avoidance and interpersonal sensitivity [5].

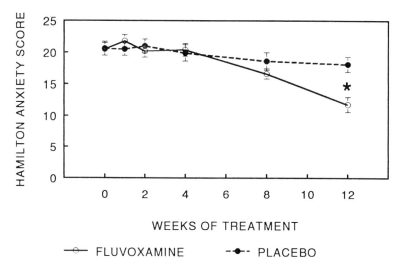

Figure 1. Mean (± sem) score on the Hamilton Anxiety Scale in patients with social phobia treated with fluvoxamine (n=15) or placebo (n=15). Fluvoxamine was superior to placebo in reducing symptoms of anxiety from week 12 of treatment.

4. Differential Effects of 5-HT Agonists in Anxiety States

Clinical evidence that 5-HT may be involved in GAD is mainly derived from studies with selective 5-HT_{1A} agonists. Studies with buspirone have revealed that this class of compounds possesses beneficial effects in GAD patients [7]. Since anxiolytic effects become apparent only after long-term treatment, it is assumed that adaptive changes of 5-HT_{1A} receptors are required to observe clinical relevant effects. Studies with buspirone in PD are either inconclusive or negative, while a preliminary study with OCD patients suggests, that it may have anticompulsive and antiobsessional effects [8,9]. Recently, we investigated the effects of flesinoxan in PD patients. Flesinoxan is a potent and selective 5-HT_{1A} agonist, surpassing buspirone in receptor affinity. Patients were treated with flesinoxan and placebo using a crossover design with a placebo run-in phase [10]. Treatment with 2.4 mg of flesinoxan, a dose that is usually well tolerated by healthy controls and patients with GAD, resulted in a substantial increase in anxiety (figure 2). In contrast to SSRIs, this deterioration did not give way to improvement when treatment was continued. In a subsequent double-blind placebo controlled dose-finding study, we found that this anxiogenic response in PD is dose-dependent; lowering the dose resulted in a behavioral response indistinguishable from placebo.

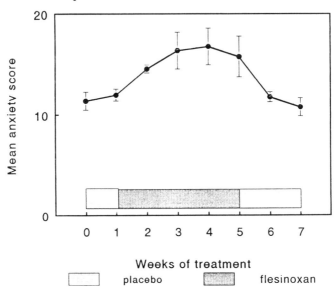

Figure 2. The mean clinical anxiety score in PD patients treated with 2.4 mg of flesinoxan and placebo. Flesinoxan caused an increase in anxiety rating in PD patients.

Its interesting to note, that flesinoxan did not have an effect on the number of panic attacks, it was anxiogenic rather than panicogenic in this sample. Several nonselective and indirect agonists, such as m-chlorophenylpiperazine (mCPP), fenfluramine, and 5-hydroxytryptophan (5-HTP), have also been tested in anxiety disorders. Challenge tests with mCPP, a nonselective 5-HT agonist, which might have preferential actions at the $5-HT_{1C}$ receptors but which also interacts with other 5-HT and nonserotonergic receptors, have demonstrated anxiogenic responses in PD, GAD, OCD, and healthy controls [7,11,12]. In addition, it causes an exacerbation of obsessions in OCD patients. The behavioral responses in patients are generally greater than in controls, suggesting that patients are hypersensitive to this stimulation. In PD, similar anxiogenic reactions were seen with fenfluramine, a compound which releases 5-HT and inhibits its uptake [13]. In contrast to mCPP, fenfluramine did not exacerbate obsessions in OCD patients [11].

5. 5-HT Antagonists in Anxiety Disorders

We studied the efficacy of ritanserin, a selective 5-HT$_{1C/2}$ antagonist, in PD patients [4]. Previous studies had revealed anxiolytic activity in GAD patients. In PD ritanserin appeared to be indistinguishable from placebo, while fluvoxamine, the reference compound, was effective. The authors concluded from this data that 5-HT$_2$ receptors are not hypersensitive in PD and that the antipanic effect of antidepressants cannot be related to their purported downregulation of the 5-HT$_2$ receptors following chronic administration.

6. Conclusions

There is a vast and expanding literature relating 5-HT to anxiety states. As of yet it is difficult to make any coherence from the collection of data. The efficacy of SSRIs in the treatment of PD and OCD remains the firmest evidence on the role of 5-HT in these conditions, whereas studies with 5-HT$_{1A}$ agonists hint at the involvement of 5-HT in GAD. Challenge studies with mCPP support the notion that 5-HT systems may be dysregulated in anxiety states, but the precise nature of the disturbance remains elusive. The paradoxical effect of the 5-HT$_{1A}$ agonist flesinoxan and the initial deterioration of SSRIs in PD are still puzzling but could be a clue to the pathogenesis of this disorder. To understand the role of 5-HT in anxiety disorders, further research is clearly required. These studies should take advantage of more selective agents and psychological strategies to characterize the behavioral responses in well-defined samples.

7. References

1. Gardner, C.R. (1985) 'Pharmacological studies of the role of serotonin in animal models of anxiety', in A.R. Green (ed.), Neuropharmacology of Serotonin, Oxford University Press, New York, pp. 281-325.
2. Weissman, M.M. (1990) 'Panic and generalized anxiety: are they separate disorders?', J. Psychiat. Res. 24, 167-162.
3. Goodman, W.K., McDougle, C.J., and Price, L.H. (1992) 'Pharmacotherapy of obsessive compulsive disorder', J. Clin. Psychiatry 53 (Suppl.), 29-37.
4. Westenberg, H.G.M. and Den Boer, J.A. (1989) 'Serotonin-influencing drugs in the treatment of panic disorder', Psychopathology 22 (Suppl. 1), 68-77.
5. Den Boer, J.A., van Vliet, I.M., and Westenberg, H.G.M. (1992) 'A double-blind study of fluvoxamine in social phobia', Clin. Neuropharmacol. 15 (suppl. 1), 38.
6. De Montigny, C. and Aghajanian, G.K. (1978) 'Tricyclic antidepressants: Long-term treatment increases responsivity of rat forebrain neurons to

serotonin, Science 202, 1303-1306.
7. Kahn, R.S., Kalus, O., Wetzler, S., and Van Praag, H.M. (1991) 'The role of serotonin in the regulation of anxiety', in, S.L. Brown and H.M. Van Praag (eds.), Brunner/Mazel, New York, pp. 129-160.
8. Pohl, R., Balon, R., Yeragani, V.K., and Gershon, S. (1989) 'Serotonergic anxiolytics in the treatment of panic disorder: a controlled study of buspirone', Psychopathology 22 (Suppl. 1), 60-67.
9. Pato, M.T., Pigott, T.A., Hill, J.L., Grover, G.N., Berstein, S., and Murphy, D.L. (1991) 'Controlled comparison of buspirone and clomipramine in obsessive-compulsive disorder', Am. J. Psychiatry 148, 127-129.
10. Westenberg, H.G.M., Van Vliet, I.M., and Den Boer, J.A. (1992) 'Flesinoxan, a selective 5-HT$_{1A}$ agonist, in the treatment of panic disorder', Clin. Neuropharmacol. 15 (Suppl. 1), 60.
11. Barr, L.C., Goodman, W.K., Price, L.H., McDougle, C.J., and Charney, D.S.(1992) 'The serotonin hypothesis of obsessive compulsive disorder: implications of pharmacological challenge studies', J. Clin. Psychiatry 53 (Suppl.), 17-28.
12. Germine, M., Goddard, A.W., Woods, S.W., Charney, D.S., and Heninger, G.R. (1992) 'Anger and anxiety responses to m-chlorophenylpiperazine in generalized anxiety disorder', Biol. Psychiatry 32, 457-461.
13. Targum, S.D. and Marshall, L.E. (1989) 'Fenfluramine provocation of anxiety in patients with panic disorder', Psychiatry Res. 28, 295-306.

ALCOHOLIC VIOLENT OFFENDERS: BEHAVIORAL, ENDOCRINE, DIURNAL RHYTHM, AND GENETIC CORRELATES OF LOW CSF 5-HIAA CONCENTRATIONS

M. LINNOILA, M. VIRKKUNEN*, R. TOKOLA*, D. NIELSEN and D. GOLDMAN
Division of Intramural Clinical and Biological Research
National Institute of Alcohol Abuse and Alcoholism
Bethesda, Maryland 20893
USA, and
*Department of Psychiatry
University of Helsinki
Helsinki 00181
Finland

ABSTRACT. In a series of studies on Finnish alcoholic violent offenders, low cerebrospinal fluid (CSF) 5-hydroxyindoleacetic acid (5-HIAA) concentrations have been associated with impulsive, unprovoked offenses committed under the influence of alcohol. In impulsive violent offenders, lowest CSF 5-HIAA concentrations are associated with a history of suicide attempts, with a family history of paternal alcoholism, and with a polymorphism of the tryptophan hydroxylase gene on chromosome 11. On the Karolinska Scales of Personality (KSP) impulsive violent offenders with low CSF 5-HIAA concentrations are characterized by traits such as impulsivity, irritability, and monotony avoidance. Their diurnal activity rhythm recordings show increased activity during night-time hours compared to healthy volunteers, and they exhibit a hypoglycemic tendency in response to an oral glucose load. Comparisons to healthy volunteers show that low CSF 5-HIAA alone is not sufficient to cause an individual to be interpersonally violent, but it is associated with irritability and impaired impulse control. High CSF free testosterone concentrations are more directly related to aggressiveness.

1. Introduction

Serotonin (5HT), one of the major monoamine neurotransmitters, affects a multitude of bodily functions, such as sleep, appetite, libido, and carbohydrate metabolism. In psychiatric literature, central serotonergic dysfunction has been postulated to play a role in anxiety disorders, mood disorders, and schizophrenia. Despite a large number of publications concerning the role of serotonin in categorically classified mental disorders only a few generally agreed upon insights have emerged from studies using this theoretical framework.

Categorically defined mental disorders typically consist of multiple behavioral aberrations grouped together as a syndrome. If clearly defined behavioral dimensions rather than syndromic mental disorders are correlated with measures of neurotransmitter function, a more coherent picture begins to emerge. This is particularly true for serotonin. The first observation conducive of the dimensional approach was made by Swedish investigators, who found that patients with unipolar depression, who had low CSF 5-HIAA concentrations, had a high incidence of suicide attempts [1]. Furthermore, these attempts were characterized by violent methods used to carry them out. This finding has stood the test of time and it has been widely replicated [2].

Brown et al. expanded upon these reports by investigating patients with a life-long pattern of repeated, aggressive behaviors. In two studies [3,4] these authors found a strong negative correlation between scores on the Brown-Goodwin aggression rating scale and CSF 5-HIAA concentrations. They suggested that reduced central serotonin turnover is conducive of a trait-like increase of risk to exhibit aggressive behavior towards others.

2. Discussion

We were initially interested in exploring whether low CSF 5-HIAA was primarily associated with increased aggressiveness or impaired impulse control. We approached this problem by investigating violent offenders and impulsive fire setters [5,6]. Based on the characteristics of the crime the violent offenders, all of whom were alcoholics, were divided into impulsive and nonimpulsive groups. Offenders who had attacked a person unknown to them, without provocation or expected monetary gain, were classified as impulsive. Impulsive fire setters were studied to provide a further test of the hypothesis, because they are thought to be relatively nonaggressive towards others while having impaired impulse control. Both impulsive violent offenders, and impulsive fire setters were found to have low CSF 5-HIAA concentrations when compared to nonimpulsive violent offenders. Therefore, the results of these studies supported the notion that reduced central serotonin turnover is primarily associated with impaired impulse control rather than increased aggressiveness.

Based on this finding and the findings of Brown et al. [3,4], we hypothesized that during a prospective followup, after release from incarceration, offenders with low CSF 5-HIAA would be more likely to commit recidivist violent crimes. We found that mild hypoglycemia after an oral glucose challenge together with low CSF 5-HIAA indeed had modest power to predict recidivist violent crimes during an average of three years of followup [7]. We observed also that offenders with alcoholic fathers, who had exhibited aggressive behaviors while intoxicated, had the lowest CSF 5-HIAA concentrations [8]. We suggested, similar to Rosenthal et al. [9], who had investigated depressed patients, that low CSF 5-HIAA may be to some extent a familial trait.

During these studies we had observed clinically that offenders, who turned out to have low CSF 5-HIAA concentrations, often experienced periods of sleeplessness while on the forensic psychiatry ward. This observation led to the hypothesis that they may exhibit temporary dysregulation of certain diurnal rhythms. This inference was bolstered by findings in animal studies demonstrating that serotonergic input to the main endogenous circadian clock, the suprachiasmatic nucleus, facilitates entrainment of diurnal rhythms to external zeitgerers [10]. Moreover, Japanese endocrinologists had observed that the suprachiasmatic nucleus also plays a role in regulating carbohydrate metabolism [11]. Based on these pieces of information we postulated a model for a possible "low central serotonin syndrome" which covers certain aspects of its pathophysiology and behavioral consequences produced by the pathophysiology (figure 1) [12].

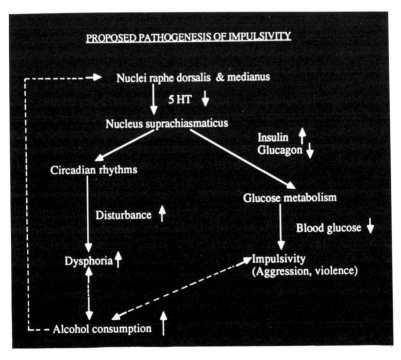

Figure 1. Proposed model to explain consequences of reduced serotonergic input to the suprachiasmatic nucleus.

In our most recent study we have tested the model and provided confirmatory evidence concerning diurnal activity rhythm disturbances in impulsive violent offenders with low CSF 5-HIAA concentrations [13]. To investigate further biochemical concentrations of impulsiveness as contrasted to aggressiveness, we computed discriminate function analyses using all behavioral and biochemical variables available to discriminate impulsive from nonimpulsive offenders and violent offenders from healthy volunteers. For the impulsiveness the only variable which contributed significantly to the discrimination was CSF 5-HIAA concentration (figure 2) concentration and for aggressiveness by far the largest contribution was made by CSF free testosterone concentration (figure 3).

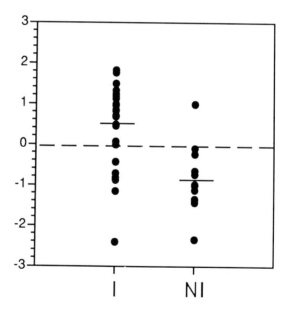

Figure 2. Results of discriminant analysis on impulsiveness. I=impulsive; NI=nonimpulsive offenders.

3. Results

Thus, the results of the study corroborate our earlier inferences concerning CSF 5-HIAA and impulsive control. Moreover, they add to the emerging picture of biochemical concomitants of violent behavior under the influence of alcohol by emphasizing an association between CSF free testosterone concentration and aggressiveness.

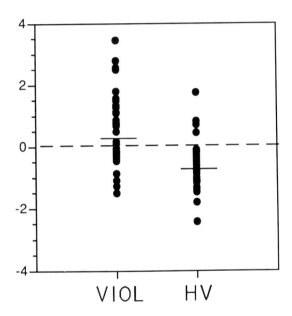

Figure 3. Results of discrimination analysis on aggressiveness. VIOL=violent offenders; HV=healthy volunteers.

Our current studies explore the neural pathways involved in impulse control related to serotonin, and molecular genetic family studies. In the first phase of the latter we found that a tryptophan hydroxylase polymorphism is highly associated with a history of suicide attempts and low CSF 5-HIAA in a group of alcoholic, impulsive, violent offenders (figure 4). The functional importance of this polymorphism is currently under investigation as well as the generalizability of this finding to other populations at high risk to attempt suicide.

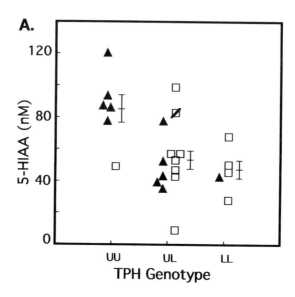

Figure 4. Tryptophan hydroxylase (TPH) genotype, CSF 5-HIAA concentration and a history positive for suicide attempts (□) in alcoholic impulsive violent offenders.

4. References

1. Asberg, M. Traskman, L., and Thoren, P. (1976) '5-HIAA in the cerebrospinal fluid: a biochemical suicide predictor?', Arch. Gen. Psychiatry 33, 1193-1197.
2. Molcho, A., Stanley, B., and Stanley, M. (1991) 'Biological studies and members in suicide and attempted suicide', Int. Clin. Psychopharmacol. 6, 77-92.
3. Brown, G.L., Goodwin, F.K., Ballenger, C., Goyer, P.F., and Major, L.F. (1979) 'Aggression in humans correlates with cerebrospinal fluid amine metabolites', Psychiatry Res. 1, 131-139.
4. Brown, G.L., Ebert, M.H., Goyer, P.F., Jimerson, D.C., Klein, W.J., Bunney, W.E., and Goodwin, F.K. (1982) 'Aggression, suicide and serotonin: relationships to CSF amine metabolites', Am. J. Psychiatry 139, 741-746.
5. Linnoila, M., Virkkunen, M., Scheinin, M., Nuutila, A., Rimon, R., and Goodwin, F.K. (1983) 'Low cerebrospinal fluid 5-hydroxyindoleacetic acid concentration differentiates impulsive from nonimpulsive behavior', Life Sci. 33, 2609-2914.
6. Virkkunen, M., Nuutila, A., Goodwin, F.K., and Linnoila, M. (1987) 'Cerebrospinal fluid monoamine metabolites in male arsonists', Arch. Gen.

Psychiatry 44, 241-247.
7. Virkkunen, M., DeJong, J., Bartko, J., Goodwin, F.K., and Linnoila, M. (1989) 'Relationship of psychobiological variables to recidivism in violent offenders and impulsive fire setters', Arch. Gen. Psychiatry, 46, 600-603.
8. Linnoila, M., DeJong, J., and Virkkunen, M. (1989) 'Family history of alcoholism in violent offenders and impulsive fire setters', Arch. Gen. Psychiatry 46, 613-616.
9. Rosenthal, N., Davenport, Y., Cowdry, R., Webster, M., and Goodwin, F. (1980) 'Monoamine metabolites in cerebrospinal fluid of depressive subgroups', Psychiatry Res. 2, 113-119.
10. Morin, L.P. and Blanchard, J. (1991) 'Depletion of brain serotonin by 5,7-DHT modifies hamster circadian rhythm response to light', Brain Res. 566, 173-185.
11. Yamamoto, H., Nagai, K., and Nagakava, H. (1984) 'Additional evidence that the suprachiasmatic nucleus is the center for regulation of insulin secretin and glucose homeostasis', Brain Res. 304, 237-241.
12. Linnoila, M., Virkkunen, M., and Roy, A. (1986) 'Biochemical aspects of aggression in man', in W.E. Bunney Jr., E. Costa, and S.G. Potkin (eds.) Clinical Neuropharmacology, Raven Press, New York, pp. 377-379.
13. Virkkunen, M., Rawlings, R., Tokola, R., Kallio, E., Poland, R.E., Guidotti, A., Nemeroff, C., Bissette, G. Kalogeras, K., Karonen, S.-L., and Linnoila, M. 'CSF biochemistries, glucose metabolism, diurnal activity rhythms, personality profiles and state aggressiveness in Finnish violent offenders', submitted for publication.

MODULATION OF DOPAMINERGIC NEUROTRANSMISSION BY 5-HT$_2$ ANTAGONISM

T.H. SVENSSON, G.G. NOMIKOS, and J.L. ANDERSSON
Department of Pharmacology
Karolinska Institute
Box 60 400
S-104 01 Stockholm
Sweden

ABSTRACT. 5-HT$_2$-antagonists cause a slight, indirect activation of midbrain DA neurons with enhanced DA release in terminal regions. Phasic rather than tonic activity seems enhanced, which may improve signal to noise ratio of DA-signalling in psychosis. When combined with some DA-D$_2$-antagonism the DA-activating effect of 5-HT$_2$-antagonism appears to preferentially involve prefrontal DA projections.

1. Introduction

The introduction of ritanserin, a rather selective 5-hydroxytryptamine (5-HT$_2$)-receptor antagonist, in the mid-eighties was of considerable clinical and conceptual interest to the medical profession. Early studies indicated a mood-elevating effect in dysthymic states, with improved energy, drive, and motivation [1]. In addition, the drug appeared to reduce negative symptoms in schizophrenia [2], as well as extrapyramidal side effects (EPS) such as parkinsonism, generally associated with treatment with classical neuroleptics [3,4]. Recent studies document augmentation of haloperidol therapy in treatment-resistant schizophrenia, in particular as regards anergia, dysphoria, and negative symptoms [4]. Such observations necessarily focused our attention to the putative effects of 5-HT$_2$-receptor antagonists on midbrain dopamine (DA) systems, in particular because of the DA hypothesis of schizophrenia as well as the critical significance of brain DA-systems for parkinsonism. The initial experiments, utilizing single cell recording technology, revealed that ritanserin causes an increased neuronal firing of midbrain DA neurons in the zona compacta, substantia nigra (SN), as well as those in the adjacent ventral tegmental area (VTA) [5]. Moreover, this effect seemed indirect, i.e. mediated through release from serotonergic inhibition, since the effect of ritanserin was blocked by pretreatment with p-chlorophenylalanine. The present paper attempts to explain, at least to some extent, how such an effect of 5-HT$_2$-antagonists may help to improve negative symptoms in schizophrenia and, perhaps, amotivational states of some drug-addictions and will also present new data on the electrophysiological and biochemical effects of 5-HT$_2$-antagonists on midbrain DA-systems, pertinent to these questions.

2. Material and Methods

2.1 ELECTROPHYSIOLOGY

Male Sprague-Dawley rats (200-300 g) were anesthetized with chloral hydrate and a tracheal cannula and a jugular vein catheter for drug administration were inserted before mounting in a stereotaxic frame. Body temperature was kept at 36-37°C. For experimental detail, see [5]. Briefly, extracellular recording electrodes were pulled from Omegadot glass tubing and filled with 2 M NaCl containing 2% Pontamine Sky Blue. Electrode impedance was 2-4 Mohms measured at 135 Hz and the electrodes were lowered through a burrhole in the skull into the brain by means of a hydraulic microdrive. Typical DA cells were found 6.5-7.5 mm from skull surface in the SN and 7.5-8.5 mm in the VTA. Cells generally fulfilled well-established criteria for midbrain DA neurons [6,7]. Action potentials were amplified, discriminated, and monitored on an oscilloscope and an audio monitor. Burst firing, firing rate, and regularity of firing were analyzed from interspike time interval histograms (ISH) created by a computer. Burst firing was measured as the percentage ratio of spikes in bursts to the total number of spikes of an ISH. Burst onset was signalled by an interspike interval less than 80 ms and burst termination by an interval greater than 160 ms [8]. Only one cell was studied in each animal. Recording sites were marked at the end of each experiment by passing a negative current (5 µA) through the recording electrode, leaving a blue spot at the tip of the electrode. Rats were subsequently perfused transcardially with 10% formalin, the brains were removed and recording sites were verified histologically. All recording sites included in this work were found within the SN or the VTA.

2.2 MICRODIALYSIS

Male Wistar rats (250-350 g) were anesthetized with barbiturates and stereotaxically implanted with vertical probes of the concentric type in either the nucleus accumbens (NAC) or the prefrontal cortex (PFC). Coordinates from bregma were AP=3.6, ML=1.4, DV=-8.2 [9] and AP=3.6, ML=0.6, DV=-6.2 [10] for the NAC and the PFC, respectively. Dialysis occurred through a 2.25 (for NAC) or a 4.0 mm (for PFC) semipermeable membrane (ID=0.24 mm, 40,000 Daltons, AN69 Hospal). All experiments were performed approximately 48 hours postsurgery in awake, freely moving animals. Microdialysis was performed using automated on-line sampling [11]. The dialysis probe was perfused with perfusion solution (147 mM NaCl, 3.0 mM KCl, 1.3 mM CaCl$_2$, 1.0 mM MgCl$_2$, and 1.0 mM sodium phosphate; pH 7.4) at a rate of 2.5 µl/min set by a microinfusion pump. Biochemical analysis was performed by HPLC-ED [11]. Separation of DA was achieved by reverse phase liquid chromatography (150x4.6 mm, Nucleosil 5 μm, C18) with mobile phase consisting of 0.055 M sodium acetate with 0.1 mM octanesulfonic acid, 0.01 mM NA_2EDTA, and 10% methanol (pH=4.1, adjusted with glacial acetic acid). The mobile phase was delivered by an HPLC pump at 0.8 ml/min and chromatograms were recorded on a two-pen chart recorder. Upon completion of the experiments, all probe placements were verified histologically. The average of two baseline samples before drug injection was defined as 100%. All subsequent measures were related to these values, and the mean percentages were calculated for each 40-minute sample across the rats in all groups. Percent changes were analyzed by analysis of variance (ANOVA) with repeated measures, followed by the Newman-Keuls test.

3. Results and Discussion

3.1 ELECTROPHYSIOLOGY

Single unit recordings from VTA DA neurons have, generally, revealed two major modes of function, e.g. single spike firing and burst firing [7,12]. The burst firing, which seems driven primarily by an excitatory amino acid (EAA) input [13], and is associated with a six-fold increased DA release [14,15] in comparison with that of single spike firing, seems largely controlled by the medial PFC in the rat [16]. Thus, local cooling of the frontal cortex was found to cause a reversible reduction in phasic, but not tonic activity in the VTA-DA cells, which displayed reversible pacemaker-like firing during this procedure. More recently, we have found that local microinfusion of lidocaine (38 nmol) into the medial PFC also caused a selective and reversible reduction in burst firing of VTA-cells, with no change in firing rate. Generally, midbrain DA cells may well influence their target neurons primarily through changes in firing pattern [17]. Moreover, VTA-DA neurons display rapid activation responses in association with goal or reward oriented behaviors [18]. Thus, a selective reduction of burst activity in the mesolimbic DA neurons, in fact parallelled by a slight but significant reduction in DA release in the accumbens nucleus [19], may suggest an impairment of the machinery for incentive behavior. Interestingly, our experiments provide experimental models for so-called hypofrontality, i.e. a reduced functional activity in prefrontal cortex, clinically observed in chronic schizophrenics and chronic alcoholics [20]. Moreover, the hypofrontality observed clinically seems to correlate to lack of goal-oriented behavior, indifference, and, generally, negative symptoms of schizophrenia. Our experiments thus suggest, that such functional deficits may relate to defective phasic, but not tonic midbrain DA activity. Interestingly, two potent and selective 5-HT_2-receptor antagonists, ritanserin [16]) and amperozide (APZ) [21], were both found to antagonize the functional deficit of VTA-DA cells caused by hypofrontality, i.e. to restore or preserve phasic activity (burst firing) without changing the firing rate.

Other experiments of ours involving schizophrenomimetic drugs such as noncompetitive NMDA-receptor antagonists indicate that DA-signalling in brain may show additional distortions correlating to psychotic symptomatology, e.g. a regional imbalance of DA systems involving DA neurons projecting to the PFC in particular, and an elevated tonic firing in spite of a reduced phasic activity [19,22]. Such a dysfunction, i.e. a reduced signal-to-noise ratio of the DA signal, may thus require not only restoration of phasic activity, but also antagonism of the elevated tonic activity, for example by means of combined DA-D_2 receptor antagonism and 5-HT_2-receptor antagonism.

Most recently, we have started to analyze the functional interplay between a 5-HT_2-receptor antagonist, ritanserin, and a selective DA-D_2-receptor antagonist, raclopride, on the VTA-DA system, utilizing single cell recording technology. Raclopride alone caused a dose-dependent increase in both burst firing and firing rate of the VTA-DA cells. An excitatory effect was seen already at 10 μg/kg IV and it was maximal at the highest dose tested, 5120 μg/kg. No cells went into depolarization inactivation regardless of the dose of raclopride. Although ritanserin alone caused slight excitation of the VTA-DA cells, the relative increase in burst firing by low doses of raclopride (10 or 20 μg/kg) was significantly larger in ritanserin pretreated rats than in controls (figure 1A). The potentiation of the effects of raclopride on burst firing was not seen with high doses of the drug. Neither was any potentiation found as regards the rate increase by raclopride in low or high doses. Thus, a selective enhancement of phasic, but not tonic DA cell activity was obtained by the drug combination, specifically involving low doses of the DA-D_2-antagonist.

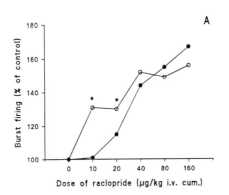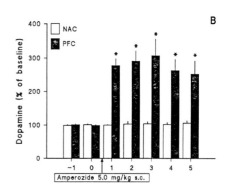

Figure 1. (A) Pretreatment with ritanserin (open circles, 1.0 mg/kg IV, 30 minutes before) selectively enabled an increased burst firing in VTA-DA neurons following administration of low, but not high doses of raclopride (solid circles) (B) Amperozide preferentially increased extracellular concentrations of dopamine in the

prefrontal cortex (PFC) as compared to the nucleus accumbens (NAC).

3.2 MICRODIALYSIS

These experiments evaluated the effects of the atypical neuroleptic APZ on extracellular concentrations of DA in the NAC and the PFC. In previous experiments APZ was found to display high affinity for 5-HT_2-receptors, but only weak affinity for DA-D_2-receptors [23], and there is accumulating behavioral and biochemical evidence that APZ acts preferentially on the limbic system [24]. Moreover, in our previous electrophysiological experiments we found that APZ exerts selective actions on DA cells projecting from the VTA [21]. Our biochemical experiments here show that APZ increases extracellular concentrations of DA in forebrain areas by a dose- and region-dependent manner. Specifically, the 5 mg/kg dose of APZ failed to significantly affect DA release in the NAC while it markedly increased extracellular DA levels in the PFC (by 207%, figure 1B). A higher dose of APZ (10 mg/kg) significantly enhanced DA concentrations in the NAC by 30% and in the PFC by 356%. Thus, these data demonstrate that APZ exerts an almost exclusively preferential action on DA release in the PFC, especially in low doses. Previous microdialysis studies have indicated that acute administration of typical and atypical antipsychotics leads to differential activation of mesotelencephalic DA systems [25]. Clozapine, in particular, seems to increase DA release in PFC much more than in either NAC or striatum. This preferential action of clozapine can probably be attributed to other actions than its DA-receptor blocking properties, since neither haloperidol nor sulpiride display a similar profile. Our recent results further indicate that the preferential effect of APZ on DA transmission in the PFC can not be directly linked to its DA-D_2-receptor antagonistic properties, since the selective DA-D_2-receptor antagonist raclopride equipotently increased DA release in both forebrain structures. Yet, administration of the relatively selective 5-HT_2-receptor antagonist ritanserin produced but a slightly more pronounced effect on DA release in the PFC compared to the NAC. It seems plausible, therefore, that the 5-HT_2-receptor antagonistic properties of APZ may act synergistically, via a dynamic interaction, with its weak DA-D_2-receptor antagonistic effects, to cause a preferential activation of DA neurons in the PFC.

4. Conclusions

Generally, 5-HT_2-antagonists such as ritanserin have been found to cause a slight, probably indirect activation of midbrain DA cell activity concomitant with a modest increase in DA release both in the NAC and the PFC. Such effects may help to explain a general improvement in energy, drive, and motivation. Yet, the classical DA-hypothesis of schizophrenia would for the same reason, if anything, predict worsening of clinical symptomatology, rather than an actual improvement. Our experiments suggest that 5-HT_2-antagonists may help to correct a distorted brain

DA-signalling in psychosis, i.e. to preferentially restore phasic rather than tonic activity of the DA neurons and, hence, improve their signal-to-noise ratio. Experiments, e.g. with APZ, suggest that small degree of DA-D_2-receptor antagonism in conjunction with 5-HT_2-antagonism may help to facilitate selectivity for the PFC DA-projections. Such PFC activation may serve to reduce hypofrontality and, secondarily, improve DA-signalling in the NAC. Thus, this type of agents should restore prefrontal cortical control over motivational processes.

5. Acknowledgements

This work was supported by the Swedish Medical Research Council (project no. 4747), Torsten and Ragnar Söderbergs Stiftelser, the Karolinska Institute and AB LEO's i Helsingborg Stiftelse för Forskning. G.G. Nomikos was supported by the Wenner-Gren Foundation. The skillful technical assistance of Mrs. M. Marcus and A. Malmerfeldt is greatfully recognized.

6. References

1. Reyntjens, A., Gelders, Y.G., Hoppenbrouwers, M.-L.J.A., and Vanden Bussche, G. (1986) 'Thymostenic effects of ritanserin (R 55667), a centrally acting serotonin-S_2 blocker', Drug Dev. Res. 8, 205-211.
2. Gelders, Y., Vanden Bussche, G., Reyntjens, A., and Janssen, P. (1986) 'Serotonin-S_2 receptor blockers in the treatment of chronic schizophrenia', Clin. Neuropharmacol. 9 (Suppl. 4), 325-327.
3. Bersani, G., Grispini, A., Marini, S., Pasini, A., Valducci, M., and Ciani, N. (1986) 'Neuroleptic-induced extrapyramidal side effects: clinical perspectives with ritanserin (R 55667), a new selective 5-HT_2 receptor blocking agent', Curr. Ther. Res. 40, 492-499.
4. Gelders, Y.G. (1989) 'Thymosthenic agents, a novel approach in the treatment of schizophrenia', Br. J. Psychiatry 155, 33-36.
5. Ugedo, L., Grenhoff, J., and Svensson, T.H. (1989) 'Ritanserin, a 5-HT_2 receptor antagonist, activates midbrain dopamine neurons by blocking serotonergic inhibition', Psychopharmacol. 98, 45-50.
6. Guyenet, P.G. and Aghajanian, G.K. (1978) 'Antidromic identification of dopaminergic and other output neurons of the rat substantia nigra', Brain Res. 150, 69-84.
7. Grace, A.A. and Bunney, B.S. (1983) 'Intracellular and extracellular electrophysiology of nigral dopaminergic neurons - 1. Identification and characterization', Neurosci. 10, 301-315.
8. Grace, A.A. and Bunney, B.S. (1984) 'The control of firing pattern in nigral dopamine neurons: burst firing', J. Neurosci. 4, 2877-2890.
9. Pelligrino, L.K., Pelligrino, A.A., and Cushman, A.J. (1979) A stereotaxic atlas of the rat brain, Plenum Press, New York.

10. Paxinos, G. and Watson, G. (1986) The rat brain in Stereotaxic Coordinates, 2nd ed., Academic Press, New York.
11. Nomikos, G.G., Damsma, G., Wenkstern, D., and Fibiger, H.S. (1989) 'Acute effects of bupropion on extracellular dopamine concentrations in rat striatum and nucleus accumbens by in vivo microdialysis', Neuropsychopharmacol. 2, 273-281.
12. Bunney, B.S., Walters, J.R., Roth, R.H., et al. (1973) 'Dopaminergic neurons: effect of antipsychotic drugs and amphetamine on single cell activity', J. Pharmacol. and Exp. Thera. 185, 560-571.
13. Grenhöff, J., Tung, C.-S., and Svensson, T.H. (1988) 'The excitatory amino acid antagonist kynurenate induces pacemaker-like firing of dopamine neurons in rat ventral tegmental area in vivo', Acta Physiol. Scand. 134, 567-568.
14. Gonon, F.G. (1988) 'Nonlinear relationship between impulse flow and dopamine released by rat midbain dopaminergic neurons as studied by *in vivo* electrochemistry', Neurosci. 24, 19-28.
15. Bean, A.J. and Roth, R.H. (1991) 'Extracellular dopamine and neurotensin in rat prefrontal corteivo: Effects of median forebrain bundle stimulation frequency, stimulation pattern, and dopamine autoreceptors', J. Neurosci. 11, (9) 2694-2702.
16. Svensson, T.H. and Tung, C.-S. (1989) 'Local cooling of prefrontal cortex induces pacemaker-like firing of dopamine neurons in rat ventral tegmental area in vivo', Acta Physiol. Scand. 136, 135-136.
17. Bunney, B.S. (1992) 'Clozapine: A hypothesised mechanism for its unique clinical profile', Br. J. Psychiatry 160 (Suppl. 17), 17-21.
18. Nishino, H., Ono, T., Muramoto, K., Fukuda, M., and Sasaki, K. (1987) 'Neuronal activity in the ventral tegmental area (VTA) during motivated bar press feeding in the monkey', Brain Res. 413, 302-313.
19. Murase, S., Mathé, J.M., Grenhoff, J., and Svensson, T.H. (1992) 'Effects of dizocilpine (MK-801) on rat midbrain dopamine cell activity: differential actions on firing pattern related to anatomical localization', J. Neural Transm., in press.
20. Ingvar, D.H. (1987) 'Evidence for frontal/prefrontal cortical dysfunction in chronic schizophrenia: the phenomenon of hypofrontality reconsidered', in H. Helmchen and F.A Henn (eds.), Biological Perspectives of Schizophrenia, John Wiley and Sons, New York, pp. 201-211.
21. Grenhoff, J., Tung, C.-S., Ugedo, L., and Svensson, T.H. (1990) 'Effects of amperozide, a putative antipsychotic drug, on rat midbrain dopamine neurons recorded in vivo', Pharmacol. & Toxicol. 66 (Suppl. 1), 29-33.
22. Pawlowski, L., Mathé, J.M., and Svensson, T.H. (1990) 'Phencyclidine activates rat A10 dopamine neurons but reduces burst activity and causes regularization of firing', Acta Physiol. Scand. 139, 529-530.
23. Meltzer, H.Y., Matsubara, S., and Lee, J.-C. (1989) 'Classification of typical

and atypical antipsychotic drugs on the basis of dopamine D-1, D-2 and serotonin$_2$ pK$_i$ values', J. Pharmacol. Exp. Ther. 251, 238-246.
24. Gustafsson, B. and Christensson, E. (1990) 'Amperozide and emotional behavior', Pharmacol. Toxicol. 66, (Suppl. 1) 34-39.
25. Moghaddam, B. and Bunney, B.S. (1990) 'Acute effects of typical and atypical antipsychotic drugs on the release of dopamine from prefrontal cortex, nucleus accumbens, and striatum of the rat: an in vivo microdialysis study', J. Neurochem. 54, 1755-1760.

COMBINED SEROTONIN 5-HT$_2$ AND DOPAMINE DA$_2$ ANTAGONIST TREATMENTS IN CHRONIC SCHIZOPHRENIA

RICHARD L. BORISON and BRUCE I. DIAMOND
Medical College of Georgia and Augusta VA Medical Center
Department of Psychiatry
1515 Pope Avenue
Augusta, Georgia 30912-3800
USA

ABSTRACT. The availability of agents with specific serotonin 5-HT$_2$ receptor blocking properties has allowed for a more thorough investigation into the contribution of serotonergic mechanisms to schizophrenic psychosis. Data is presented on the clinical actions of risperidone, a specific and potent blocker of both 5-HT$_2$ and DA$_2$ receptors, in schizophrenicin-patients. In comparison to placebo, risperidone is an efficacious antipsychotic with a rapid onset of action, whereas in comparison to haloperidol, it has a quicker onset of therapeutic action, there is a trend for overall therapeutic superiority, and it produces significantly fewer neurological side-effects. These data, and a review of the data of others, suggests that 5-HT$_2$ receptor blockade significantly enhances the therapeutic efficacy of antipsychotic drugs while reducing associated extrapyramidal side-effects.

1. Introduction

The role of serotonin (5-HT) in schizophrenia has been unclear, in part, due to the fact that the ability to test its central pharmacology has been difficult in the absence of specific and potent serotonergic receptor blockers that are available for clinical trials. This obstacle was overcome, and the possible contributory role of serotonin to the schizophrenic psychosis was elucidated when the specific 5-HT$_2$ receptor blocking agent ritanserin was introduced. This drug is not only specific for the 5-HT$_2$ receptor, but is quite potent, with an IC$_{50}$ of 0.30 nM [1].

In clinical trials, ritanserin was used as an adjunct treatment with the specific dopamine-2 (DA$_2$) antagonist, haloperidol. These trials demonstrated that not only did ritanserin improve the antipsychotic efficacy of haloperidol, with particular benefit in the negative symptoms and affective symptoms, but that it also greatly reduced the extrapyramidal side-effects (EPS) produced by haloperidol [2]. As combination treatment is not always practical, ultimately a new compound, risperidone, was synthesized, which combined the 5-HT$_2$ blocking properties of ritanserin and the DA$_2$ blocking properties of halperidol [1]. In this paper we will present individual site data from the first multicenter, double-blind, placebo-controlled trial of risperidone in schizophrenic patients.

2. Methods

The study enrolled male and female patients, diagnosed by DSM-IIIR criteria with schizophrenia, from whom written informed consent had been obtained. Subjects had to be in an acute exacerbation of their illness, and at baseline required a minimum total Brief Psychiatric Rating Scale (BPRS) score of 30, with a score of moderate or greater on at least two of the positive symptom BPRS items (conceptual disorganization, unusual thought content, suspiciousness, and hallucinatory behavior).

At the initial study visit, patient health was monitored via medical and psychiatric histories, a physical examination, clinical laboratory testing, and an electrocardiogram. Subjects with other Axis I DSM-IIIR psychiatric disorders were excluded from entry, as well as excluding patients with clinically significant or unstable concurrent medical illness. Women of child-bearing potential were not allowed entry into the study.

At the screening visit, and at subsequent visits, the major measures for efficacy and side-effects were administered, namely the BPRS, the Scale for the Assessment of Negative Symptoms (SANS), the Clinical Global Improvement (CGI) scale, the Extrapyramidal Side-Effect Rating Scale (ESRS), and the Abnormal Involuntary Movement Scale (AIMS).

All subjects received a three-day placebo washout period (or a two-week washout for depot neuroleptics), prior to being randomized to double-blind study medication treatment with either risperidone, haloperidol, or placebo. during the first 18 days of double-blind treatment, dosage was flexible, and could be increased every two days by increments of 1 mg for risperidone or 2 mg of haloperidol, up to a maximum of 10 tablets daily of either placebo, risperidone (total of 10 mg/day), or haloperidol (total of 20 mg/day). During the subsequent four weeks the dosage was fixed, and on a weekly basis efficacy and side-effects were assessed.

3. Results

There were 36 subjects who participated in the trial. Our results showed that patients receiving haloperidol received an average dose of 18.0 ± 1.44 mg/day, whereas those receiving risperidone were treated with an average dose of 9.67 ± 0.33 mg/day. The efficacy of risperidone was manifest after the first week of treatment, with a significant fall in BPRS score ($p < 0.01$), with a further efficacy demonstrated by the end of the second week of treatment ($p < 0.001$) which was sustained throughout the study. By comparison, haloperidol first produced a significant lowering of BPRS score (p $<$ 0.028) only after the third week of treatment. Although both risperidone and haloperidol were superior to placebo treatment in reducing BPRS score, there was a trend for risperidone to be superior to haloperidol treatment. In using the criterion of a medication responder as a 20% or greater reduction in BPRS, there were greater than double the number of risperidone as haloperidol responders. Similar results were obtained for CGI data. In contrast to these results, our measurement of negative symptoms via the SANS failed to detect any difference between risperidone, haloperidol, or placebo.

When investigating EPS, haloperidol produced a significant increase in EPS in comparison to either placebo or risperidone treatment, whereas there was no statistical difference between risperidone and placebo in the production of EPS.

There were no significant adverse experiences noted with risperidone treatment in regard to vital signs, clinical laboratories, or electrocardiograms.

4. Discussion

The role of 5-HT_2 receptor blockade in the antipsychotic actions of neuroleptic medication remains speculative. It has been suggested that the antiserotonergic actions of clozapine may account for its greater therapeutic action in the treatment of refractory schizophrenia, and putatively its efficacy in treating the negative symptoms of schizophrenia [3]. Apart from these indirect clinical observations, it has otherwise been difficult to establish the exact contribution of serotonergic dysfunction in the clinical picture of schizophrenia. An analysis of postmortem brains of schizophrenics for 5-HT receptors has produced contrasting results, with authors reporting increased 5-HT_{1A} activity [4], whereas others have measured decreased 5-HT_2 receptor density [5,6,7]; however, these findings have not been replicated by other [8,9].

Attempts to ascertain abnormal 5-HT function in schizophrenics has also been investigated via neuroendocrinologic strategies. The most frequent agent used in these studies has been meta-chlorophenylpiperazine (MCPP) which has a complex interaction with various 5-HT receptor subtypes, namely agonistic properties at 5-HT_{1C} and 5-HT_{1D} receptors, and antagonistic properties at 5-HT_2 and 5-HT_3 receptors. In a study of MCPP in schizophrenics, when compared to controls [10], it was found that MCPP (0.35 mg/kg) increased body temperature and BPRS scores in controls, whereas it failed to affect blood pressure and decreased BPRS scores, particularly the positive symptoms, in schizophrenics; actions of MCPP on adrenocorticotropic hormone and prolactin did not differentiate controls from schizophrenics. Other studies have suggested MCPP induced a blunted response of cortisol and prolactin [11], and other studies have reported a worsening [12] or no change in psychosis [13] after the intravenous infusion of MCPP. The greater relative specificity of ritanserin and risperidone for 5-HT_2 receptors makes these agents better pharmacologic tools for determining the influence of serotonin on schizophrenic psychosis.

Initially, risperidone was used in an acute open label treatment of schizophrenics with positive therapeutic results [14]. This study reported that 11/17 psychotic men showed a beneficial response to an average dose of 4.5 mg/day of risperidone. In another open trial of risperidone in 17 psychotic patients [15], a curvilinear dose-response relationship was observed, with decreased efficacy noted in doses greater that 15 mg/day. This study indicated that the negative symptoms of schizophrenia showed particular benefit from risperidone therapy. In another flexible dosing trial [16], with an average dose of 4.6 mg/day, risperidone was shown to be an effective antipsychotic agent in 20 schizophrenic subjects. The largest open trial of risperidone [17], conducted in 83 subjects, in which flexible dosing led to an average dose of 4.3-5.7 mg/day, there was a highly statistically significant improvement in BPRS score with a major improvement in the negative symptom subscale. Risperidone has also been used as an adjunctive treatment with haloperidol, in which compared to placebo, it lowered BPRS and SANS scores significantly [18].

Long-term trials with risperidone have yielded results similar to acute studies. In the first longitudinal study of 111 schizophrenics up to one year [19], the antipsychotic actions of risperidone became manifest after the first two weeks. After this rapid improvement, the rate of therapeutic gain slowed, but nonetheless continued until the seventh month of treatment, at which time it plateaued. The endpoint results of this study showed that 37% of patients had a 75% or greater improvement in BPRS score, whereas 62% had a 50% or greater lessening of scores. In this study efficacy was demonstrated for both the positive and negative symptoms of schizophrenia. Although somewhat less that 20% of patients required antiparkinsonian medication during the trial, in contrast, over one-third had required concomitant antiparkisonian medication while receiving their previous medications. Moreover, by the end of the trial, all EPS had resolved.

In a large multicenter, double-blind, dose-response study of risperidone, it appeared that 6 mg/day produced maximal efficacy, with significant improvement in both positive and negative symptoms, with EPS no greater than that observed with placebo treatment. In contrast, higher doses were less efficacious and EPS showed a proportional linear increase with dose (R. Meibach, Ph.D., personal communication). These results suggest that trials that failed to show efficacy for negative symptoms (such as our own) or differences in EPS may have done so due to inappropriately high doses of risperidone. This dose-response curve also suggests that a maximal balance between sertonergic and dopaminergic blockade is needed for the successful treatment of negative symptoms and the reduction of EPS. At present, it is unclear as to which constellation of neurochemical mechanisms may underlie the psychopathology associated with negative symptoms. In contrast, it appears that 5-HT_2 receptor blockade may lessen DA_2 receptor blockade in striatal tissue [20], thereby accounting for the lessening of EPS observed with risperidone treatment relative to haloperidol.

In conclusion, the potent 5-HT_2 blocking properties of risperidone confer on this compound properties unique from typical neuroleptics, namely in its efficacy in treating the negative symptoms of schizophrenia, and in reducing the incidence of EPS. As such, risperidone appears to present an important advance in the treatment of schizophrenia.

5. References

1. Leysen, J.C., Gommeren, W., Eens, A., dechaffoy de Courcelles, D., Stoof, J.C., and Janssen P.A.J. (1988) 'Biochemical profile of risperidone, a new antipsychotic', J. Pharmacol. Exp. Thera. 247, 661-670.
2. Bersani, G., Grispini, A., Marini, S., Pasini, A., Valducci, M., and Ciani, N. (1986) 'Neuroleptic-induced extrapyramidal side-effects: clinical perspectives with ritanserin (R 55667), a new selective $5\text{-}HT_2$ receptor blocking agent', Curr. Ther. Res. 40, 492-499.
3. Meltzer, H.Y. (1989) 'Clinical studies on the mechanism of action of clozapine: the dopamine-serotonin hypothesis', Psychopharmacology 99, S18-S27.
4. Hashimoto, T., Nishino, N., Nakai, H., and Tanaka C. (1991) 'Increase in serotonin $5\text{-}HT_{1A}$ receptors in prefrontal and temporal cortices of brains from patients with chronic schizophrenia', Life Sciences 48, 355-363.
5. Bennet, J.P., Enna, S.J., Bylund, D., Gillin, J.C., and Snyder, S.H. (1979) 'Neurotransmitter receptors in frontal cortex of schizophrenics', Arch. Gen. Psychiatry 37, 927-934.
6. Mita, T., Hanada, S., Nishino, N., Kuno, T., Nakai, H., Yamadori, T., Mizoi, Y., and Tanaka, C. (1986) 'Decreased serotonin S2 and increased dopamine D2 receptors in chronic schizophrenics', Biol. Psychiatry 21, 1407-1414.
7. Kleinman, J.E., Laurelle, M., Casanova, M.F., Toti, R., Weinberger, D.R., and Abi-Dargham, A. (1991) 'Frontal cortex 5-HT receptors in schizophrenia', Proc. Am. Psych. Assoc., p. 182.
8. Whitaker, P.M., Crow, T.J., and Ferrier, I.N. (1981) 'Triated LSD binding in frontal cortex in schizophrenia', Arch. Gen. Psychiatry 38, 278-280.
9. Reynolds, G.P., Rossor, M.N., and Iverson, J.L. (1983) 'Preliminary studies of human $5\text{-}HT_2$ receptors in their involvement in schizophrenia and neuroleptic drug action', J. Neural Transmission 18 (Suppl.), 278-280.
10. Kahn, R.S., Siever, L.J., Gabriel, S., Amin, F., Stern, R.G., DuMont, K., Apter, S., and Davidson, M. (1992) 'Serotonin function in schizophrenia: Effects of metachlorophenylpiperazine in schizophrenic patients and health subjects', Psychiat. Res. 43, 1-12.
11. Iqbal, N., Asnis, G.M., Wetzler, S., Kahn, R.S., Kay, S., and Van Praag, H.M. (1991) 'The MCPP challenge test in schizophrenia: Hormonal and behavioral responses', Biol. Psychiatry 30, 770-778.
12. Krystal, J.H., Seibyl, J.P., Price, L.P., Woods, S.W., Heninger, G.R., and Charney, D.S. (1991) 'MCPP effects on schizophrenic patients before and

after typical and atypical neuroleptic treatment', Schiz. Res. 4, 350.
13. Owen, R.R., Gutierrez, L., Hadd, K.I., Benkelfat, C., and Murphy, D.L. (1990) 'Serotonergic responsibility in schizophrenia', New Research Abstracts, Proc. Am. Psych. Assoc., NR235.
14. Roose, K., Gelders, Y., and Heylen, S. (1988) 'Risperidone (R 64766) in psychotic patients', Acta psychiat. belg. 88, 233-241.
15. Mesotten, F., Pietquin, M., Burton, P., Heylen, S., and Gelders, Y. (1989) 'Therapeutic effect and safety of increasing doses of risperidone (R 64766) in psychotic patients', Psychopharmacology 99, 445-449.
16. Castelao, J.R., Ferreira, L., Gelders, Y.G., and Heylen, S.L.E. (1989) 'The efficacy of the D-2 and 5-HT_2 antagonists risperidone (R 64766) in the treatment of chronic psychosis', Schizophrenia Research 2, 411-415.
17. Ishigooka, J. (1991) 'Clinical experience with risperidone in Japan - a multicentre early phase II trial', in J.M. Kane (ed.), Risperidone: Major Progress in Antipsychotic Treatment, Oxford Clinical Communications, Oxford, pp. 40-43.
18. Meco, G., Bedini, L., Bonifati, V., and Sonsini, U. (1989) 'Risperidone in the treatment of chronic schizophrenia with tardive dyskinesia', Curr. Ther. Res. 46, 876-884.
19. Mertens, C. (1991) 'Long-term treatment of chronic schizophrenic patients with risperidone', in J.M. Kane (ed.), Risperidone: Major Progress in Antipsychotic Treatment, Oxford Clinical Communications, Oxford, pp. 44-48.
20. Saller, C.F., Czupryna, M.J., and Salama A.I. (1990) '5-HT_2-receptor blockade by ICI 169,369 and other 5-HT_2 antagonists modulates the effects of D2-dopamine receptor blockade,' J. Pharmacol. Exp. Ther. 253, 1162-1170.

IS SEROTONIN INVOLVED IN THE PATHOGENESIS OF SCHIZOPHRENIA?

H.M. VAN PRAAG
Professor and Chairman
Academic Psychiatric Center
University of Limburg
Maastricht, P. O. Box 1918
The Netherlands

ABSTRACT. Though one of the first hypotheses in modern biological psychiatry proposed a relationship between serotonin (5-hydroxytryptamine; 5-HT) dysfunction in the brain and schizophrenia, work in this field was kept simmering once the neuroleptics with their strong impact on dopaminergic functioning were introduced. Recently, however, evidence has been found that blockage of certain populations of 5-HT receptors may result in antipsychotic effects. Ritanserin, a 5-HT$_2$ receptor blocker was reported to possess antipsychotic properties. Risperidone and clozapine, combined 5-HT$_2$ and dopamine (DA)2 receptor blockers, are probably superior to classical neuroleptics. The antipsychotic potential of 5-HT$_3$ receptor antagonists was inferred from animal experiments and the first human data are encouraging. These findings justify the question whether in schizophrenic patients 5-HT disturbances do occur.

1. Early Hypotheses on Serotonin and Schizophrenia

The dawn of biological psychiatry is marked by the hypothesis of Woolley and Shaw [1] launched in the early fifties that the then just discovered hallucinogenic effect of LSD was related to the drug's ability to bind to serotonin (5-hydroxytryptamine, 5-HT) receptors, thus causing either excess of attenuated activity in 5-HTergic systems. Characteristic of the LSD psychosis is that severe disturbances in perception and cognition occur while consciousness remains clear. In this respect the LSD psychosis is unlike other toxic psychosis and resembles the group of the schizophrenic psychoses. Accordingly, the LSD/5-HT hypothesis was widened to include the notion that 5-HTergic dysfunctions might be involved in the pathogenesis of schizophrenia.

These hypothesis did not receive the attention they deserved for two reasons. First of all, LSD fell into discredit and was for all intents and purposes barred from both biological experimentation and studies exploring its value as a catalyst of explorative psychotherapies. Scientific considerations played a lesser role in the promulgation of this ban than political and emotional ones did. In the cultural revolution that swept Western Europe and the United States in the mid-1960s and 1970s, LSD became a symbol of the flower-power generation dreaded and shunned by the establishment at that time.

A second reason why the LSD/schizophrenia 5-HT-hypothesis was prematurely discarded, was the discovery that drugs with antipsychotic activity share the ability to block dopaminergic (DAergic) activity in the brain. Thus it came to pass that DA for many years firmly held center stage in biological schizophrenia research.

2. Revival of Interest for Serotonin in Schizophrenia

A few years ago, however, the interest in the possible role of 5-HT in schizophrenia surged, mainly due to a number of psychopharmacological discoveries. Ritanserin, an antagonist of the 5-HT_2 receptor and the closely linked 5-HT_{1C} receptor, was reported to have a therapeutic effect in schizophrenia, in particular against negative symptoms [2]. Moreover, clozapine was reported to have a unique therapeutic profile in chronic schizophrenia, showing efficacy in patients refractory for classical neuroleptics [3]. Clozapine is a strong antagonist of 5-HT_2 and 5-HT_{1C} receptors and to a lesser extent of 5-HT_3 receptors. Possibly, the ratio of DA/5-HT receptor blockage is linked to its unusual therapeutic profile [4] though there are other explanations as well [5,6]. Risperidone, likewise a substance with pronounced DA_2 and 5-HT_2 receptor antagonism, is also showing powerful therapeutic effects in chronic schizophrenia with negative symptoms [7]. Finally, drugs were developed that block 5-HT_3 receptors selectively and animal data identified them as potential antipsychotics [8]. Controlled clinical trials have not yet unambiguously demonstrated that that prediction bears out, nor did uncontrolled trials (e.g. [9]).

Conversely, a number of drugs with hallucinogenic potential, most notably LSD, bind strongly to 5-HT_2 and 5-HT_{1C} receptors; this effect highly correlates with their psychotogenic potency. In rats, these drugs induce the so-called 5-HT syndrome, a syndrome that is characteristic for drugs that mimic the action of 5-HT. This syndrome can be blocked by 5-HT_2 receptor antagonists [10].

3. Central Serotonin Disturbances in Schizophrenics

Assuming that blockade of certain subpopulations of 5-HT receptors reduces psychotic symptomatology and that, conversely, activating those receptors will induce or aggravate psychotic symptomatology, one may raise the question whether in schizophrenia 5-HT functioning is disturbed and if so, with what psychopathological dimensions are the 5-HT disturbances, if any, correlated. Elsewhere we reviewed the entire literature [11,13]. Here I restrict myself mainly to a brief discussion of the data we collected over the years, using three strategies considered to provide the most direct information on 5-HT in the living human brain, i.e.:

1. Measurement of the 5-hydroxyindoleacetic acid (5-HIAA)

concentration in cerebrospinal fluid (CSF) before and after administration of probenecid. 5-HIAA is the major degradation product of 5-HT. Probenecid blocks the efflux of 5-HIAA from the central nervous system (CNS), including the CSF, to the blood stream. The probenecid-induced accumulation of CSF 5-HIAA is a crude indicator of the metabolism of 5-HT in the CNS [14].

    2.   Manipulation of central 5-HT and subsequent study of the behavioral effects in schizophrenic patients and normals.

    3.   Challenge test with compounds that exert a more or less selective effect on central 5-HT systems.

3.1   CSF 5-HIAA MEASUREMENT

The data on baseline and postprobenecid concentrations of 5-HIAA in CSF are inconsistent. Both increased, decreased, and normal values have been reported [12]. Our data are negative [15,16], in that we found no consistent abnormalities, in whatever way we classified the group of 68 patients with schizophrenic psychoses we studied. They were first-break patients (n=29) and patients who had suffered 1-5 psychotic episodes in the past but who were able to maintain themselves in society between episodes. They were diagnosed as suffering from acute and chronically relapsing schizophrenia.

Comparing the CSF 5-HIAA data in the patient group versus controls (figure 1) and in the patient group subdivided according to DSM-III-R criteria (figure 2), according to the Kraepelinian subgroupings (figure 3), and according to the first-break/multiple-break criterium (figure 4), no significant subgroup differences were established.

In chronic schizophrenia several authors found a relation between lowered CSF 5-HIAA and enlarged cerebral ventricles, an observation that possibly reflects loss of 5-HTergic cells [17]. The lowered concentration of CSF-5-HIAA found in suicidal patients with depression [18,19] and personality disorders [20], moreover, was also observed in suicidal schizophrenic patients [21-24]. This observation reflects probably the relation that exists between 5-HT dysfunctions and disturbed aggression (or impulse) regulation, rather than being specific for schizophrenia [25,26].

In summary, the CSF 5-HIAA studies have so far failed to reveal evidence for disturbances in central 5-HT metabolism, related to the schizophrenic psychosis as such.

3.2   MANIPULATION OF CENTRAL 5-HT

There is little data on the behavioral effects of increasing the availability of central 5-HT in schizophrenia. Massive increases probably exert psychotogenic effects. Ad-

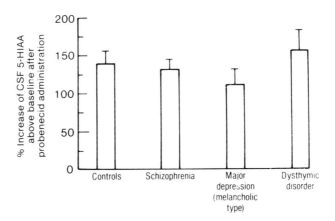

Figure 1. Postprobenecid CSF 5-HIAA in schizophrenic patients compared with other diagnostic categories.

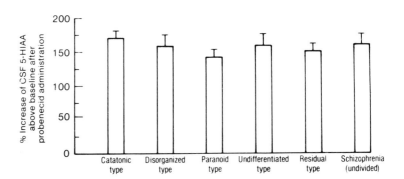

Figure 2. Postprobenecid CSF 5-HIAA in schizophrenic patients subdivided according to DSM-III-R criteria.

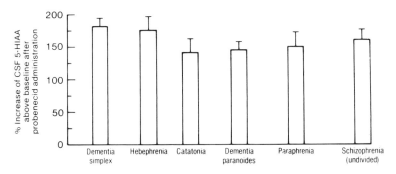

Figure 3. Postprobenecid CSF 5-HIAA in schizophrenic patients subdivided according to Kraepelinian criteria.

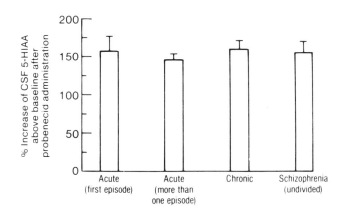

Figure 4. Postprobenecid CSF 5-HIAA in schizophrenic patients subdivided according to the first-break/multiple-break criterium.

Administration of the 5-HT precursor, 5-hydroxytryptophan (5-HTP), in combination with a peripheral decarboxylase inhibitor made the condition of schizophrenic patients worse [27,28]. I found that the combination of the monoamineoxidase inhibitor iproniazid (150 mg dd for two weeks) and the 5-HT precursor 1-tryptophan (in a single oral dose of 5 g) led to transient LSD-like psychotic phenomena in 7 out of 8 normal volunteers [29].

The selective 5-HT uptake inhibitors, though they have not been systematically studied in schizophrenic psychoses, have only in sporadic cases been reported to induce or enhance psychotic phenomena [30,31]. The data on the effect of the 5-HT releaser fenfluramine in schizophrenic psychoses are inconsistent [11]. However, the dl-form was used, and the racemic mixture of fenfluramine is not 5-HT-selective but increases the availability of 5-HT as well as noradrenalin in the brain [32].

In summary, the available data suggest that massive increases of 5-HT availability in the brain exerts psychogenic effects, while a more modest overall increase is generally tolerated without compromising the stability of the psychic apparatus.

3.3 HORMONAL CHALLENGE TESTS

In this domain, too, data are scarce and controversial. Until recently, however, the challengers used, i.e. the 5-HT precursors 1-tryptophan and 1-5-HTP, and the 5-HT releaser fenfluramine, were not 5-HT selective. These substances exert a pronounced effect on catecholaminergic systems as well [25]. Our own data using 1-tryptophan (figure 5) and 1-5-HTP (figure 6) are negative. We could not ascertain any differences in 5-HT receptor sensitivity between (subgroups of) schizophrenic patients and controls [16]. Recently a more selective challenger of 5-HT receptors has been introduced, i.e. m-chlorophenylpiperazine (MCPP). MCPP binds onto all 5-HT receptors, though in the human brain it has a certain preference for the $5-HT_{1C}$ and the $5-HT_3$ subtypes [33].

Iqbal et al. [34,35] reported about a study of 7 chronically relapsing, though still functional, schizophrenics and 8 normal controls. They were challenged with 0.25 mg/kg MCPP orally or with placebo. This dose of MCPP is the lowest effective one in releasing prolactine and ACTH/cortisol, hormones the release of which is under 5-HTergic control. The patients had been drug-free for at least 2 weeks and their behavior was assessed with the positive and negative symptom scale (PANSS) [36].

The major findings were twofold. First we observed that the patients, after a single MCPP administration showed a significant worsening of the Total Psychopathology Scale (i.e. the sum of all PANSS symptoms), due to an increase in positive symptoms and in excitement and tension. MCPP is an anxiogenic substance [37,38] and it is as yet unknown whether the increase in psychotic symptoms is a direct effect of MCPP or one mediated via a primary increase in anxiety.

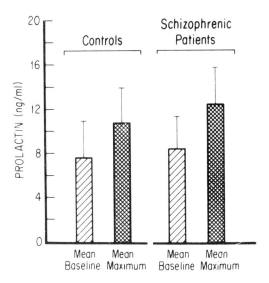

Figure 5. Prolactin response to an oral 1-tryptophan load (5 g) in a group of drug-free patients with relapsing schizophrenic psychoses.

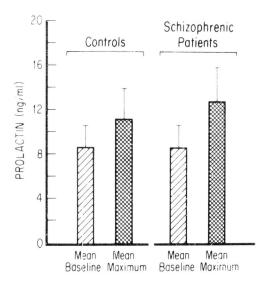

Figure 6. Prolactin response to an oral load of 1-5-hydroxytryptophan (200 mg) after 3-days treatment with the peripheral decarboxylase inhibitor carbidopa in a group of drug-free patients with relapsing schizophrenic psychoses.

Secondly, 5-HT receptor sensitivity differed in schizophrenic patients as compared to controls. In the former group the placebo-corrected increase in plasma cortisol, usually seen after MCPP, was blunted. This phenomenon is suggestive of hyposensitivity of (certain) 5-HT receptors. The behavioral effects, on the other hand suggest 5-HT receptor hypersensitivity.

4. Multiple 5-HT Receptor Disturbances in Schizophrenia?

This dissociation of behavioral and hormonal effects of MCPP implies that the alleged 5-HT disturbance in schizophrenia involves multiple receptor systems, some being hyposensitive, some hypersensitive. Since MCPP binds to all 5-HT receptor subsystems, this drug is well-suited to reveal a multiplicity of 5-HT receptor disturbances.

Evidence supporting the concept of multiple, though opposite, 5-HT receptor dysfunctions is provided by the Kobe group in Japan. In the prefrontal and temporal cortex of schizophrenic patients they found an increased number of 5-HT_{1A} binding sites [39], combined with a decrease in number of 5-HT_2 binding sites [40].

Recently [38] reported observations also suggestive of a complex 5-HT disturbance in schizophrenia. MCPP raised body temperature in control subjects, but not in patients. Hormonal responses to MCPP did not differ between patients and controls, while psychotic symptoms decreased after the MCPP challenge. The type of patients they studied was different from our research subjects. Our patients could still maintain themselves in society. The subjects of Kahn et al. [38] had been chronically ill and had long histories of institutionalization. Another study confirmed our finding that MCPP worsens psychotic symptoms in schizophrenia [41].

5. Are the 5-HT Disturbances Specific for Schizophrenia?

A dissociation of hormonal and behavioral responses to MCPP similar to the one we found in schizophrenia, Zohar et al. [42] observed in obsessive compulsive disorder (OCD). After a MCPP challenge psychopathological symptoms worsened, in particular obsessions, while the hormonal response to MCPP was blunted. The similarity of MCPP responses in schizophrenia and OCD, suggests that the MCPP findings are not specific to either one of the nosological entities but rather to a psychopathological phenomenon that is common to both disorders. In other words, I hypothesize that the observed 5-HT disturbance is not syndromally or nosologically but functionally specific. The functional approach, searching for specific relations between biological dysfunctions and psychological dysfunctions has been much more fruitful than the search for "markers" or causes of psychopathological syndromes or psychiatric disorders [13,26,32,43].

A possible common psychopathological denominator for the 5-HT disturbances in schizophrenia and OCD could be sought in the domain of thought disorders. A central feature of OCD is the intrusion of senseless ideas increasingly preoccupying the patient who is however unable to eliminate or correct them. The senseless idea may become an overvalued idea, with a concomitant decrease of the realization of the absurdity of the thought. Overvalued ideas are close relatives of delusions. A delusion can be defined as an overvalued idea, combined with a complete loss of critical function, that is the ability to test the probability of an idea and to drop it if it is in disparity with perceived reality or with previous experiences.

A common cognitive disturbance in schizophrenia and OCD is the inability to correct and thus interrupt reasonable beliefs that enter consciousness. I submit the hypothesis that this cognitive disturbance is the behavioral correlate of 5-HT disturbances observed in both conditions; a hypothesis that is speculative but testable.

6. Summary and Conclusions

Suppression of certain 5-HTergic circuits seems to exert a therapeutic effect in schizophrenic psychoses. There is some evidence that in particular the negative symptom cluster is responsive but the observation needs confirmation. To prove this contention studies are needed in which selective 5-HT receptor blockers are compared with selective DA receptor blockers as to their effects on the spectrum of schizophrenic phenomena. It is unknown whether the alleged antipsychotic effects of 5-HT receptor blockers is related to a direct effect on the 5-HT system or is mediated via alterations in the DAergic system.

Until recently, few data supported the notion that 5-HTergic dysregulation is involved in the pathogenesis of schizophrenic psychoses. Challenge tests with the relative selective 5-HT receptor agonist MCPP revealed for the first time tentative evidence of the existence of central 5-HT dysfunctions in schizophrenia. It was hypothesized that the 5-HT disturbances observed in schizophrenia relate to particular cognitive disturbances occurring in schizophrenia but not specific for that disorder. In other words, the 5-HT disturbances might be functionally specific while being nonspecific on the syndromal and nosological level.

7. References

1. Woolley, D.W. and Shaw, E. (1954) 'A biochemical and pharmacological suggestion about certain mental disorders. Proceedings of the National Academy of Sciences 40, 228-231.
2. Gelders, Y., Ceulemans, D.L.S., Hoppenbrouwers, M.L., Reyntjens, A., and Mesotten, F. (1985) 'Ritanserin, a selective serotonin antagonist in chronic schizophrenia', Proceedings of the Fourth World Congress of Biological Psychiatry, Philadelphia, p. 338.
3. Kane, J., Honigfield, G., Singer, J., Meltzer, H., and Clozaril Collaborative

Study Group (1988) 'Clozapine for the treatment-resistant schizophrenic', Arch. Gen. Psych. 45, 798-796.
4. Meltzer, H.Y., Matsubara, S., and Lee, J.C. (1989) 'Classification of typical and atypical antipsychotic drugs on the basis of dopamine D-1, D-2 and serotonin 2 pk_i values', J. Pharm. Exper. Ther. 251, 238-246.
5. van Praag, H.M. (1992) 'Serotonergic mechanisms in the pathogenesis of schizophrenia', in J.P. Lindemayer and S.R. Kay (eds.), New Biological Vistas on Schizophrenia, Brunner Mazel, New York.
6. Chen, J., Paredes, W., van Praag, H.M., and Gardner, E.L. (1992) 'Serotonin denervation enhances responsiveness of presynaptic dopamine efflux to acute clozapine in nucleus accumbens but not in caudate-putamen', Brain Res., 582, 173-178.
7. Claus, A., Bollen, J., De Cuyper, H., Eneman, M., Malfroid, M., Peuskens, J., and Heylen, S. (1992) 'Risperidone versus haloperidol in the treatment of chronic schizophrenic in patients: a multicentre double-blind comparative study' Acta Psych. Scand. 85, 295-305.
8. Costall, B., Domeney, A.M., Naylor, R.J., and Tyers, M.B. (1987) 'Effects of the 5-HT3 receptor antagonist, GR38032F, on raised dopaminergic activity in the mesolimbic system of the rat and marmoset brain', Br. J. Pharm. 92, 881-894.
9. Newcombes, J.W., Faustman, W.J., Zipursky, R.B., and Csernansky, J.G. (1992) 'Zacopride in schizophrenia: a fragile blood serotonin type 3 antagonist trial', Arch. Gen. Psych. 49, 751.
10. Jacobs, B.L. (1987) 'How hallucinogenic drugs work', American Scientist 75, 386-392.
11. Bleich, A., Brown, S.L., Kahn, R.S., and van Praag, H.M. (1988) 'The role of serotonin in schizophrenia', Schizophrenia Bull. 14, 297-315.
12. Bleich, A., Brown, S.L., and van Praag, H.M. (1991) 'A serotonergic theory of schizophrenia', in S.L. Brown and H.M. van Praag (eds.), The role of serotonin in psychiatric disorders, Brunner Mazel, New York, pp. 183-214.
13. van Praag, H.M. (1992) Make Believes in Psychiatry or The Perils of Progress, Brunner Mazel, New York.
14. Korf, J. and van Praag, H.M. (1971) 'Amine metabolism in de human brain: further evaluation of the probenecid test', Brian Res. 35, 221-230.
15. van Praag, H.M. (1977) 'The significance of dopamine for the mode of action of neuroleptics and the pathogenesis of schizophrenia', Br. J. Psych. 130, 463-474.
16. van Praag, H.M. (1982) 'Neurotransmitters and schizophrenia. Part A: Catecholamines and schizophrenia', in P.J.V. Beumont and G.D. Burrows (eds.), Handbook of Psychiatry and Endocrinology, Elsevier, Amsterdam.
17. Potkin, S.G., Weinberger, D.R., Linnoila, M., and Wyatt, R.J. (1983) 'Low CSF 5-hydroxyindoleacetic acid in schizophrenic patients with enlarged cerebral ventricles', Am. J. Psych. 140, 21-25.

18. Asberg, M., Traskman, L., and Thoren, P. (1976) '5-HIAA in the cerebrospinal fluid: a biochemical suicide predictor?', Arch. Gen. Psych. 33, 1193-1197.
19. van Praag, H.M. (1982) 'Depression, suicide and the metabolism of serotonin in the brian', J. Affective Disorders 4, 275-290.
20. Traskman, L., Asberg, M., Bertilsson, L., and Sgostrand, L. (1981) 'Monoamine metabolites in CSF and suicidal behavior', Arch. Gen. Psych. 38, 631-636.
21. van Praag, H.M. (1983) 'CSF 5-HIAA and suicide in non-depressed schizophrenics', Lancet 2, 977-978.
22. Ninan, P.T., van Kammen, D.P., Scheinin, M. Linnoila, M., Bunney Jr., W.E., and Goodwin, F.K. (1984) 'CSF 5-hydroxyindoleacetic acid levels in suicidal schizophrenic patients', Am. J. Psych. 141, 566-569.
23. Banki, C.M., Arato, M. Papp, Z., and Kurcz, M. (1984) 'Cerebrospinal fluid amine metabolites and neuroendocrine charges in psychoses and suicide', in E. Usdin, A. Carlsson and A. Dahlstrom (eds.), Catecholamines: Neuropharmacology and central nervous system-therapeutic aspects, Alan Liss, New York, pp. 153-159.
24. Cooper, S.J. Kelly, C.B., and King, D.J. (1992) '5-hydroxyindoleacetic acid in cerebrospinal fluid and prediction of suicidal behavior in schizophrenia', Lancet 340, 940-941.
25. van Praag, H.M., Lemus, C., and Kahn, R. (1987) 'Hormonal probes of central serotonergic activity. Do they really exist?', Biol. Psych. 22, 86-98.
26. van Praag, H.M., Asnis, G.M., Brown, S.L., and Korn, M. (1990) 'Monoamines and abnormal behavior. A multidimensional perspective, Br. J. Psych. 157 723-734.
27. Wyatt, R.J., Vaughan, T., Galanter, M., Kaplan, J., and Green, R. (1972) 'Behavioral changes of chronic schizophrenic patients given 1-hydroxytryptophan', Science 177, 1124-1126.
28. Bigelow, L., Walls, P., Gillin, J.C., and Wyatt, R.J. (1979) 'Clinical effects of 1-5-Hydroxytryptophan in chronic schizophrenic patients', Biol. Psych. 14, 53-67.
29. Praag, van H.M. (1962), 'A critical investigation of the importance of monoamine oxidase inhibition as a therapeutic principle in the treatment of depression' (Dutch), Thesis, Utrecht.
30. Mandalos, G.E. and Szarek, B.L. (1990) 'Dose-related paranoid reaction associated with fluxotine', Journal of Nervous and Mental Disease 178, 57-58.
31. Lindenmayer, J.P., Vakharia, M., and Kanofsky, D. (1990) 'Fluoxetine in chronic schizophrenia', Clinical Psychopharm. 10, 76.
32. van Praag, H.M., Kahn, R., Asnis, G.M., Wetzler, S., Brown, S., Bleich, A., and Korn, M. (1987) 'Denosologization of biological psychiatry or the specificity of 5-HT disturbances in psychiatric disorders', Journal of Affective Disorders 13, 108.

33. Glennon, R.A., Ismael, A., McCarthy, B., and Peroudka, S.J. (1989) 'Binding of azylpiperazines to 5-HT$_3$ serotonin receptors: results of a structure-affinity study', Eur. J. Pharmacol. 168, 387-392.
34. Iqbal, N., Asnis, G.M., Wetzler, S., Kahn, R.S., Kay, S.R., and van Praag, H.M. (1991) 'The role of serotonin in schizophrenia: new findings', Schizophr. Res. 5, 515-519.
35. Iqbal, N., Asnis, G.M., Wetzler, S., Kahn, R.S., Kay, S.R., and van Praag, H.M. (1991), 'The MCPP challenge test in schizophrenia: Hormonal and behavioral responses', Biological Psychiatry 30, 770-78.
36. Kay, S.R. (1991) 'Positive and negative syndromes in schizophrenia. Assessment and research. Brunner Mazel, New York.
37. Charney, D.S., Words, S.W., Goodman, W.K., and Henniger, G.R. (1987) 'Serotonin function in anxiety. II Effects of the serotonin agonist MCPP in panic disorder patients and healthy subjects', Psychopharmacology 92, 14-24.
38. Kahn, R.S., Siever, L.J., Gabriel, S., Amin, F., Stern, R.J., Dumont, K., Apter, S., and Davidson, M. (1992) 'Serotonin function in schizophrenia: effects of metachlorophenylpiperazine in schizophrenic patients and healthy subjects', Psychiat. Res. 43, 1-12.
39. Hashimoto, T., Nishino, N., Nakai, H., and Tanaka, C. (1991) 'Increase in serotonin 5-HT1A receptors in prefrontal and temporal cortices of brains from patients with chronic schizophrenia', Life Sciences 48, 355-363.
40. Mita, T., Hanada, S., Nishino, N., Kuno, T., Nakai, H., Yamadori, T., Mizoi, Y., and Tanaka, C. (1986) 'Decreased serotonin (S$_2$) and increased dopamine (D$_2$) receptors in chronic schizophrenics', Biological Psychiatry 21, 1407-1414.
41. Krystal, J.H., Seibyl, J.P., Price, L.P., Woods, S.W., Henniger, G.R., and Charney, D.S. (1991) 'MCPP effects in schizophrenic patients before and after typical and atypical neuroleptic treatment', Schizophrenia Res. 4, 350.
42. Zohar, J., Mueller, E.A., Insel, T.R., Zohar-Kadouch, R.C., and Murphy, D.L. (1987) 'Serotonergic responsivity in obsessive-compulsive disorder', Archives of General Psychiatry 44, 946-951.
43. van Praag H.M., Korf, J., Lakke, J.P.W.F., and Schut, T. (1975) 'Dopamine metabolism in depression, psychoses and parkinson's disease: The problem of the specificity of biological variables in behavior disorders', Psychol. Med. 5, 138-146.

SEROTONIN ANTAGONISTS AS ANTIPSYCHOTICS

BECKY L. KINKEAD, MICHAEL J. OWENS, and
CHARLES B. NEMEROFF
Laboratory of Neuropsychopharmacology
Department of Psychiatry and Behavioral Sciences
Emory University School of Medicine
Atlanta, Georgia 30322
USA

ABSTRACT. Recent evidence has indicated that antagonism of serotonin receptor subtypes may, in part, be responsible for the atypical profile of some antipsychotic drugs. In the past, our lab has investigated the neurobiology of the tridecapeptide neurotransmitter neurotensin (NT). NT has been implicated in the mechanism of action of antipsychotic drugs and the pathophysiology of schizophrenia. Of particular interest is the finding that acute and chronic treatment with clinically efficacious typical antipsychotic drugs selectively increase NT concentrations in the nucleus accumbens and caudate nucleus of rats, while treatment with the atypical antipsychotic clozapine increases NT concentrations in the nucleus accumbens only. Recently, our lab has examined the effects of the putative atypical antipsychotic, sertindole. Sertindole is a potent antagonist of 5-HT_2, D_2, and α_1-adrenoreceptors and has a high 5-HT_2/D_2 ratio similar to a number of other atypical antipsychotic drugs. Sertindole mimicked the changes in NT concentration produced by the atypical antipsychotic clozapine. These findings support the hypothesis that sertindole may be an atypical antipsychotic and provide further evidence for the role of serotonin receptor antagonism in the mechanism of action of atypical antipsychotic drugs.

1. Background

The widely held dopamine (DA) hypothesis of the pathophysiology of schizophrenia is based on two lines of evidence: 1) almost all known antipsychotic drugs, haloperidol, chlorpromazine, etc., are DA receptor antagonists and 2) DA agonists, drugs that increase the synaptic availability of DA, such as d-amphetamine can produce psychosis indistinguishable from paranoid schizophrenia. Indeed, the ability of neuroleptic drugs to produce an antipsychotic action, as well as extrapyramidal symptoms (EPS), has been mainly attributed to their ability to block the DA D_2 receptor subtype in the mesolimbocortical and nigrostriatal dopamine systems, respectively. Although the inhibitory effects of these neuroleptics on classical D_2 receptors appears to correlate very well with clinical antipsychotic potency, this neurochemical effect alone cannot explain all the clinically relevant differences between these drugs.

The main impetus for changing our thinking about how antipsychotic drugs act comes from experience with clozapine. Clozapine is a dibenzodiazepine which was synthesized in the early 1960s based upon its structural similarity to tricyclic antidepressants. Early trials with clozapine demonstrated its efficacy for both positive and negative symptoms of schizophrenia, as well as its utility in treatment-resistant schizophrenia. It produced little or no EPS and tardive dyskinesia. Unlike other "typical" antipsychotics, clozapine was found to be a relatively weak D_2 antagonist but possessed high affinity for 5-HT_2, histamine, muscarinic and α-adrenergic receptors. The inability to completely differentiate typical from atypical antipsychotics on the basis of D_2 binding has been interpreted to indicate that D_2 antagonism alone is insufficient to account for the atypical profile of clozapine, and that another pharmacological property in combination with D_2 antagonism must subserve these differences.

Several hypotheses have been promulgated to explain the differences between typical and atypical antipsychotics. One hypothesis postulates that clozapine may bind to different D_2 receptor isoforms and/or other DA receptors subtypes, which then results in an atypical profile. Clozapine is known to bind with higher affinity to D_1 and D_4 receptors than it does to D_2 receptors; in contrast, haloperidol and other typical antipsychotics have a higher binding affinity for D_2 receptors.

Others have suggested that clozapine may preferentially bind to particular brain regions, different from the pattern of binding produced by typical antipsychotics. Indeed, there is evidence from positron-emission tomography (PET) studies that at clinically relevant doses, clozapine occupies only 40-50% of D_2 receptors in the striatum whereas 80-90% D_2 occupancy is observed in limbic areas [1].

In recent years, a third hypothesis has been put forth: that clozapine's atypical action may be due at least in part, to its action at other neurotransmitter receptors such as certain 5-HT receptor subtypes. The involvment of serotonin (5-HT) neural circuits in the mechanism of action of atypical antipsychotic drugs was postulated partly because of the fact that 5-HT is known to exert a regulatory action on DA neurons. Relatively recent neurochemical studies suggest that 5-HT projections tonically inhibit mesolimbic and nigroneostriatal dopaminergic activity. Moreover, 5-HT may directly inhibit DA release from striatal nerve terminals [2,3]. These findings led to the hypothesis that 5-HT_2 antagonists might decrease the inhibition of DA activity produced by chronic neuroleptics, functionally increasing DA activity in the striatum, cortical and limbic areas somewhat above the level produced during neuroleptic-induced depolarization block. This suggests that there may be relative differences in the optimal magnitude of dopaminergic blockade needed in different brain areas. Moreover, complete dopaminergic blockade may not be beneficial for negative symptoms and may result in EPS.

It has been reported that the 5-HT_2 antagonist, ritanserin, when coadministered with typical neuroleptics decreases EPS [4,5,6]. Also, risperidone, which is both a 5-HT_2 and D_2 antagonist is effective in treating both the positive and negative symptoms of schizophrenia as well as producing nominal EPS [7].

Meltzer and colleagues [8] have examined the receptor binding profile of a large series of antipsychotic drugs. They noted that "typical" and "atypical" antipsychotics can be distinguished on the basis of lower D_2 and higher 5-HT_2 pK_i values of the atypical compounds. Absolute potency at the 5-HT_2 receptor alone is not the determining factor, but atypical antipsychotics appear to have 5-HT_2/D_2 pK_i ratios >1.1. Likewise, Seeman and colleagues [9] have reported that those antipsychotics with less propensity to cause rigidity have higher 5-HT_2/D_2 ratios. 5-HT_2 blockade, however, cannot account for the atypical actions of remoxipride, raclopride, or thioridazine, because their 5-HT_2/D_2 blocking profile is similar to that of the classical antipsychotics.

It has been suggested that 5-HT_2 antagonists be co-administered with a typical antipsychotic (D_2 antagonist) to determine if this would produce a clozapine-like profile. However, even low doses of the typical neuroleptics produce 80-90% D_2 receptor blockade which makes it difficult to produce the relatively greater 5-HT_2 to D_2 receptor occupancy rates as that observed with clozapine (40-50% D_2 occupancy).

In addition to the 5-HT_2 receptor, there is evidence that other serotonin receptor subtypes may play a role in the action of atypical antipsychotics. The 5-HT_{1C} receptor has been implicated in the mechanism of action of atypical antipsychotics because clozapine has a higher affinity for 5-HT_{1C} receptors than for DA D_2 receptors [10]. Also, *in situ* hybridization studies have shown that there are high densities of 5-HT_{1C} messenger RNA in the substantia nigra and nucleus accumbens, while the striatum and ventral tegmental area have lower densities [11]. However, chlorpromazine has a higher affinity for 5-HT_{1C} receptors than D_2 receptors. Indeed, most typical and atypical antipsychotics bind weakly to 5-HT_{1C} receptors labeled with [^{3}H]-mesulergine [10].

5-HT_3 antagonists have also been proposed as potential antipsychotics. Unlike other 5-HT receptors, the 5-HT_3 receptor is a component of a ligand-gated cation channel. Following binding, 5-HT rapidly increases membrane sodium and potassium conductance resulting in depolarization. 5-HT_3 receptors are found in the area postrema where they are likely responsible for the antiemetic properties of these compounds. Receptors are also found in the nucleus of the solitary tract, entorhinal cortex, and limbic regions [12]. Activation of these receptors leads to an increase in DA release in the mesolimbic and perhaps the nigrostriatal pathway. It is also known that clozapine binds with moderate affinity ($K_d \approx 100nM$) to the 5-HT_3 receptor while the typical antipsychotics spiperone, haloperidol, fluphenazine, and (-)sulpiride are inactive [13,14]. However, loxapine, which is also a typical antipsychotic, binds to 5-HT_3 receptors with the same affinity as clozapine [15]. There is also some evidence, that the human nucleus accumbens and striatum may not have 5-HT$_3$ receptors, indicating that 5-HT$_3$ binding may be irrelevant to the mechanism of action of atypical antipsychotic drugs [16]. In a recent review article by Reynolds [17], he concluded that in several clinical trials the 5-HT$_3$ antagonist, ondansetron, was not clinically effective in the treatment of schizophrenia. However, this may have been due to the nature of the experimental design of these studies, and pharmacokinetic properties of the drug more than a lack of intrinsic efficacy.

Conflicting evidence has also been generated from electrophysiological studies examining the effect of 5-HT$_3$ antagonists on DA neuron firing rates. In a study by Sorensen et al. [18], comparing MDL 73,147EF (a 5-HT$_3$ antagonist) and haloperidol, acute haloperidol administration increased DA firing rates in the substantia nigra and ventral tegmental area (VTA) while acute MDL 73,147EF had no effect. However, chronic treatment with both haloperidol and MDL 73,147EF decreased the firing rate of neurons in both the substantia nigra and VTA, results consistent with evidence of depolarization block and likely antipsychotic efficacy. However, *in vivo* studies using the 5-HT$_3$ antagonist, BRL 43694, showed no effect on the firing rate of dopamine neurons acutely or chronically in either brain region [19]. Recently, Blackburn and colleagues [20] reported that the potent and selective 5-HT$_3$ antagonist BRL 46470A, selectively reduced the number of spontaneously active DA neurons in the A$_{10}$ area of the rat brain while having no effect on the A$_9$ region, an action shared by clozapine. This finding suggests that BRL 46470A may possess an atypical antipsychotic profile.

For a number of years, our lab has studied the neurobiology of the tridecapeptide neurotransmitter neurotensin (NT). Several lines of evidence have implicated central nervous system NT containing neurons in the pathophysiology of schizophrenia and in the mechanism of action of antipsychotic drugs. Numerous anatomical, neurochemical, electrophysiological, and pharmacological findings have revealed interactions between CNS NT and DA systems. Of particular interest, is the observation that the pharmacologic effects of centrally administered NT resemble those of peripherally administered clinically efficacious antipsychotics. These effects include decreased avoidance but not escape responding in a discrete trial conditioned avoidance paradigm, antagonism of amphetamine (and other psychostimulant)-induced locomotor activity and induction of hypothermia [21]. Furthermore, the concentration of NT in cerebrospinal fluid of drug-free schizophrenic patients has been shown repeatedly to be decreased when compared to age- and sex-matched controls and has been observed to increase after neuroleptic drug treatment [22].

Of particular interest is the finding that both acute and chronic treatment with clinically efficacious antipsychotic drugs, such as haloperidol, chlorpromazine and pimozide, produce dose-dependent increases in NT concentrations in both the nucleus accumbens and the caudate nucleus [23,24,25]. In contrast, phenothiazines without antipsychotic efficacy fail to increase neurotensin concentrations in the rat brain. We and others have shown that these changes are not related to anticholinergic, antihistaminergic, or antiadrenergic properties of these compounds.

Moreover, no other class of psychotropic drugs tested thus far, including antidepressants, anxiolytics or anticonvulsants, alters NT concentrations. Work from Dorsa and colleagues has shown that these increases in NT concentrations are preceded by increases in the expression of neurotensin mRNA in these same areas [26].

In contrast to the alterations in NT concentrations produced by typical antipsychotics, the atypical antipsychotic clozapine, produced increases in NT concentrations in the nucleus accumbens but not in the caudate nucleus [27]. Similarly, Merchant and colleagues [28], reported that while a series of typical and atypical antipsychotics increase NT mRNA expression, and presumably NT concentrations, in the nucleus accumbens, the atypical antipsychotic clozapine (and other atypical antipsychotics) does not alter NT mRNA expression in the caudate.

2. Methods and Results

Recently we have examined the effects of the putative atypical antipsychotic sertindole on NT concentrations in the rat brain. Sertindole is a potent antagonist at $5\text{-}HT_2$, D_2 and α_1-adrenoreceptors and has a high $5\text{-}HT_2/D_2$ binding ratio similar to a number of other atypical antipsychotics (*vide supra*). In an acute study, rats received either sertindole (0.05, 0.5, or 5 mg/kg) or haloperidol (1mg/kg). Controls received vehicle (1.0 ml/kg polyethylene glycol 400:0.3% tartaric acid [3:1]). As previously described [25,27,29], haloperidol produced increases in NT concentrations in the nucleus accumbens, anterior and posterior caudate. The lowest dose of sertindole produced no changes in NT concentration in any brain region, while the two higher doses of sertindole produced an increase in NT concentrations in the nucleus accumbens. Sertindole did not alter NT concentrations in the anterior or posterior caudate at any dose studied. A decrease in NT concentration ($p < 0.01$) was seen in the hippocampus following treatment with all three doses of sertindole and after haloperidol as well. These changes in NT concentrations observed after sertindole administration mimic the changes produced by the atypical antipsychotic clozapine (i.e. increased NT concentration in the nucleus accumbens with no change in NT concentration in the caudate). These effects on NT concentration support the hypothesis that sertindole itself may indeed have an atypical antipsychotic profile. These findings also provide further evidence for the possible role of antagonism of 5-HT receptors, $5\text{-}HT_2$ in the case of sertindole, in the mechanism(s) of action of atypical antipsychotic drugs.

We are currently assessing the effects of atypical antipsychotic candidates such as sertindole and novel $5\text{-}HT_3$ antagonists on NT concentrations, as well as using PCR techniques to measure NT and NT receptor mRNA expression, in order to further determine whether measures of NT neuronal activity may differentiate between typical and atypical antipsychotics.

3. Acknowledgments

Supported by NIMH MH-39415.

4. References

1. Farde, L., Nordstrom, A.-L., Weisel, F.-A., and Sedvall, G. (1989) 'D1- and D2 dopamine occupancy during treatment with conventional and atypical neuroleptics', Psychopharmacology 99 (Suppl.), S28-S31.
2. Dray, A., Gonye, T.J., Oakley, N.R., and Tanner, T. (1976) 'Evidence for the existence of a raphe projection to the substantia nigra in rat', Brain Res. 113, 45-57.
3. Soubrie, P., Reisine, T.D., and Glowinski, J. (1984) 'Functional aspects of serotonin transmission in the basal ganglia: a review and an *in vivo* approach using the push-pull cannula technique', Neuroscience 131, 615-624.
4. Castelao, J.F., Ferreira, L., Gelders, Y.G., and Heylen, S.L.E. (1989) 'The efficacy of the D_2 and 5-HT2 antagonist risperidone (R 64 766) in the treatment of chronic psychosis. An open dose-finding study', Schizophrenia Res. 2, 411-415.
5. Monfort, M., Manus, A., Bourgvignon, J., and Bouhours, P. (1989) 'Risperidone: treatment of schizophrenic patients with negative symptoms', Abstr. VIII World Congress Psychiat., Athens, Greece, Excerpta Medica International Congress Series 899, Elsevier, Amsterdam, p. 440.
6. Borison, R.L., Diamond, B.I., Pathiraja, A. and Meibach, R.C. (1992) 'Clinical overview of Risperidone', in H.Y. Meltzer (ed.), Novel Antipsychotic Drugs, Raven Press, Ltd., New York, pp. 233-239
7. Gelders, Y.G. (1989) 'Thymosthenic agents, a novel approach in the treatment of schizophrenia', Br. J, Psychiatry 155 (Suppl. 5), 33-36.
8. Meltzer, H.Y., Matsubara, S., and Lee, J.-C. (1989) 'Classification of typical and atypical antipsychotic drugs on the basis of dopamine D-1 and D-2 and serotonin$_2$ pK_i values', J. Pharmacol. Exp. Ther. 251(1), 238-246.
9. Seeman, P. (1992) 'Receptor selectivities of atypical neuroleptics', in H.Y. Meltzer (ed.), Novel Antipsychotic Drugs, Raven Press Ltd., New York.
10. Canton, H., Verriele, L., and Colpaert, F.C. (1990) 'Binding of typical and atypical antipsychotics to 5-HT_{1C} and 5-HT_2 sites: clozapine potently interacts with 5-HT_{1C} sites', Eur. J. Pharmacol. 191, 93-96.
11. Molineaux, S.M., Jessell, T.M., Axel, R., and Julius, D. (1989) '5-HT_{1C} receptor is a prominent serotonin receptor subtype in the central nervous system', Proc. Natl. Acad. Sci. USA 86, 6793-6797.
12. Kilpatrick, G.J., Jones, B.J., and Tyers, M.B. (1987) 'Identification and distribution of 5-HT_3 receptors in rat brain using radioligand binding', Nature 330, 746-748.
13. Bolanos, F.J., Schechter, L.E., Miquel, M.C., Emerit, M.B., Rumigny, J.F.,

Hamon, M., and Gozlan, H. (1990) 'Common pharmacological and physicochemical properties of 5-HT$_3$ binding sites in the rat cerebral cortex and NG 108-15 clonal cells', Biochem. Pharmacol. 40, 1541-1550.
14. Watling, K.J., Beer, M.S., Stanton, J.A., and Newberry, N.R. (1990) 'Interaction of the atypical neuroleptic clozapine with 5-HT$_3$ receptors in the cerebral cortex and superior ganglion of the rat', Eur. J. Pharmacol. 182, 465-472.
15. Hoyer, D., Gozlan, H., Bolanos, F., Schechter, L.E., and Hamon, M. (1989) 'Interaction of psychotropic drugs with central 5-HT$_3$ recognition sites: fact or fiction?', Eur. J. Pharmacol. 171, 137-139.
16. Waeber, C., Hoyer, D., and Palacios, J.M. (1989) '5-Hydroxytryptamine$_3$ receptors in the human brain: autoradiographic visualization using [^{3}H]ICS 205-930', Neuroscience 31, 393-400.
17. Reynolds, G.P. (1992) 'Developments in the drug treatment of schizophrenia', Trends Pharmacol. Sci. 13, 116-121.
18. Sorensen, S.M., Humphreys, T.M., and Palfreyman, M.F. (1989) 'Effects of acute and chronic MDL 73,147, a 5-HT$_3$ receptor antagonist, on A9 and A10 dopamine neurons', Eur. J. Pharmacol. 163, 115-120.
19. Ashby Jr., C.R., Jiang, L.H., Kasser, R.J., and Wang, R.Y. (1990) 'Electrophysiological characterization of 5-hydroxytryptamine$_3$ receptors in rat medial prefrontal cortex', J.Pharmacol. Exp. Ther. 251, 171-178.
20. Blackburn, T.P., Kennet, G., Ashby, C.R., Minabe, Jr. Y., Wang, R.Y., Domeney, A.M., and Costall, B. (1992) 'BRL 46470A: A novel, potent 5-HT$_3$ receptor antagonist with atypical antipsychotic properties', in press.
21. Levant, B., Bissette, G., and Nemeroff, C.B. (1991) 'Neurotensin', in C.B. Nemeroff (ed.), Neuropeptides in Psychiatry, APA Press, Washington, pp. 149-168.
22. Nemeroff, C.B. (1980) 'Neurotensin: Perchance an endogenous neuroleptic?', Biol. Psychiatry 15, 283-302.
23. Govoni, S., Hong, J.S., Yang, H.-Y.T., and Costa, E. (1980) 'Increases of neurotensin content elicited by neuroleptics in nucleus accumbens', J. Pharmacol. Exp. Ther. 215, 413-417.
24. Frey, P., Lis, M., and Coward, D.M. (1988) 'Neurotensin concentrations in rat striatum and nucleus accumbens: further studies of their regulation', Neurochem. Int. 12, 33-38.
25. Myers, B., Levant, B., Bissette, G., and Nemeroff, C.B. (1992) 'Pharmacological specificity of the increase in neurotensin concentrations after antipsychotic drug treatment', Brain Res. 573, 325-328.
26. Merchant, K.M., Miller, M.A., Ashleigh, E.A., and Dorsa, D.M. (1991) 'Haloperidol rapidly increases the number of neurotensin mRNA- expressing neurons in neostriatum of the rat brain', Brain Res. 540, 311-314.
27. Kilts, C.D., Anderson, C.M., Bissette, G., Ely, T.D., and Nemeroff, C.B. (1988) 'Differential effects of antipsychotic drugs on the neurotensin

concentration of discrete rat brain nuclei', Biochem. Pharmacol. 37, 1547-1554.
28. Merchant, K.M., Dobner, P.R., and Dorsa, D.M. (1992) 'Differential effects of haloperidol and clozapine on neurotensin gene transcription in rat neostriatum', J. Neuroscience 12, 652-663.
29. Nemeroff, C.B., Kilts, C.D., Levant, B., Bissette, G., Campbell, A., and Baldessarini, R.J. (1991) 'Effects of the isomers of N-n-propylnorapomorphine and haloperidol on regional concentrations of neurotensin in rat brain', Neuropsychopharmacology 4, 27-33.

PHARMACOLOGICAL ANALYSIS OF 5-HT RECEPTORS IN HUMAN SMALL CORONARY ARTERIES

J.A. ANGUS, *T.M. COCKS, and *M. ROSS-SMITH
Department of Pharmacology
The University of Melbourne
Parkville, Victoria
Australia, and
*Baker Medical Research Institute
P.O. Box 348
Prahran, Victoria
Australia

ABSTRACT. We analyzed the reactivity of human isolated small coronary arteries in response to 5-HT, the 5-HT_1-like receptor agonist sumatriptan, and ergometrine. The arteries (100-300 μm diameter) were removed from the atrial appendage of patients undergoing heart-lung bypass surgery, and were mounted on wires in a myograph for simultaneous measurement of force and membrane potential. Only 39% (n=7) of all arteries contract to 5-HT (0.01-10 μM) and the maximum was only 43 $\pm$ 14% of the depolarization to K^+ (124 mM). Sumatriptan (0.01-10 μM) and ergometrine (0.001-1 μM) did not contract these vessels. In contrast acetylcholine (0.01-10 μM) consistently contracted these small arteries to $\approx$80% of K^+ and caused an unusual oscillation in membrane potential. Thus human small coronary arteries (over the diameter range 100-300 μm) appear to have a sparse population of 5-HT receptors and probably do not respond to 5-HT if released upstream from aggregating platelets.

1. Introduction

Serotonin (5-hydroxytryptamine, 5-HT) has attracted attention as a potential candidate for the cause of coronary vasospasm because it is released during platelet aggregation and it contracts large coronary arteries [1,2]. Ergometrine (ergonovine) has been used as a diagnostic agent in patients with variant angina and recent evidence suggests it has affinity for 5-HT receptors in a variety of arteries in addition to α-adrenoceptors [3,4]. In patients with angina, intracoronary infusions of 5-HT constrict epicardial large coronary arteries especially in the presence of atherosclerosis [5,6,7]. 5-HT_2-receptors probably mediate the major contribution of

5-HT-induced larger artery constriction since ketanserin abolished the response [5,7]. In some patients the smaller distal arteries still contracted to 5-HT after ketanserin and one conclusion was that non-5-HT_2 receptors may be involved [7]. In isolated ring segments of large coronary arteries the 5-HT concentration-response curve was antagonized by ketanserin [1,2] and the selective 5-HT_1-like receptor agonist, sumatriptan, was a relatively weak agonist contracting epicardial arteries to approximately 30% of the maximum to 5-HT [1,2,8].

Since nothing is known of the reactivity of human small coronary arteries to 5-HT, 5-HT-receptor subtype selective agents, or ergometrine, we developed techniques to study isometric force and membrane potential in vessels of 100-300 μm in diameter taken from the right atrial appendage and mounted on wires in the controlled environment of the myograph chamber. These small arteries contribute to the coronary resistance and can be studied here without the influences of local factors generated by cardiac metabolism.

2. Methods

The tip of the right atrial appendage was removed from patients undergoing bypass surgery for mitral valve replacement or coronary graft surgery. The tissue was immediately placed in cool physiological salt solution (PSS) saturated with 5% CO_2 in O_2 [9]. It was pinned out in a dissecting dish and with the aid of a cool light source and microscope a small coronary artery about 2 mm long was removed from the muscle. The vessel was mounted on 40 μm diameter wires in a Mulvany-Halpern myograph and heated to 37° C. Thirty minutes later the artery was stretched to set the resting force at a normalized level of 0.9L_{100}, where L_{100} is the circumference of the artery when distended at a transmural pressure of 100 mmHg [9]. Maximum contractions were determined by exposing the artery twice to K^+ (124 mmHg, KSS) for two minutes. Cumulative concentration-response curves were constructed to 5-HT creatinine sulphate, sumatriptan (GR43175, Glaxo Group Research, Ware, UK), ergometrine maleate (David Bull Labs, Mulgrave, Australia), and acetylcholine bromide. In some experiments, membrane potential (E_m) was recorded with conventional glass electrodes simultaneously with the measurement of wall force. These techniques have been published in detail [9].

3. Results

3.1 SELECTION OF VESSELS

Only arteries that contracted to KSS with an equivalent transmural pressure (Δ P) of $\rangle$ 40 mmHg were used in the analysis. From the Laplace relationship, Δ P = internal radius/Δ T, where T is wall tension. There were some vessels that had been damaged during the tissue removal or were too short in length. Table 1 shows the contractile responses to 18 arteries, one per patient, divided retrospectively into responders and nonresponders, depending upon their response to 5-HT (10 μm). By comparison, the two groups contracted similarly to KSS and to acetylcholine (ACH, 10 μM) and were of similar internal diameter.

TABLE 1. A comparison of contraction responses to KSS, 5-HT, and acetylcholine in small coronary arteries.

	Diameter μm	KSS mmHG	5-HT % of KSS	ACH % of KSS
		Maximum responses		
Responders (n=7)	160 ± 14	76 ± 15	43 ± 14	82 ± 20
Nonresponders (n=11)	199 ± 18	83 ± 90	0	78 ± 9
t-test	ns	ns	--	ns

These results show that only 39% of all arteries contracted to 5-HT and the maximum contraction (E_{max}) was only 43% of the depolarization induced by KSS. In contrast, all the arteries contracted to acetylcholine with an E_{max} about double that to 5-HT in the responders (figure 1). In these seven arteries, the maximum response ranged from 8% to 92% of the E_{max} to KSS. Logistic curves were satisfactorily computer fitted to 5 of the 7 vessels, giving an EC_{50} for 5-HT of 6.73 ± 0.20 (mean ± sem, -logM).

3.2 SUMATRIPTAN

Out of 11 patients (one artery from each that satisfied the inclusion on technical grounds), only one artery contracted to sumatriptan (0.01-1 μM) and that was to 12% of KSS at a concentration of 1 μM. Of these 11 arteries, 5-HT was tested in 8 and only 3 responded to 5-HT (1 μM) with contractions of 9, 27, and 92% of KSS. In the artery that contracted to 92%, ketanserin (0.1 μM) abolished the response.

3.3 ERGOMETRINE

In 7 arteries, ergometrine (1-1000 nM) was without effect while 5-HT (10 μM) contracted 3 of these 7 by 9, 27, and 92% in the same vessels as for the sumatriptan assays. In 5 of these 7 vessels sumatriptan was tested up to 1 µM without effect.

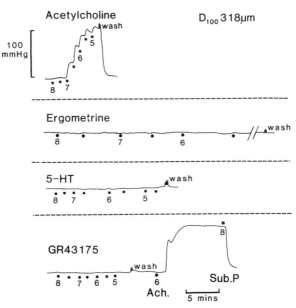

Figure 1. Chart record of force changes in a human small coronary artery (318 µm diameter) in response to acetylcholine, ergometrine, 5-HT, and sumatriptan (GR43175). Endothelium was intact given the relaxation to substance P. Concentrations are -logM at 0.5 unit steps. Force change is in active pressure units ($\Delta P = \Delta$tension/radius).

3.4 SYNERGY

In 5 arteries, acetylcholine (0.3 µM) was applied to cause a steady small contraction of 9.7 ± 2.2% of KSS maximum (range 6-17%). Applying sumatriptan at 3 µM in the presence of this contraction caused the force to rise in 3 out of 5 arteries to 16.6 ± 4.3% (n=5) of KSS (range 6-31%). In one vessel that did not respond to sumatriptan at 3 µM, we applied 10 µM and this contracted the artery to 76% of KSS maximum in the presence of the acetylcholine. In 3 arteries, only one contracted to ergometrine (0.1 µM) to 100% of KSS and this was the same vessel that contracted to sumatriptan (76%) in the presence of acetylcholine.

3.5 OTHER ARTERIES

In two arteries with larger internal diameters (760 µm and 435 µm) taken from the ventricle of an explant heart with cardiomyopathy but no atherosclerosis, neither sumatriptan nor 5-HT (from 0.01-30 μM) caused any contraction. From the same heart an atrial artery (351 μm) and a vessel from the ventricle (783 μm) were submaximally contracted by endothelin (10 nM) to 65% and 63% respectively of KSS. 5-HT (10 μM) had no effect on these contracted arteries.

3.6   5-HT$_4$ RECEPTORS

In 4 atrial arteries (198-220 μm) from 2 patients, the 5-HT$_2$ and 5-HT$_1$-like receptors were antagonized by ketanserin (1 μM) and methiothepin (0.1 μM), respectively. Exposure of the arteries to K$^+$ (25-35 mM) caused a steady submaximum contraction to 55-65% of KSS. Neither 5-HT nor the selective 5-HT$_4$-receptor agonist BMIU8 [10] (0.01-30 μM) had any effect.

3.7   ELECTROPHYSIOLOGY

Human atrial small arteries mounted in the wire myograph had a resting membrane potential (E$_m$) of -62±3 mV (n = 24 arteries, first cell) [11]. KSS infusion depolarized the artery to ≈ -10 mV and simultaneously increased the isometric force (figure 2). Note the marked depolarization before the force began to rise. In the same artery (153 μm diameter), acetylcholine infusion (1 μM) caused a rapid depolarization of 30-40 mV followed by repolarization to near the resting E$_m$. This pattern of oscillation continued as the acetylcholine was infused and was accompanied by a sustained contraction greater than that to KSS. This sustained E$_m$ oscillation at a frequency of 0.54 ± 0.08 Hz was observed in ≈ 40% of arteries. In the remainder of arteries, the oscillations ceased after 4-12 cycles in the continued presence of acetylcholine [11]. In arteries with sustained E$_m$ oscillations, the infusion of substance P (30 nM) abruptly repolarized the artery and inhibited the contraction to acetylcholine. Similar oscillations in E$_m$ were observed to other constrictor agents of the human small coronary arteries such as endothelin (10 nM), noradrenaline (5 μM), and with 5-HT (10 μM).

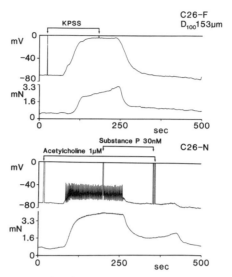

Figure 2. Simultaneous records of membrane potential (Em, mV, top traces) and active force (mN, bottom traces) recorded from the same human small coronary artery (153 µm diameter) in response to K$^+$ (124 mM, KPSS, top) and acetylcholine, with Substance P (bottom). The bar indicates the infusion period and there was a 60-second delay between the pump switching and the drug change in the myograph chamber.

4. Discussion

These studies show that the majority ($\rangle$ 60%) of human small coronary arteries form the atrial appendage fail to contract to 5-HT compared with conduit arteries from explant hearts [1,2,8]. In arteries where 5-HT was effective, the maximum contraction was similar for the large and small arteries, i.e. 40-50% of the contraction to KSS. The lack of response to 5-HT in the majority of small arteries was not due to damage since every artery contracted to near KSS maximum to acetylcholine, and relaxed to the endothelium-dependent relaxant, substance P. One explanation for the lack of contraction to 5-HT in the small arteries is that subtype receptors mediating relaxation could be present that would functionally antagonize any contractile response. This is unlikely since the 5-HT$_1$-like constrictor agonist sumatriptan, and the mixed 5-HT$_1$- and α-adrenoceptor agonist ergometrine were without effect on these small arteries. In addition, we could find no evidence for the novel 5-HT$_4$ receptors that mediate positive inotropism in the human atrial appendage [13] and relaxation of the sheep pulmonary vein [14].

There is always the potential for interactions between different receptor agonists resulting in synergy. We used the constrictor acetylcholine at a concentration that caused a steady level of force before applying the 5-HT analogues. Acetylcholine may have no role in the coronary vasculature *in vivo*. But apart from K^+, these small atrial arteries did not appear to react to the thromboxane mimetic U46619, nor to other amines such as histamine. Thus testing for synergy was confined to acetylcholine and 5-HT receptor agonists. Nevertheless, in the occasional artery we did see some response to high concentrations of ergometrine and sumatriptan in the presence of acetylcholine. Thus it would appear that the human small coronary artery is generally unresponsive to 5-HT and especially to sumatriptan. Our results also suggest that ergometrine and sumatriptan have similar properties in these small arteries and that the 5-HT_1-like receptors are sparce in these vessels. If the results from the small arteries taken from the ventricle are generally applicable, then small arteries ($\langle 800 \ \mu m$) from the human heart are relatively unresponsive to 5-HT receptor activation.

The electrophysiology shows that these small coronary arteries behave very differently from small coronary arteries from other species and from small arteries from noncardiac tissue. The oscillations of E_m to constrictor agents have not been observed in atrial vessels from dog or pig nor from human buttock skin small arteries in response to noradrenaline [11,12]. Not only are there muscarinic receptors on smooth muscle cells in the human small coronary artery, but the E_m oscillation is probably mediated by the release of intracellular calcium since the consistent E_m response to acetylcholine was blocked by ryanodine (Angus & Broughton, unpublished). The paucity of arteries responding to 5-HT precluded significant E_m studies with this amine.

In conclusion, comparative *in vitro* studies suggest that the human small coronary vessels are relatively unreactive to 5-HT and the specific 5-HT_1-like agonist sumatriptan. Therefore, it is unlikely that these arteries play a significant role in the response to 5-HT if released upstream from aggregating platelets at sites of atherosclerosis.

5. Acknowledgements

This work was supported by an Institute Grant form the National Health and Medical Research Council of Australia and by Glaxo Australia.

6. References

1. Connor, H.E., Feniuk, W., and Humphrey, P.P.A. (1989) '5-hydroxytryptamine contracts human coronary arteries predominantly via 5-HT_2 receptor activation', Eur. J. of Pharmacol. 161, 91-94.
2. Chester, A.H., Martin, G.R., Bodelsson, M., Arneklo-Nobin, B., Tadjkarimi, S., Tornebrandt, K., and Yacoub, M.H. (1990) '5-hydroxytryptamine receptor profile in healthy and diseased human epicardial coronary arteries',

Cardiovascular Res. 24, 92-93.
3. Brazenor, R.M. and Angus, J.A. (1981) 'Ergometrine contracts canine coronary arteries by a serotonergic mechanism: no role for alpha-adrenoceptors', J. Pharmacol. Exp. Ther. 218, 530-536.
4. Feniuk, W., Humphrey, P.P.A., and Perren, M.J. (1989) 'The selective carotid arterial vasoconstrictor action of GR43175 in anaesthetized dogs', Br. J. Pharmacol. 96, 83-90.
5. Golino, P., Piscione, F., Willerson, J.T., Cappelli-Bigazzi, M., Focaccio, A., Villari, B., Indolfi, C., Russolillo, E., Condorelli, M., and Chiariello, M. (1991) 'Divergent effects of serotonin on coronary-artery dimensions and blood flow in patients with coronary atherosclerosis and control patients', N. Engl. J. Med. 324, 641-645.
6. McFadden, E.P., Clarke, J.G., Davies, G.J., Kaski, J.C., Haider, A.W., and Maseri, A. (1991) 'Effect of intracoronary serotonin on coronary vessels in patients with stable angina and patients with variant angina', N. Engl. J. Med. 324, 648-654.
7. McFadden, E.P., Bauters, C., Lablanche, J.M., Leroy, F., Clarke, J.G., Henry, M., Schandrin, C., Davies, G.J., Maseri, A., and Bertrand, M.E. (1992) 'Effect of ketanserin on proximal and distal coronary constrictor responses to intracoronary infusion of serotonin in patients with stable angina, patients with variant angina, and control patients', Circulation 86, 187-195.
8. Cocks, T.M., Kemp, B.K., Pruneau, D., and Angus, J.A. (1992) 'Comparison of contractile responses to 5-hydroxytryptamine (5-HT) and sumatriptan in human isolated coronary artery: synergy with the thromboxane A2-receptor agonist, U46619', (in preparation).
9. Angus, J.A., Broughton, A., and Mulvany, M.J. (1988) 'Role of alpha-adrenoceptors in constrictor responses of rat, guinea pig and rabbit small arteries to neural activation', J. Physiol. (London) 403, 495-510.
10. Schiantarelli, P., Bockaert, J., Cesana, R., Donetti, A., Dumuis, A., Giraldo, E., Ladinsky, H., Monferini, E., Rizzi, C., Sagrada, A., and Schiavone, A. (1990) 'The prokinetic properties of new benzimidazolone derivatives are related to 5-HT_4 agonism', Pharmacol. Res. 22, 453.
11. Angus, J.A., Broughton, A., and McPherson, G.A. (1991) 'Membrane potential and contractility responses to acetylcholine and other vasoconstrictor stimuli in human small coronary arteries', in M.J. Mulvany, C. Aalkjaer, A.M. Heagerty, N.C.B. Nyborg, and S. Trandgaard (eds.), Resistance Arteries, Structure and Function, Elsevier (Int. Congress Series), Amsterdam, pp. 255-260.
12. Angus, J.A., Broughton, A., Cocks, T.M., and McPherson, G.A. (1991) 'The acetylcholine paradox--a constrictor of human small coronary arteries even in the presence of endothelium', Clin. Exp. Pharmacol. Physiol. 18, 33-36.
13. Kaumann, A.J., Sanders, L., Brown, A.M., Murray, J.J., and Brown, M.J.

(1991) 'A 5-HT$_4$-like receptor in human right atrium', Naunyn-Schmiedeberg's Arch. Pharmacol. 344, 150-159.
14. Cocks, T.M. and Arnold, P.J. (1992) '5-hydroxytryptamine (5-HT) mediates potent relaxation in the sheep isolated pulmonary vein via activation of 5-HT$_4$ receptors', Br. J. Pharmacol., in press.

PLATELETS, PLATELET MEDIATORS, AND UNSTABLE CORONARY ARTERY DISEASE SYNDROMES

JAMES T. WILLERSON
The University of Texas Medical School and
Texas Heart Institute at Houston
6431 Fannin, Room 1.150
Houston, Texas 77030
USA

ABSTRACT. We have speculated previously that the abrupt conversion from chronic stable to unstable angina and the continuum to acute myocardial infarction may result from myocardial ischemia caused by progressive platelet aggregation and dynamic vasoconstriction themselves caused by local increases in thromboxane A_2, serotonin, adenosine diphosphate (ADP), thrombin, and platelet activating factor (PAF) at sites of coronary artery stenosis and endothelial injury. Platelet aggregation and dynamic coronary artery vasoconstriction probably result from the local accumulation of thromboxane A_2, serotonin, ADP, thrombin, and PAF and also relative decrease in the local concentrations of endothelially derived vasodilators and inhibitors of platelet aggregation, such as endothelium-derived relaxing factor (EDRF) and prostacyclin. With severe reductions in coronary blood flow caused by these mechanisms, platelet aggregates increase, and an occlusive thrombus composed of platelets and white and red blood cells in a fibrin mesh develops. When coronary arteries are occluded or narrowed for a sufficient period of time by these mechanisms, myocardial necrosis, electrical instability, or sudden death may occur. We believe that unstable angina and acute myocardial infarction are a continuum in relation to the process of coronary artery thrombosis and vasoconstriction. When the period of platelet aggregation or dynamic vasoconstriction at sites of endothelial injury and coronary artery stenosis is brief, unstable angina or non-Q-wave infarction occur. When the period of reduction in blood flow is prolonged, Q-wave infarction results.

1. Introduction

Acute coronary heart disease syndromes, including unstable angina, variant angina (Prinzmetal's angina), and acute myocardial infarction are usually caused by a primary decrease in myocardial oxygen delivery [1-15]. Unstable angina occurs in some patients with endothelial injury or ulceration at the site of a coronary artery stenosis [6]; some patients also have an intraluminal coronary artery thrombus [4-6,9-11]. Specifically, Falk [7] and Davies and Thomas [8] have suggested that atherosclerotic plaque rupture or fissuring may lead to coronary arterial thrombi and the development of unstable angina, acute myocardial infarction, or sudden death. Patients with Q-wave (usually transmural) myocardial infarcts often have ulcerated or fissured atherosclerotic plaques and the subsequent development of occlusive coronary artery thrombi [2,3]. A similarity exists in the coronary arteriographic appearance in the relevant coronary artery in patients with unstable angina and in those patients who develop acute myocardial infarction [15]. Patients with non-Q-wave infarcts (usually subendocardial infarcts) much less commonly have coronary artery thrombi that are permanently occlusive [3], but these patients have transient reductions in coronary blood flow followed by reperfusion [16]. The transient reductions in coronary blood flow are most likely related to intermittent platelet aggregations and dynamic vasoconstriction occurring at sites of coronary artery stenosis and endothelial injury [4,5].

We have suggested that unstable angina, non-Q-wave myocardial infarction, and Q-wave myocardial infarcts represent a continuum, such that transient reductions in coronary blood flow associated with platelet aggregation and dynamic vasoconstriction at sites of coronary artery stenosis and endothelial injury lead to the abrupt development of crescendo or unstable angina [4,5]. If important reductions in coronary blood flow and oxygen delivery are sustained for 20 minutes to 2 hours, a non-Q-wave myocardial infarction may occur, and if the period of important reduction in myocardial oxygen delivery persists for more than 2 hours, a Q-wave myocardial infarction usually occurs [4,5]. In this suggested pathophysiologic scheme, factors responsible for the conversion from chronic to acute coronary heart disease syndromes include endothelial injury at sites of coronary artery stenosis [4,5]; the endothelial injury could be, among other factors, the result of plaque fissuring or ulceration, [7,8] hemodynamic factors, systemic arterial hypertension, infection, smoking, catheter-induced injury, or balloon angioplasty.

With endothelial injury, platelets attach to the exposed subendothelium and collagen, and the platelet aggregation quickly leads to the release and local accumulation of thromboxane A_2, ADP, thrombin, and serotonin that promote further plate aggregation, dynamic coronary artery vasoconstriction, and consequent reduction in coronary blood flow [4,5,10-14,17-28] (figure 1). Platelet aggregation and dynamic coronary artery vasoconstriction probably result from a combination of the local arterial accumulation of thromboxane A_2, thrombin, PAF, serotonin, and adenosine diphosphate (ADP) (and possible other mediators) and relative or absolute decreases in local arterial concentrations of endothelially derived vasodilators and inhibitors of platelet aggregation, such as endothelially derived relaxing factors (EDRFs) and prostacyclin [20,29,30] (figure 1).

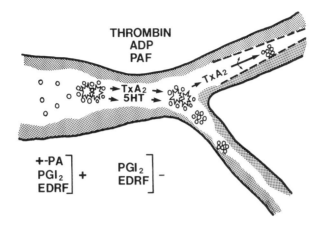

Figure 1. Schematic diagram indicating the possible mechanisms by which thromboxane A_2 and serotonin promote platelet aggregation and decrease coronary blood flow in the patient with unstable angina. Aggregating platelets (stars) release thromboxane A_2 and serotonin at sites of coronary artery injury, which cause further platelet aggregation at that site and downstream, dynamic coronary vasoconstriction, and partial or total coronary artery thrombosis and reductions in endothelially derived relaxing factor (EDRF) and prostacyclin (PGI_2) at vascular sites with endothelial injury probably contribute to the development of vasoconstriction and thrombosis. See text for details. Modified and reproduced from Willerson et al. [29] by permission.

Chronic endothelial injury at sites of coronary artery stenosis lasting from several hours to days is associated with the accumulation of platelets, white and red blood cells, and a fibrin mesh, and potentially, this injury may be associated with the accumulation of other mediators contributing to platelet aggregation and dynamic vasoconstriction, including platelet activator factor, selected leukotrienes (LTC_4 and LTD_4), histamine, prostaglandin D_2, and possibly endothelin [4,5,31].

2. Unstable Angina Pectoris

Unstable angina pectoris is defined as angina that increases in frequency with progressively less effort, and often, it occurs at rest. Any factor that results in abrupt or progressive coronary artery luminal diameter narrowing may cause diminished myocardial perfusion with consequent myocardial ischemia and unstable angina. Table 1 lists several possible causes of unstable angina. In some patients, the syndrome may result from the progression of severe, multifocal coronary atherosclerosis. However, in many patients, we believe that other mechanisms are involved because a good correlation often does not exist between acute coronary heart disease and the anatomic extent and severity of coronary atherosclerosis.

TABLE 1. Potential Causes of Unstable Angina Pectoris

- Progressive coronary artery narrowing from atherosclerosis
- Platelet aggregation and white blood cell adhesion and platelet- or white blood cell-derived mediator release at the site of endothelial injury and coronary artery stenosis leading to increased luminal narrowing from anatomic obstruction and dynamic increases in coronary vascular resistance
- Any combination of plaque fissure, hemorrhage, or thrombosis with progressive coronary artery narrowing
- Coronary artery spasm

3. Mechanisms Potentially Responsible for the Development of Sustainment of Unstable Angina

Our hypothesis that unstable angina is associated with increases in transmyocardial thromboxane A_2 concentrations in patients was tested from 1979 to 1980 [10]. In 60 patients undergoing cardiac catheterization, blood samples were obtained from the coronary sinus and ascending aorta, and prostaglandins were measured by radioimmunoassay [10].

Our data showed a temporal relation between continuing unstable angina and increases in transcardiac thromboxane concentrations, and they were consistent with the hypothesis that thromboxane A_2 accumulation and consequent platelet aggregation and coronary vasoconstriction are important factors in the pathogenesis of unstable angina in patients [10]. However, the data did not prove a causal relation.

Subsequently, Fitzgerald et al. [11] at Vanderbilt and Hamm et al. [12] have shown that many patients with unstable angina have elevated urinary concentrations of a major thromboxane metabolite, that is, 2-3 dinor-thromboxane B_2 and that in some patients, further increases occur in the thromboxane urinary metabolite with new episodes of chest pain.

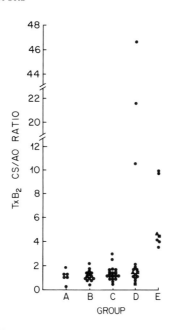

Figure 2. Plot of ratios of thromboxane B_2 (TxB$_2$) in the coronary sinus (CS) and ascending aorta (AO) in the five groups of patients (n=60) studied. Each point represents the data from one patient. ■, patients who received a cyclooxygenase inhibitor within 5 days of study; ▲, patients with coronary artery spasm. In group A (valvular and congenital nonischemic heart disease), B (chest pain syndrome without ischemic heart disease), and C (ischemic heart disease without chest pain for at least 96 hours), the patients had thromboxane B_2 coronary sinus aorta ratios of 3.1 or less. Group D patients (unstable angina pectoris with chest pain 24-96 hours before study) had a bimodal distribution; 12 patients had low ratios, whereas three had very high ratios. Group E patients (unstable angina pectoris with chest pain within 24 hours before study) had ratios that were higher than those in patients in group A, B, and C ($p < 0.05$). Reprinted from Hirsh et al. [10] by permission.

4. Evaluation of the Physiologic Importance of Thromboxane Accumulation

4.1 EXPERIMENTAL MODELS

In the canine heart, a severe proximal coronary artery stenosis with associated endothelial injury causes cyclical coronary blood flow reductions as originally described by Folts et al. (figure 3) [17]. In this model, endothelial injury is caused by the application of an external constrictor and by gentle stroking of the left anterior descending coronary artery with cloth-covered forceps. The cyclic coronary flow alterations are related to platelet aggregation and leukocyte and red blood cell accumulation at the stenotic site. Dazoxiben (UK-37-248, Pfizer Pharmaceuticals, New York, New York), a thromboxane synthetase inhibitor, [18] or SQ29,548 (Squibb Pharmaceuticals, Princeton, New Jersey), or a thromboxane receptor antagonist, [19] abolishes or significantly attenuates the frequency of cyclic flow reductions in approximately 70% of treated animals (figure 3). Thromboxane B_2 concentrations in the distal portion of the narrowed coronary artery and at the site of the coronary stenosis and endothelial injury are increased during cyclic coronary flow reductions and are reduced to control values after administration of dazoxiben [18,20]. In contrast, 6-keto-prostaglandin $F_{1\alpha}$ (the inactive metabolite of prostacyclin) generation is reduced at the site of coronary artery constriction compared with generation in the nonconstricted, noninjured coronary artery [20]. The 6-keto-prostaglandin $F_{1\alpha}$ levels distal to the narrowed coronary artery increase significantly during cyclic flow reductions and remain elevated after the administration of dazoxiben [18]. In the canine model, systemically administered dazoxiben (2.5 mg/kg body wt IV) suppresses arachidonic acid-induced thromboxane A_2 (but not prostaglandin E_2) production by platelets, but it does not significantly influence prostacyclin synthesis by coronary artery rings [18]. Furthermore, the topical application or intra-atrial administration of a thromboxane-mimetic (U46619) generally restores cyclic coronary flows in this model after they are abolished by a thromboxane synthesis inhibitor [18]. Folts et al. [17] have shown that aspirin often eliminates cyclic coronary artery flow reductions in the same experimental model.

Therefore, thromboxane concentration increases at the site of a coronary artery stenosis and endothelial injury in the canine model, and the narrowed, injured artery makes substantially more thromboxane when arachidonic acid is added *in vitro* [18,20]. The administration of a thromboxane synthetase inhibitor or receptor antagonist usually abolishes cyclic flow alterations in this model [17-19]. More recently, we have shown that dynamic coronary vasoconstriction occurs adjacent to the site of the stenosis in this canine model and that thromboxane and serotonin receptor antagonists prevent the dynamic vasoconstriction [21]. Thromboxane synthesis inhibitors and receptor antagonists are also protective against the development of *in situ* platelet aggregation [22] in this experimental model.

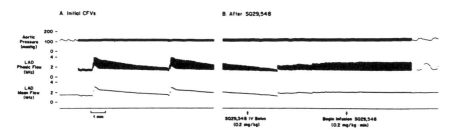

Figure 3. Representative recording from a dog with cyclic coronary flow variations (CFVs) and abolition of the cyclic flows with SQ29,548, a thromboxane receptor antagonist from Squibb Pharmaceuticals (Princeton, N.J.). Panel A: Initial CFVs. Note the reoccurring pattern of decline and restoration of mean and phasic left anterior descending coronary artery (LAD) flow. Panel B: Normal pattern of flow restored minutes after administration of SQ29,548. CFVs occur in the canine model with a tight proximal coronary artery stenosis and endothelial injury as previously reported (see references [17-26]). Reproduced from Ashton et al. [19] by permission.

5. Additional Humoral Mediators Influencing Platelet Aggregation and Dynamic Changes in Coronary Artery Tone

5.1 SEROTONIN

In our initial studies, we found that thromboxane synthesis inhibitor or receptor antagonist was protective and abolished cyclical alterations in coronary blood flow in approximately 70% of the treated animals. Thus, another mediator(s) appeared to contribute to the development of cyclic flow alterations in this canine model. Subsequent studies showed that serotonin concentration increases by at least 18-fold at the site of a coronary artery stenosis and endothelial injury when cyclic coronary flow reductions occur [23]. Ketanserin, a serotonin receptor antagonist (Janssen Pharmaceuticals, Beerse, Belgium), usually abolishes cyclic flow reductions [23,24] and works in essentially every instance in which a thromboxane synthetase inhibitor or receptor antagonist, or both, have failed (figure 4) [25]. Other serotonin receptor antagonists without α-adrenergic antagonist properties at the concentrations at which they were used also abolish cyclic coronary flow alterations in approximately 90% of animals and are effective when thromboxane synthetase inhibitors or receptor antagonists have failed [25].

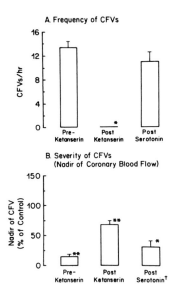

Figure 4. Panel A: Bar graph of number of cyclic coronary artery flow variations (CFVs) per hour in dogs during a control period, after abolition of cyclic coronary flows by ketanserin (a serotonin receptor antagonist), and after reestablishment of cyclic coronary artery flows by the infusion of serotonin (1-3 μg/min) directly into the left atrium. Panel B: Bar graph of severity of CFVs, obtained by averaging the two lowest mean flow velocity values (nadirs) as a percentage centage of controlled unconstricted blood flow velocity throughout a 30-minute interval, is shown during the control period of cyclic flows and after the serotonin receptor antagonist ketanserin, had abolished cyclic alterations in coronary blood flow. The severity of cyclic coronary flows after their restoration by serotonin administration is shown in the far right. The mean flow velocity nadir in serotonin-restored CFVs was taken as a percentage of mean flow velocity after infusion of serotonin but before the restoration of CFVs. *, significant reduction in the frequency of cyclic flow variations per hour by the serotonin receptor antagonist, ketanserin. †, significantly different from coronary blood flow velocity following ketanserin. **$p<0.05$, significantly different from coronary blood flow velocity during infusion of serotonin and before CFVs were restored. Reprinted from Ashton et al. [23] by permission.

After cyclic flows are abolished by serotonin receptor antagonists, they may be restored by the intra-atrial administration of serotonin [23]. There is also an amplification of thromboxane and serotonin's effects on cyclic coronary flow variations, such that interfering with the effect of either is generally protective in eliminating cyclic flow alterations [25]. Thus, both thromboxane and serotonin are important in initiating or sustaining cyclic flow reductions in this experimental model [25]. If the contribution from both thromboxane and serotonin receptor stimulation is eliminated, cyclic coronary flow reductions are nearly always abolished in this experimental model [25] and are very difficult to restore even if systemic epinephrine concentrations are markedly increased [26].

Recently, subsets of patients with limiting angina and significant coronary heart disease were shown to have increased transcardiac serotonin concentrations [13]. Moreover, subsets of patients with limiting angina have a vasoconstrictor substance in their coronary sinus effluents whose vasoconstrictor effect is blocked by a serotonin receptor antagonist [14].

5.2 ADP

ADP increases at sites of coronary artery narrowing and endothelial injury and ADP antagonists, such as apyrase, inhibit cyclic coronary blood flow in approximately eighty percent of animals [27].

5.3 THROMBIN

Thrombin antagonists inhibit cyclic coronary blood flows as do PAF antagonists when cyclic flow alterations have been in progress for 6-8 hours in the canine model [28,29].

5.4 EDRF

Promoters of EDRP release, such as L-carnitine, the substrate for EDRP formation, abolish cyclic flow variations when they have been caused by inhibitors of EDRP formation [30].

6. Coronary Artery Stenosis and Endothelial Injury in Closed-Chest, Awake, Unsedated Dogs

One to three weeks after the application of the coronary artery constrictor and endothelial injury and after frequent cyclic coronary flow alterations, chronically instrumented dogs develop marked intimal proliferation that converts a moderately severe coronary artery stenosis to a more critical one [31,32]. Similar intimal proliferation has been described previously as occurring in endothelially injured carotid arteries in the rat [33] and in the patient who develops restenosis after coronary artery angioplasty [34]. We and others have suggested that in the stenosed coronary artery with endothelial injury and chronic cyclic coronary flow alterations in the awake, unsedated dog and in patients with restenosis after coronary artery angioplasty, the marked intimal proliferative responses found may be caused by the accumulation of selected growth factors from aggregating platelets or infiltrating white blood cells and activated T cells, including one or more of the following: platelet-derived growth factors, transforming growth factors, epidermal growth factor, fibroblast growth factor, serotonin, endothelin, or a heparin-associated growth factor [35-40]. Further studies are needed to test this suggestion and to identify which growth factor(s) and which cell population(s) may contribute to the fibroproliferative reaction. However, we have shown that intense therapy with combined thromboxane A_2 and serotonin antagonists together markedly attenuate the neointimal proliferation in this experimental model [32].

7. References

1. Maseri, A., L'Abbate, A., Pesola, A., Marzilli, M., Severi, S., Parodi, O., Ballestra, A.M., Maltini, G., DeNes, D.M., and Biagini, A. (1977) 'Coronary vasospasm in angina pectoris', Lancet 1, 713-717.
2. DeWood, M.A., Spores, J., Notske, R., Mouser, L.T., Burroughs R., Golden, M.S., and Lang, H.T. (1980) 'Prevalence of total coronary occlusion during the early hours of transmural myocardial infarction', N. Engl. J. Med. 303, 897-902.
3. Buja, L.M. and Willerson, J.T. (1981) 'Clinicopathologic correlates of acute ischemic heart disease syndromes', Am. J. Cardiol. 47, 343-256.
4. Willerson, J.T., Campbell, W.B., Winniford, M.D., Schmitz, J., Apprill, P., Firth, B., Ashton, J., Smitherman, T., Bush, L., and Buja, L.M. (1984) 'Conversion from chronic to acute coronary artery disease, Speculation regarding mechanisms', Am. J. Cardiol. 54, 1349-1354.
5. Willerson, J.T., Hillis, L.D., Winniford, M.D., and Buja, L.M. (1986) 'Speculations regarding mechanisms responsible for acute ischemic heart disease syndromes', J. Am. Coll. Cardiol. 8, 245-250.
6. Sherman, C.T., Litvak, F., Grundfest, W., Lee, M., Hickey, A., Chaux, A., Kass, R., Blanche, C., Matloff, J., Morgenstern, L., Ganz, W., Swan, H.J.C., and Forrester, J. (1986) 'Coronary angioscopy in patients with unstable angina pectoris', N. Engl. J. Med. 315, 913-919.
7. Falk, E. (1983) 'Plaque rupture with severe pre-existing stenosis precipitating coronary thrombosis: Characteristics of coronary atherosclerotic plaques', Br. Heart J. 50, 127-134.
8. Davies, M.J. and Thomas, A.C. (1985) 'Plaque fissuring--The cause of acute myocardial infarction, sudden ischemic death, and crescendo angina', Br. Heart J, 53, 363-373.
9. Gotoh, K., Minamino, T., Katoh, O., Hamano, Y., Fukui, S., Hori, M.,

Kusuoka, H., Mishima, M., Inoue, M., and Kamada T. (1988) 'The role of intracoronary thrombus in unstable angina: Angiographic assessment and thrombolytic therapy during ongoing anginal attacks', Circulation 77, 526-534.
10. Hirsh, P.D., Hillis, L.D., Campbell, W.B., Firth, B.G., and Willerson J.T. (1981) 'Release of prostaglandins and thromboxane into the coronary circulation in patients with ischemic heart disease', N. Engl. J. Med. 304, 685-691.
11. Fitzgerald, D.J., Roy, L., Catella, F., and Fitzgerald, G.A. (1986) 'Platelet activation in unstable coronary disease', N. Engl. J. Med. 315, 983-989.
12. Hamm, C.W., Lorenz, R., Bleifeld, W., Kupper, W., Wober, W., and Weber, P. (1987) 'Biochemical evidence of platelet activation in patients with persistent unstable angina', J. Am. Coll. Cardiol. 9, 998-1004.
13. Van den Berg, E.K., Schmitz, J.M., Benedict, C.R., Malloy, C.R., Willerson, J.T., and Dehmer G.J. (1989) 'Transcardiac serotonin concentration is increased in selected patients with limiting angina and complex coronary lesion morphology', Circulation 79, 116-124.
14. Rubanyi, G.M., Frye, R.L., Holmes, D.R., and Vanhoutte, P.M. (1987) 'Vasoconstrictor activity of coronary sinus plasma from patients with coronary artery disease', J. Am. Coll. Cardiol. 9, 1243-1249.
15. Ambrose, J.A., Winters, S.L., Arora, R., Eng, A., Riccio, A., Gorlin, R., and Fuster, V. (1986) 'Angiographic evolution of coronary artery morphology in unstable angina', J. Am. Coll. Cardiol. 7, 472-478.
16. Gibson, R.S., Beller, G.A., Gheroghiade, M., Nygaard, T.W., Watson, D.D., Huey, B.L., Sayre, S.L., and Kaiser D.L. (1986) 'The prevalence and clinical significance of residual myocardial ischemia 2 weeks after uncomplicated non-Q-wave infarction, A prospective natural history study', Circulation 73, 1186-1198.
17. Folts, J.D., Crowell Jr., E.B., and Rowe, C.G. (1976) 'Platelet aggregation in partially obstructed vessels and its elimination by aspirin', Circulation 54, 365-370.
18. Bush, L.R., Campbell, W.B., Tilton, G.D., Buja, L.M., and Willerson J.T. (1984) 'Effects of the selective thromboxane synthetase inhibitor, dazoxiben, on variations in cyclic blood flow in stenosed canine coronary arteries', Circulation 69, 1161-1170.
19. Ashton, J.H., Schmitz, J.M., Campbell, W.B., Ogletree, M.L., Raheja, S., Taylor, A.L., Fitzgerald, C., Buja, L.M., and Willerson J.T. (1986) 'Inhibition of cyclic flow variations in stenosed canine coronary arteries by thromboxane A_2/prostaglandin H_2 receptor antagonists', Circ. Res. 59, 568-578.
20. Schmitz, J., Apprill, P., Buja, L.M., Willerson, J.T., and Campbell, W.B. (1985) 'Vascular prostaglandin and thromboxane production in a canine model of myocardial ischemia', Circ. Res. 57, 223-231.
21. Golino, P., Ashton, J.H., Buja, L.M., Taylor, A.L., McNatt, J., Rosolowsky, M., Campbell, W.B., and Willerson, J.T. (1989) 'Local platelet activation

causes vasoconstriction of large epicardial canine coronary arteries in vivo, Thromboxane A_2 and serotonin are possible mediators', Circulation 79, 154-166.

22. Golino, P., Buja, L.M., Ashton, J.H., Kulkarni, P., Taylor, A., and Willerson, J.T. (1988) 'Effect of thromboxane and serotonin receptor antagonists on intracoronary platelet deposition in dogs with experimentally stenosed coronary arteries', Circulation 78, 701-711.

23. Ashton, J.H., Benedict, C.R., Fitzgerlad, C., Raheja, S., Taylor, A., Campbell, W., Buja, L.M., and Willerson, J.T. (1986) 'Serotonin is a mediator of cyclic flow variations in stenosed canine coronary arteries', Circulation 73, 572-578.

24. Bush, L.R., Campbell, W.B., Kern, K., Tilton, G.D., Apprill, P., Ashton, J., Schmitz, J., Buja, L.M., and Willerson, J.T. (1984) 'The effects of alpha$_2$-adrenergic and serotonergic receptor antagonists on cyclic blood flow alterations in stenosed canine coronary arteries', Circ. Res. 55, 642-652.

25. Ashton, J.H., Ogletree, M.L., Michel, I.M., Taylor, A.L., Raheja, S., Schmitz, J., Buja, L.M., Campbell, W.B., and Willerson, J.T. (1987) 'Cooperative mediation by serotonin S_2 and thromboxane A_2/prostaglandin H_2 receptor activation of cyclic flow variations in dogs with severe coronary artery stenoses', Circulation 76, 952-959.

26. Ashton, J.H., Golino, P., McNatt, J.M., Buja, L.M., and Willerson, J.T. (1989) 'Serotonin S_2 receptor blockade (LY53857) combined with thromboxane A_2/prostaglandin H_2 receptor blockade (SQ29,548) provide protection against epinephrine-induced intermittent thrombosis in severely narrowed canine coronary arteries', J. Am. Coll. Cardiol. 13, 755-763.

27. Yao, S.K., Ober, J.C., McNatt, J., Benedict, C.R., Rosolowsky, M., Anderson, H.V., Cui, K., Maffrand, J.P., Campbell, W.B., Buja, L.M., and Willerson, J.T. (1992) 'ADP plays an important role in mediating platelet aggregation and cyclic flow variations in vivo in stenosed and endothelium-injured canine coronary arteries', Circ. Res. 70, 39-48.

28. Eidt, J.F., Allison, P., Nobel, S., Ashton, J., Golino, P., McNatt, J., Pastor, P., Buja, L.M., and Willerson, J.T. (1989) 'Thrombin is an important mediator of platelet aggregation in stenosed and endothelially-injured canine coronary arteries', J. Clin. Invest. 84, 18-27.

29. Willerson, J.T., Yao, S.K., Ferguson, J.J., Anderson, H.V., Golino, P., and Buja, L.M. (1991) 'Unstable angina pectoris and the progression to acute myocardial infarction: Role of platelets and platelet-derived mediators', Journal of the Texas Heart Institute 18, 243-237.

30. Yao, S.K., Ober, J., Krishnaswami, A., Ferguson, J.J., Anderson, H.V., Golino, P., Buja, L.M., and Willerson, J.T. (1992) 'Endogenous endothelium-derived relaxing factor protects against platelet aggregation and cyclic flow variations in stenosed and endothelium-injured arteries', Circulation, in press.

31. Eidt, J.F., Ashton, J., Golino, P., McNatt, J., and Buja, L.M. (1988) 'Flow variations in stenosed coronary arteries are mediated by serotonin and thromboxane A_2 in awake, closed chest dogs (abstract)', Clin. Res. 36, 603A.
32. Willerson, J.T., Yao, S.K., McNatt, J., Benedict, C.R., Anderson, H.V., Golino, P., Murphree, S.S., and Buja, L.M. (1991) 'Frequency and severity of cyclic flow alterations and platelet aggregation predict the severity of neointimal proliferation following experimental coronary atenosis and endothelial injury', Proc. Nat. Acad. Sci. 88, 10624-10628.
33. Fishman, J.A., Ryan, G.B., and Karnovsky, M.J. (1975) 'Endothelial regeneration in the rat carotid artery and the significance of endothelial denudation in the pathogenesis of myointimal thickening', Lab. Invest. 32, 339-351.
34. Austin, G.E., Ratliff, N.B., Hollman, J., Tabei, S., and Phillips, D.F. (1985) 'Intimal proliferation of smooth muscle cells as an explanation for recurrent coronary artery stenosis after percutaneous transluminal coronary angioplasty', J. Am. Coll. Cardiol. 6, 369-375.
35. Jonasson, L., Holm, J., and Hansson, G.K. (1988) 'Cyclosporin A inhibits smooth muscle proliferation in the vascular response to injury', Proc. Natl Acad. Sci. USA 85, 2302-2306.
36. Berk, B.C., Alexander, R.W., Brock, T.A., et al. (1986) 'Vasoconstriction, A New activity for platelet-derived growth factor', Science 232, 87.
37. Waterfield, M.D., Scrace, G.T., Whittle, N., et al. (1983) 'Platelet-derived growth factor is structurally related to the putative transforming protein p28^{cis} of simian sarcoma virus', Nature (Lond.) 304, 35-39.
38. Williams, L.T., Tremble, P.M., Lavin, M.R., and Sunday, M.E. (1984) 'Platelet-derived growth factor receptors form a high affinity state in membrane preparations, Kinetics and affinity cross-linking studies', J. Biol. Chem. 259, 5287-5294.
39. Roberts, A.B., Sporn, M.B., Assoian, R.K., et al. (1986) 'Transforming growth factor type beta: Rapid induction of fibrosis and angiogenesis *in vivo* and stimulation of collagen formation *in vitro*', Proc. Natl Acad. Sci. USA 83, 4167-4171.
40. Akhurst, R.J., Fee, F., and Balmain, A. (1988) 'Localized production of TGF-beta mRNA in tumour promoter-stimulated mouse epidermis', Nature 331, 363-364.

ABNORMAL CORONARY VASOCONSTRICTOR RESPONSES TO SEROTONIN IN PATIENTS WITH ISCHEMIC HEART DISEASE

C.J. VRINTS, J. BOSMANS, H. BULT[*], A.G. HERMAN[*], and J.P. SNOECK
Department of Cardiology, University Hospital of Antwerp and
[*]*Division of Pharmacology, University of Antwerp (UIA)*
Wilrijkstraat 10
2590 Edegem
Belgium

ABSTRACT. The coronary vasomotor responses to acetylcholine, serotonin (5-HT) and to isosorbide dinitrate were assessed by quantitative coronary arteriography in 31 patients.
  Acetylcholine induced constrictions in diseased coronary arteries (n=15). In the patients with smooth normal coronary arteriograms, acetylcholine caused dilatations in 7 patients (smooth dilators) and constrictions in 9 patients (smooth constrictors). 5-HT evoked no significant diameter changes in the smooth dilators (+1.4 ± 4.1%), but in the smooth constrictors and in the patients with diseased coronary arteries 5-HT caused potent constrictions. The vasoconstrictor responses to 5-HT were similar in patients with minimal (-26.5 ± 4.7%) and with advanced atherosclerosis (-30.9 ± 5.3%). Isosorbide dinitrate reversed rapidly the vasoconstriction induced by 5-HT.
  Both in early and advanced atherosclerosis, impairment of endothelium-dependent vasodilation causes a hypersensitivity of the coronary artery to the vasoconstrictor stimulus of 5-HT.

1. Introduction

In normal coronary arteries the vascular endothelium constitutes an active defense mechanism against platelet aggregation and abnormal vasoconstriction [1-3]. The adenine nucleotides and 5-HT released by aggregating platelets stimulate the endothelial cells to release nitric oxide which completely counteracts the direct contracting effects of 5-HT and thromboxane on the vascular smooth muscle cells. In endothelium-denuded arteries this protective function of the endothelium is lost and aggregating platelets cause vasoconstriction. This has led to the hypothesis that endothelial dysfunction in diseased coronary arteries may lead to augmented vasoconstrictor responses to 5-HT released by aggregating platelets [2]. These abnormal vasoconstrictions could play an important role in the pathogenesis of myocardial ischemia in patients with unstable coronary artery syndromes [4].

Studies in several animal models have shown that atherosclerosis leads to a major impairment of endothelium-dependent vasodilation which results in a potentiation of the vasoconstrictor responses to ergonovine and to 5-HT [5-10]. The results of several clinical studies with intracoronary infusion of acetylcholine suggest that an impairment endothelium-dependent vasodilation is also present in patients with coronary artery disease [11-14]. However, the impaired cholinergic responses observed in atherosclerotic coronary arteries could be related to a selective dysfunction of the endothelial muscarinic receptors as was suggested by some findings in isolated human coronary arteries [8]. Moreover, several studies with isolated human coronary arteries could not demonstrate an important endothelial-dependent vasodilator response to 5-HT in apparently normal coronary arteries [15-18]. Although one study showed an attenuation by the endothelium of the contractions of isolated normal human coronary arteries to 5-HT [16], it could be that the endothelium-dependent vasodilator effect to 5-HT is less important in human coronary arteries than that observed in other species. The results with intracoronary infusion of acetylcholine can therefore not directly be extrapolated to 5-HT. The purpose of this study was therefore to examine whether the impairment of endothelium-dependent vasodilation in patients with coronary artery disease also leads to a change in the coronary vasomotor responses to 5-HT.

2. Materials and Methods

2.1 PATIENTS

Coronary vascular responses to acetylcholine, 5-HT and to isosorbide dinitrate were assessed by quantitative coronary arteriography in 31 patients. The patients were classified in two groups based on the presence or absence of coronary atherosclerosis as observed on the coronary arteriogram.

2.1.1 Group 1. Sixteen patients had angiographically smooth coronary arteries. These patients were further divided in two subgroups. Group 1a comprised seven patients with a nonsignificant or a mild vasodilator response to acetylcholine and with atypical symptoms (smooth dilators). Most of these patients showed no objective evidence of myocardial ischemia and may be considered to represent normal patients. Group 1b consisted of nine patients with a vasoconstrictor response (> 5% diameter decrease) to acetylcholine (smooth constrictors). Most of them presented typical anginal pain symptoms and were therefore suspected to have syndrome X.

2.1.2 Group 2. Fifteen patients with overt coronary artery disease were further subdivided in two subgroups based on the degree of atherosclerosis present in the studied artery. Group 2a consisted of eleven patients with mild lesions (< 30% stenosis) in the studied artery. Group 2b comprised five patients with moderate lesions ranging from 50 to 75% luminal diameter narrowing.

Mean age and serum cholesterol levels were not significantly different and all other coronary risk factors were equally distributed in the different patients groups.

2.2 STUDY PROTOCOL

All vasoactive medication was withheld for at least 14 hours before cardiac catheterization. A 3 French infusion catheter was advanced into the proximal segment of a major coronary artery. Serial intracoronary infusions were performed each during two minutes at a constant infusion rate of 2 ml/min in the following sequence: 1) 10^{-5} M acetylcholine; 2-4) 5-HT with stepwise increasing concentrations (10^{-7} M, 10^{-6} M, and finally 10^{-5} M); 5) 10^{-4} M acetylcholine if no angina or excessive vasoconstriction was induced by 5-HT. At the end of the protocol or earlier if a marked vasoconstriction accompanied by angina or ischemic repolarization changes occurred 1 mg of isosorbide dinitrate was injected intracoronary. Coronary arteriograms were obtained under control conditions, at the end of each infusion of acetylcholine and 5-HT and 30 seconds after the intracoronary administration of isosorbide dinitrate.

3. Results

3.1 RESPONSES IN PATIENTS WITH NORMAL CORONARY ARTERIOGRAMS (GROUP 1)

In the patients with angiographically smooth normal coronary arteries acetylcholine and 5-HT evoked nonsignificant or small vasodilator changes in group 1a or marked vasoconstrictions in group 1b.

3.1.1 *Group 1a, smooth vasodilators.* In this group of patients who to some extent can be regarded as normals, 10^{-5} M acetylcholine induced a mild vasodilation. Infusion of 5-HT evoked no vasoconstrictor responses in this subgroup. A slight but nonsignificant diameter increase was observed in response to 5-HT in the distal coronary artery segments. Subsequent infusion of 10^{-4} M acetylcholine induced no significant changes. Isosorbide dinitrate evoked a marked vasodilation.

3.1.2 *Group 1b; smooth vasoconstrictors.* In this subgroup of patients with smooth normal coronary arteriograms in whom 10^{-5} M acetylcholine evoked vasoconstrictions, infusion of 5-HT also caused vasoconstrictor responses. Infusion of 10^{-6} M and 10^{-5} M 5-HT caused marked vasoconstrictions both in the mid and distal coronary artery segments. Infusion of 10^{-4} M acetylcholine increased further the vasoconstriction induced by 5-HT. In several patients (5 of 10) the induced vasoconstriction caused anginal chest pain and ischemic ECG changes. Isosorbide dinitrate rapidly relieved the anginal pain. After the administration of the nitrate the coronary diameters returned to their baseline diameter in the mid segments, whereas a significant diameter increase occurred in the distal coronary artery segments.

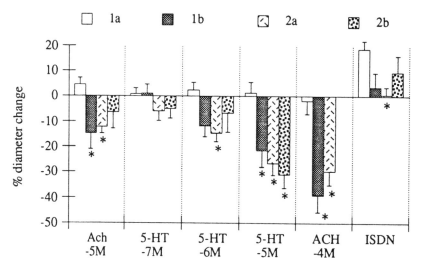

Figure 1. Diameter changes of the mid coronary artery segments after selective intracoronary infusions of 10^{-5} M acetylcholine (Ach), graded concentrations ($10^{-7} \rightarrow 10^{-5}$ M) of 5-HT, 10^{-4} M acetylcholine and after the intracoronary administration of 1 mg of isosorbide dinitrate (ISDN). Values are shown as means ± sem; * denotes a significant difference as compared to the patients of subgroup 1a.

3.2 RESPONSES IN PATIENTS WITH CORONARY ARTERY DISEASE (GROUP 2)

In the patients with coronary artery disease 10^{-5} M acetylcholine induced a mild vasoconstriction of all coronary artery segments. Infusion of 5-HT caused dose-dependent vasoconstrictions in all patients. The responses to 5-HT were similar in the patients with mild (group 2a) or more advanced coronary lesions (group 2b). At low concentrations the vasoconstrictions were limited to the diseased segments whereas at high concentrations the more normally appearing segments also showed a constrictor response. In the patients in whom 10^{-4} M acetylcholine was subsequently infused, marked constrictions were observed which were somewhat more pronounced than those observed with 10^{-5} M 5-HT. During the vasoconstriction induced by 10^{-5} M 5-HT and by 10^{-4} M acetylcholine several patients developed anginal chest pain and ischemic repolarization changes. Intracoronary administration of isosorbide dinitrate reversed the vasoconstriction and relieved rapidly the anginal pain in all patients except in one in whom 10^{-5} M 5-HT caused a temporary thrombotic occlusion of the distal left anterior descending artery (LAD).

4. Discussion

The present study confirms prior clinical studies with intracoronary infusion of acetylcholine which have demonstrated that endothelium-dependent vasodilation is impaired in diseased coronary arteries [11, 12, 14]. This impairment starts relatively early in the evolution of coronary artery disease and can already be observed at a stage when atherosclerotic lesions cannot yet be detected by coronary arteriography [13, 14].

The major finding of the present study is the observation of paradoxical vasoconstrictions to low concentrations of 5-HT in patients with coronary artery disease. Furthermore, the changes in the responses to 5-HT exactly mirror those observed with acetylcholine: in patients with normal coronary arteries in whom acetylcholine induces a small dilator response, 5-HT causes a nonsignificant change or a mild vasodilation; whereas the patients with early or advanced atherosclerosis respond with potent vasoconstriction to both acetylcholine and 5-HT. The close alignment of the changes in coronary vascular reactivity to acetylcholine and 5-HT suggests therefore that the impairment of endothelium-dependent vasodilation in diseased coronary arteries is not limited to the muscarinic receptors but may involve the endothelial serotonergic receptors as well. As also observed with acetylcholine, 5-HT caused already marked constrictions in arteries with only mild atherosclerosis and also in the angiographically normal-appearing adjacent and distal coronary artery segments. Moreover, 5-HT also caused potent vasoconstrictor responses in patients with smooth coronary arteries but with symptoms of angina which were suspected to have syndrome X. It appears therefore that the impairment of endothelium-dependent vasodilation to 5-HT occurs as early in the evolution of coronary atherosclerosis as that observed with acetylcholine and may already be present in angiographically normal coronary arteries.

The present study corroborates the results of two other studies with intracoronary infusion of 5-HT which also demonstrated vasoconstrictor responses in patients with coronary artery disease whereas mild vasodilations or nonsignificant changes were observed in patients without coronary atherosclerosis [19, 20]. The patients with stable angina showed a marked but nonocclusive vasoconstrictor response to low concentrations of 5-HT. However, in patients with variant angina, the vasoconstrictor responses to 5-HT were more marked than in patients with stable angina and a complete occlusive spasm was observed [20]. It appears therefore that the coronary arteries of patients with variant angina are not only hypersensitive but also hyperreactive to the vasoconstrictor stimulus of 5-HT whereas the patients with stable angina only are hypersensitive. A similar difference can be observed with ergonovine [21]. The hyperreactivity of patients with variant angina can probably not solely be explained by the unbalance between impaired endothelium-dependent vasodilation and the direct contracting effect of 5-HT. Potential mechanisms which also could be involved in the hyperreactivity in variant angina are: geometrical magnification of the constriction at the site of a coronary stenosis, increased production of endothelium-derived contracting factors in response to 5-HT or an increased contractility of the smooth muscle cells.

Clinical studies with ketanserin suggest that the abnormal vasoconstrictor responses to 5-HT in coronary artery disease are mediated by both $5-HT_1$-like and $5-HT_2$ receptors [19, 22]. These findings are consistent with earlier studies in canine [23] and in isolated human coronary arteries [16, 17]. Ketanserin potentiates the vasodilations of normal coronary arteries and inhibits the vasoconstrictions of normal distal coronary segments induced by high concentrations of 5-HT. However, in the patients with variant angina and in the distal coronary artery segments of patients with stable angina, ketanserin has no inhibitory effect. It appears therefore that in coronary artery disease $5-HT_1$-like receptors are the predominant receptor type involved in the abnormal vasoconstrictions to 5-HT.

The abnormal vasoconstrictor responses to 5-HT in atherosclerotic coronary arteries play probably an important role in the pathogenesis of myocardial ischemia in acute coronary artery syndromes. In unstable angina, aggregation of platelet at the site of a ruptured atherosclerotic plaque may lead to a local accumulation of 5-HT [24]. As shown in this study, this will lead to a dynamic increase in coronary artery tone not only at the level of the endothelium-denuded proximal lesion but also in the distal coronary artery segments where the endothelium-dependent vasodilation is also impaired. Myocardial ischemia may result from an obstructive vasospasm at the level of the proximal stenosis. The diffuse constrictions induced by 5-HT in the distal vessels may also contribute to the development of ischemia according to the recently introduced concept on the role of distal coronary artery constriction [25].

5. References

1. Houston, D.S., Shepherd, J.T., and Vanhoutte, P.M. (1985) 'Adenine nucleotides, serotonin, and endothelium-dependent relaxations to platelets', Am. J. Physiol. 248, H389-H395.
2. Vanhoutte, P.M. and Houston, D.S. (1985) 'Platelets, endothelium, and vasospasm', Circulation 72, 728-734.
3. Furchgott, R.F. and Vanhoutte, P.M. (1989) 'Endothelium-derived relaxing

and contracting factors.', FASEB J 3, 2007-2018.
4. Willerson, J.T., Golino, P., Eidt, J., Campbell, W.B., and Buja, L.M. (1989) 'Specific platelet mediators and unstable coronary artery lesions. Experimental evidence and potential clinical implications', Circulation 80, 198-205.
5. Henry, P.D. and Yokoyama, M. (1980) 'Supersensitivity of atherosclerotic rabbit aorta to ergonovine. Mediation by a serotonergic mechanism', J. Clin. Invest. 66, 306-313.
6. Verbeuren, T.J., Jordaens, F.H., Zonnekeyn, L.L., Van Hove, C.E., Coene, M.C., and Herman, A.G. (1986) 'Effect of hypercholesterolemia on vascular reactivity in the rabbit. I. Endothelium-dependent and endothelium-independent contractions and relaxations in isolated arteries of control and hypercholesterolemic rabbits', Circ. Res. 58, 552-564.
7. Heistad, D.D., Armstrong, M.L., Marcus, M.L., Piegors, D.J., and Mark, A.L. (1986) 'Potentiation of vasoconstrictor responses to serotonin in the limb of atherosclerotic monkeys', J. Hypertension 4, S17-S21.
8. Bossaller, C., Habib, G.B., Yamamoto, H., Williams, C., Wells, S., and Henry, P.D. (1987) 'Impaired muscarinic endothelium-dependent relaxation and cyclic guanosine 5'-monophosphate formation in atherosclerotic human coronary artery and rabbit aorta', J. Clin. Invest. 79, 170-174.
9. Shimokawa, H. and Vanhoutte, P.M. (1989) 'Impaired endothelium-dependent relaxation to aggregating platelets and related vasoactive substances in porcine coronary arteries in hypercholesterolemia and atherosclerosis', Circ. Res. 64, 900-914.
10. Vrints, C., Verbeuren, T.J., Snoeck, J., and Herman, A.G. (1990) 'Effects of hypercholesterolemia on coronary vascular reactivity', in Rubanyi, G. M. and Vanhoutte, P. M. (eds.), Endothelium-derived contracting factors, Karger, Basel, pp. 162-168.
11. Ludmer, P.L., Selwyn, A.P., Shook, T.L., Wayne, R.R., Mudge, G.H., Alexander, R.W., and Ganz, P. (1986) 'Paradoxical vasoconstriction induced by acetylcholine in atherosclerotic coronary arteries', N. Engl. J. Med. 315, 1046-1051.
12. Zeiher, A.M., Drexler, H., Wollschlaeger, H., Saurbier, B., and Just, H. (1989) 'Coronary vasomotion in response to sympathetic stimulation in humans: importance of the functional integrity of the endothelium', J Am Coll Cardiol 14, 1181-1190.
13. Vita, J.A., Treasure, C.B., Nabel, E.G., McLenachan, J.M., Fish, R.D., Yeung, A.C., Vekshtein, V.I., Selwyn, A.P., and Ganz, P. (1990) 'Coronary vasomotor response to acetylcholine relates to risk factors for coronary artery disease', Circulation 81, 491-497.
14. Vrints, C.J.M., Hitter, E., Bult, H., Herman, A.G., and Snoeck, J.P. (1992) 'Impaired endothelium-dependent cholinergic coronary vasodilation in patients with angina and normal coronary arteriograms', J. Am. Coll. Cardiol.

19, 21-31.
15. Förstermann, U., Mügge, A., Bode, S.M., and Frölich, J.C. (1988) 'Response of human coronary arteries to aggregating platelets: importance of endothelium-derived relaxing factor and prostanoids', Circ. Res. 63, 306-312.
16. Berkenboom, G., Unger, P., Ying Fang, Z., and Degre, S. (1989) 'Comparison of responses to acetylcholine and serotonin on isolated canine and human coronary arteries', Cardiovasc. Res. 23, 780-787.
17. Chester, A.H., Martin, G.R., Bodelsson, M., Arneklo-Nobin, B., Tadjkarimi, S., Tornebrandt, K., and Yacoub, M.H. (1990) '5-Hydroxytryptamine receptor profile in healthy and diseased human epicardial coronary arteries', Cardiovasc Res 24, 932-937.
18. Toda, N. and Okamura, T. (1990) 'Comparison of the response to 5-carboxamidotryptamine and serotonin in isolated human, monkey and dog coronary arteries', J. Pharmacol. Exp. Therap. 253, 676-682.
19. Golino, P., Piscione, F., Willerson, J.T., Capelli-Bigazzi, M., Focaccio, A., Villari, B., Indolfi, G., Russolillo, E., Condorelli, M., and Chiariello, M. (1991) 'Divergent effects of serotonin on coronary artery dimensions and blood flow in patients with coronary atherosclerosis and control patients.', N Engl. J. Med. 324, 641-648.
20. McFadden, E.P., Clarke, J.G., Davies, G.J., Kaski, J.C., Haider, A.W., and Maseri, A. (1991) 'Effect of intracoronary serotonin on coronary vessels in patients with stable angina and patients with variant angina', N. Engl. J. Med. 324, 648-654.
21. Kaski, J.C., Crea, F., Meran, D., Rodriguez, L., Araujo, L., Chierchia, S., Davies, G., and Maseri, A. (1986) 'Local coronary supersensitivity to diverse vasoconstrictive stimuli in patients with variant angina', Circulation 74, 1255-1265.
22. McFadden, E.P., Bauters, C., Lablanche, J.M., Leroy, F., Clarke, J.G., Henry, M., Schandrin, C., Davies, G.J., Maseri, A., and Bertrand, M.E. (1992) 'Effect of ketanserin on proximal and distal coronary constrictor responses to intracoronary infusion of serotonin in patients with stable angina, patients with variant angina, and control patients', Circulation 86, 187-195.
23. Houston, D.S. and Vanhoutte, P.M. (1988) 'Comparison of serotonergic receptor subtypes on the smooth muscle and endothelium of the canine coronary artery', J. Pharmacol. Exp. Ther. 244, 1-10.
24. Van den Berg, E.K., Schmitz, J.M., Benedict, C.R., Malloy, C.R., Willerson, J.T., and Dehmer, G.J. (1989) 'Transcardiac serotonin concentration is increased in selected patients with limiting angina and complex coronary lesion morphology', Circulation 79, 116-124.
25. Pupita, G., Maseri, A., Kaski, J.C., Galassi, A.R., Gavrielides, S., Davies, G., and Crea, F. (1990) 'Myocardial ischemia caused by distal coronary constriction in stable angina pectoris', N. Engl. J. Med. 323, 514-520.

5-HT AND THE IMMUNE SYSTEM

G. FILLION, M.-P. FILLION, I. CLOEZ-TAYARANI,
H. SARHAN, F. HAOUR, and F. BOLANOS
Unité de Pharmacologie Neuro-Immuno-Endocrinienne,
Institut Pasteur
28, rue du Dr Roux
F75015 Paris
France

ABSTRACT. Functional interactions between the central nervous system and the immune system occur through as a complex network involving neurotransmitters, hormones, and cytokines. The serotoninergic system likely plays a role in these interactions.

1. Introduction

The reciprocal interactions between the Central Nervous Systems (CNS) and the Immune System (IS) were suggested long ago [1], but it was only in the eighties that the field of "Psychoneuroimmunology" retained the interest of scientists [2] and then was largely studied [3,4,5]. Parallelly, the interest for 5-hydroxytryptamine started in the early fifties with the discovery of this amine and was renewed in the seventies with that of the receptors which progressively increased in number and complexity.

The role of 5-HT in the immune response was suggested early on [6] but it was only in the last decade that experimental studies supported this hypothesis. Alterations of the 5-HT system in the CNS have been shown to affect the immune response [7]. The antibody production induced by sheep red blood cells (SRBC) in rats is enhanced by inhibition of 5-HT synthesis [8,9] and conversely the increase in 5-HT level, using 5-HT itself or its precursor 5-HTP, decreases the immune response [8]. Accordingly, lesion of the raphe area (where all the serotonergic cellular bodies are located) leads to an increase of the response [9], but surprisingly the use of 5,7-DHT, a specific serotonergic neurotoxic, does not affect the response [8]. Furthermore, the fact that 5-HT inhibits the *in vitro* proliferation of human lymphocytes [10] and that recognition sites for 5-HT are present on macrophages [11,12] and lymphocytes [13] are in favor of a direct effect of 5-HT on immunocompetent cells (this hypothesis was also proposed by van Loveren et al. [14]. More recently, functional serotonergic effects on lymphocytes involving specific receptors were demonstrated, i.e. 5-HT is able to modulate voltage-gated potassium conductance through 5-HT$_1$ and 5-HT$_3$ receptors [15]. In addition, Hellstrand and Hermodsson [16,17 and this meeting] convincingly proposed that 5-HT regulates NK-cell responsivness to interleukin-2 (IL-2) via 5-HT$_{1A}$ receptors. It is also known that 5-HT may indirectly affect the immune response through the pituitary-adrenocortical function [18].

The reciprocal interaction of the immune activity on the cerebral 5-HT system is only suggested on the basis of a few indications. The group of S. Carlson studied 5-HT levels in various brain areas in mice following immunization with SRBC and reported a significant increase (nucleus tractus solitarius) or decrease (paraventricular nucleus, supraoptic nucleus, hippocampus) in 5-HT content 2 or 4 days after primary immunization [19]. It has also been shown that ICV injection of interleukin-1 (IL-1) in rats is able to induce a rapid increase in extracellular 5-hydroxy indole acetic acid, a metabolite of 5-HT, in the anterior hypothalamus. This effect is reversed by a specific antagonist, and presumably corresponds to a direct activation of the serotonergic system by IL-1 [20].

We have examined the presence of potential 5-HT receptors on immunocompetent tissues and the effect of modifications of the immune activity on the central serotonergic system.

2. Existence of Potential 5-HT Receptors in Immunocempetent Cells

2.1 QUANTITATIVE AUTORADIOGRAPHY

Using tritiated 5-HT at low concentrations (2-3 nM) the reversible, high-affinity binding for 5-HT was examined in various tissues containing immunocompetent cells. The results were obtained in mouse, rat, and human tissues. They show that high-affinity binding sites for [^{3}H]5-HT are present in mice thymus, in rat spleen, in human touncils, and also in sections of pellets of human lymphocytes obtained by centrifugation. These results favor the hypothesis of the existence in these tissues of serotonergic receptors having a nanomolar affinity for 5-HT. However, the fact that the pharmacological profile of the binding was not defined and the functions of the various sites were still unknown, does not allow us to conclude that these sites could correspond to receptors, i.e. 5-HT$_1$-like.

The binding of [^{3}H]5-HT to human lymphocytes was more closely studied. Bonnet et al. [21] had already observed a low-affinity binding (Kd = 198 nM) on lymphocytes; this binding presumably corresponds to the uptake of the amine since it was observed on whole cells and was inhibited by imipramine, a drug known as an uptake inhibitor. In our assays, human lymphocyte membranes were obtained using the method described by Boyum [22]. Saturation curves obtained at concentrations of [^{3}H]5-HT in the nanomolar range showed a saturable, reversible, high-affinity binding. The affinity constant (Kd) varied from 2 to 8 nM and the total number of sites ranged from 40 to 110 fmol/mg of protein depending on the preparation. Competing experiments with serotonergic ligands exhibited several populations of sites which could not be pharmacologically defined under our experimental conditions.

In order to further characterize the potential receptor sites present on immunocompetent cells, studies of the expression of gene coding for particular 5-HT receptors were performed. Using the technique of reverse transcriptase-polymerase chain reaction (RT-PCR), in which a particular messenger RNA is transcribed to cDNA, then amplified using oligonucleotide primers specific of nucleotidic sequence of a given 5-HT receptor and, finally, the amplified product after transfer on a nylon membrane is identified by hybridization with a specific labeled probe, we were able to demonstrate, in mouse thymus, the presence of mRNAs corresponding to $5-HT_{1A}$, $5-HT_{1B}$, and $5-HT_2$ receptors; preliminary assays have also shown the presence of mRNAs corresponding to $5-HT_{1D}$ receptors. These results indicate that in mouse thymus, gene coding for $5-HT_{1A}$, $5-HT_{1B}$, $5-HT_{1D}$, and $5-HT_2$ receptors are expressed. Therefore, the corresponding protein could be produced in the tissue. Although not demonstrated, the high-affinity binding sites observed in this tissue presumably correspond to the traduction of these mRNAs as receptor proteins. That $5-HT_{1A}$ mRNA are found in thymus is in good agreement with the results showing functional properties of $5-HT_{1A}$ receptors in thymocyte-derived cells (Hellstrand et al., this meeting).

3. Effect of Modulations of the Immune Activity on the Serotonergic Cerebral System

Modulations of the immune activity in mice or rats were obtained using three different experimental approaches: immunization with SRBC, pharmacological agents stimulating or depressing the immune response, and viral infection.

3.1 IMMUNIZATION OF MICE WITH SHEEP RED BLOOD CELLS (SRBC)

Immunization of mice (C_3H/OU, male, 8 weeks old) was obtained with SRBC. Brains of mice were collected 16 hours and 4 days after immunization. The functional activity of hippocampal $5-HT_{1A}$ receptors was examined by measuring the effect of 8-OH-DPAT (1 μM) to inhibit the adenylyl cyclase activity stimulated by forskolin (10 μM). Preliminary results show that the inhibitory activity of 8-OH-DPAT was reduced in immunized animals (19.9 ± 1% in control animals, 13.5 ± 0.6% after 16 hours, and 11.8 ± 0.7% 4 days after immunization). This effect was not observed in hypothalamus possibly because the inhibitory effect of 8-OH-DPAT on adenylyl cyclase activity is weak in this area (maximal inhibition = 10.9 ± 1%).

Parallelly, the number of $5-HT_{1A}$ receptor sites was examined by quantitative autoradiography. Thus, in the hippocampus [^{3}H]8-OH-DPAT binding was not significantly modified in stratum radiatum and lacunosum in CA_1 and CA_3, slightly but significantly increased in stratum Oriens in CA_3 (18%) after 16 hours but not after 4 days, and no significant changes were observed in dentate gyrus for the same periods. These results suggest that functional regulations at the 5-HT_{1A} receptors do not parallel changes in the number of sites.

The specific radioligand binding for 5-HT_{1B} (in subtantia nigra and subiculum, [^{125}I-Cyanopindolol), 5-HT_{1D} (in basal ganglia and substantia nigra, [^{3}H]-5HT in the presence of drugs masking 5-HT_{1A}, 5-HT_{1B}, 5-HT_{1C}), 5-HT_{1E} ([^{3}H]5-HT in the presence of 5-CT), and 5-HT_{2} (in frontal cortex [layer IV] [^{3}H]ketanserin) receptors was also examined. Nonsignificant variations were observed between control and immunized animals. These results suggest that either the latter receptors are not modified by immunization or they are, as 5-HT_{1A} receptors, functionally modulated.

3.2 EFFECT OF IMMUNOMODULATORS

Muramyl dipeptide (MDP) (15 mg/kg ip) injected 2 hours before sacrifice was used to stimulate the immune response in rat. The function of the heterologous presynaptic receptors (5-$HT_{1B/D}$), was examined *in vitro* by measuring the capacity of the nonselective serotonergic agonist TFMPP, to inhibit the K^+ evoked release of acetylcholine from synaptosomal preparations isolated from cortex, hippocampus, or striatum. MDP induced a significant decrease of the inhibitory activity of TFMPP in cortex (-38 ± 12 %) but not in hippocampus or striatum.

Cyclosporine (10 mg/kg per day 3 times) was used to depress the immune activity in rat. Under these conditions, the effect of TFMPP was markedly reduced in cortex (-40 ± 17 %), slightly and not significantly in striatum and not modified in hippocampus.

These results suggest that the modification of the immune activity may alter the serotonergic function particularly in the cortex.

3.3 VIRUS INFECTION

Rabies virus was administered to rats through bilateral intramasseter injections. Binding of [^{3}H]5-HT to 5-HT_{1} receptors was examined in cortex, hippocampus, and striatum, on day 5 after infection. The corresponding binding decreased significantly in cortex, modestly in striatum, and was not modified in hippocampus. The 5-HT_{1D} specific binding was examined in the cortex every day after infection. It progressively diminished to 50% of the control during day 1-2-3 and then remained at this value up to day 6 when the animal died.

The cellular function of 5-$HT_{1B/1D}$ receptors as presynaptic modulatory receptors, was examined. At day 5 after infection, the inhibitory activity of TFMPP on the K^+-evoked release of ACh was markedly reduced in synaptosomes from cortex, slightly and nonsignificantly in striatum whereas it was not affected in hippocampus. At day 4, no significant modification was observed.

Therefore, these results strongly suggest an early modification of 5-HT_{1D} receptors density in cortex induced by the infection. This effect might be in relation with a primary immune response since it is not a local effect of the virus; indeed viral particles have not invaded the cortical cerebral tissue at that time. In contrast, the functional modifications on presynaptic regulatory serotonergic receptors which occur only at day 5 after infection are presumably directly related to the presence of the virus in the nervous tissue.

4. Conclusion

It is likely that the reciprocal interactions occurring between CNS and IS involve the serotonergic system. In fact, immunocompetent tissues appear to possess recognition sites for 5-HT which may represent functional receptors, presumably similar to those existing in the brain. Therefore, the 5-HT system may use many different receptor types to control the immune response. As in the case of CNS, this large variety of receptors would favor the selectivity of the serotonergic activity on various tissues or cell types and allow a fine tuning in the control of the immune activity. However, at the present time, the cellular functions of these receptors are not yet known except in a few cases [15,17, and see introduction]. It is not yet known whether this function is directly related to the CNS activity or to that of a peripheral 5-HT system, indirectly regulated via neuronal activity, endocrine secretions, or other signals [18,23, and this meeting].

Furthermore, alterations of the cerebral serotonergic system are induced by changes in the immune activity. The mechanisms are not yet understood; however, it is likely that immune signalling in the brain is involved. The presence of IL-1 receptors in hippocampus and, to a less extent, in cortex, has been demonstrated in mouse brain [24,25,26] suggesting that brain activity might be modulated by the cytokines. Moreover, the fact that ICV administration of IL-1 enhances the release of 5-HT [20] strongly suggests the implication of the 5-HT system in the CNS response to an immune stimulus.

Better knowledge of these reciprocal interactions will undoubtedly be the basis for improvements in the field of psychiatry as well as that of cancer therapy.

5. References

1. Metalnikov, S. (1931) 'Role du système nerveux et des réflexes conditionnels dans l'immunité', Ann. Institut Pasteur 46, 137-168.
2. Ader, R. (ed.) (1981) Psychoneuroimmunology, Academic Press, New York.
3. Weigent, D.A. and Blalock, J.E. (1987) 'Interactions between the neuroendocrine and immune systems: common hormones and receptors', Immunological Rev. 100, 79.
4. Besedovsky, H.O., del Rey, A., and Sorkin, E. (1985)'Immunoneuroendocrine interactions', J. Immun. 135, 750s-754s.
5. Ader, R., Felten, D., and Cohen, N. (1990) 'Interactions between the brain and the immune system', Annu. Rev. Pharmacol. Toxicol. 30, 561-602.

6. Davis, R.B. (1968) 'Discussion of the role of 5-hydroxyindoles in the carcinoid syndrome', Advances in Pharmacol. 6, 146-149.
7. Boranic, M., Pericic, D., Poljak-Blazi, M., Manev, H., Sverko, V., Gabrilovac, J., Radacic, M., Pivac, N., and Miljenovic, G. (1990) 'Immune response of stressed rats treated with drugs affecting serotoninergic and adrenergic transmission', Biomed. & Pharmacother. 44, 381-387.
8. Jackson, J.C., Cross, R.J., Walker, R.F., Markesbery, W.R., Brooks, W.H., and Roszman, T.L. (1985) 'Influence of serotonin on the immune response', Immunology 54, 505-512.
9. Devoino, L., Morozova, N., and Cheido, M. (1988) 'Participation of serotoninergic system in neuroimmunomodulation: intraimmune mechanisms and the pathways providing an inhibitory effect', Int. J. Neurosci. 40, 111-128.
10. Slauson, D.O., Walker, C., Kristensen, F., Wang, Y., and de Weck, A.L. (1984) 'Mechanisms of serotonin-induced lymphocyte proliferation inhibition', Cell. Immunol. 1984, 84, 240-252.
11. Silverman, D.H.S., Wu, H., and Karnovsky, M.L. (1985) 'Muramyl peptides and serotonin interact at specific binding sites on macrophages and enhance superoxide release', Biochem. Biophys. Res. Com. 131 (3), 1160-1167.
12. Silverman, D.H.S., Krueger, J.M., and Karnovsky, M.L. (1986) 'Specific binding sites for muramyl peptides on murine macrophages', J. Immunol. 136, 2195-2198.
13. Bonnet, M., Lespinats, G., and Burtin, C. (1987) 'Evidence for serotonin (5HT) binding sites on murine lymphocytes'. Int. J. Immunopharmac. 9, 551-556.
14. Van Loveren, H., Den Otter, W., Meade, R., Terheggen, P.M.A.B., and Askenase, P.W. (1985) 'A role for mast cells and the vasoactive amine serotonin in T cell-dependent immunity to tumors', J. Immunol. 134 (2), 1292-1299.
15. Choquet, D. and Korn, H. (1988) 'Dual effects of serotonin on a voltage-gated conductance in lymphocytes', Proc. Nat. Acad. Sci. 85, 4557-4561.
16. Hellstrand, K. and Hermodsson, S. (1987) 'Role of serotonin in the regulation of human natural killer cell cytotoxicity', J. Immunol. 139 (3), 869-875.
17. Hellstrand, K. and Hermodsson, S. (1990) 'Monocyte-mediated suppression of IL-2-induced NK-cell activation. Regulation by 5-HT1A-type serotonin receptors', Scand. J. Immunol. 32, 183-192.
18. Fuller, R. (1992) 'The involvement of serotonine in regulation of pituitary adreno-cortical function', Frontiers in Neuroendocrin. 3, 250-270.
19. Carlson, S.L., Felten, D.L., Livnat, S., and Felten, S.Y. (1987) 'Alterations of monoamines in specific central autonomic nuclei following immunization in mice', Brain, Behavior, and Immunity 1, 52-63.
20. Gemma, C., Ghezzi, P., and De Simoni, M.G. (1991) 'Activation of the

hypothalamic serotoninergic system by central interleukin-1', Eur. J. Pharmacology 209, 139-140.
21. Bonnet, M., Lespinats, G., and Burtin, C. (1984) 'Histamine and serotonin suppression of lymphocyte response to phytohemagglutinin and allogeneic cells', Cellular Immunology 83, 280-291.
22. Boyum, A. (1968) 'Isolation of mononuclear cells and granulocytes', Scan. J. Clin. Lab. Invest. 21 (Suppl.), 77-99.
23. Lorens, A.S., Hata, N., Handa, R.J., Van de Kar, L.D., Guschwan, M., Goral, J., Lee, M.J., Hamilton, M.E., Bethea, C.L., and Clancy Jr., J. (1990) 'Neurochemical, endocrine and immunological responses to stress in young and old fischer 344 male rats', Neurobiol. Aging 11, 139-150.
24. Haour, F., Ban, E., Milon, G., Baran, D., and Fillion, G. (1990) 'Brain interleukin-1 receptors: characterization and modulation after lipopolysaccharide injection', Progress in NeuroEndocrinImmunology 3(3), 1-4.
25. Ban, E., Milon, G., Prudhomme, N., Fillion, G., and Haour, F. (1991) 'Receptors for interleukin-1 (α and ß) in mouse brain: mapping and neuronal localization in hippocampus', Neuroscience 43, 21-31.
26. Cunningham, E.T., Jr., Wada, E., Carter, D.B., Tracey, D.E., Battey, J.F., and De Souza, E.B. (1992) 'In situ histochemical localization of Type I Interleukin-1 receptor messenger RNA in the central nervous system, pituitary, and adrenal gland of the mouse', J. Neurosci. 12 (3), 1101-1114.

THE STIMULATORY EFFECTS OF D-FENFLURAMINE (d-FEN) ON BLOOD AND SPLENIC IMMUNE FUNCTIONS IN THE FISCHER 344 RAT ARE AGE AND SEX DEPENDENT

S.A. LORENS, L. PETROVIC, G. HEJNA, X.W. DONG, and
J. CLANCY, JR.
Departments of Pharmacology and Cell Biology,
Neurobiology and Anatomy
Loyola University Chicago Medical Center
2160 S. First Avenue
Maywood, Illinois 60153
USA

ABSTRACT. In order to study the effects of serotonin (5-HT) on splenic and blood immune functions, male and female F344 rats (5-6 and 21-23 mo old) received the selective 5-HT releaser and reuptake inhibitor, d-FEN (0.6-1.8 mg/kg/day, p.o.) for 30-44 days. d-FEN significantly increased the relative level of splenic as well as peripheral blood large granular lymphocytes (LGL) in not only old females (1.6x) but also young males (2.3x). In addition, d-FEN significantly increased base-line and recombinant interleukin-2 (rIL-2) stimulated YAC-1 killing (NK activity) by young male (1.4x) and old female (1.8x) spleen ocytes. d-FEN only increased (1.7-2.4x) NK cytotoxicity in the old females. The data indicate that in the F344 rat 5-HT modulates the NK component of the immune system in an age and sex dependent manner. The closer correlation of LGL elevation and functional NK activity in the spleen versus peripheral blood may indicate that d-FEN induces a greater compartmental increase in mature LGLs in the spleen than in blood. High concentrations of 5-HT and its metabolite 5-HIAA, 5-HT immunoreactive processes, as well as 5-HT recognition and reuptake sites have been found in the spleen. Morphological and biochemical evidence thus supports the view that 5-HT can play an important role in the regulation of NK-mediated immune functions.

1. Introduction

Several lines of evidence suggest that 5-HT plays an important role in the modulation of immunological processes (see Clancy et al., Fillion et al., and Hellstrand et al., this volume). Recently, we [1,2] demonstrated that subchronic administration of d-FEN enhanced the percentage of splenic large granular lymphocytes (LGL) and of natural killer cell (NK) cytotoxicity in young (5 mo) male and old (21 mo) female F344 rats. d-FEN also increased splenic Con A-stimulated T-cell blastogenesis in young males and lipopolysaccharide (LPS)-induced B-cell proliferation in old females. Old male and young female rats failed to evidence any immunological effects of d-FEN. The presence or absence of d-FEN-induced changes could not be ascribed to group differences in splenic d-FEN or metabolite levels. Thus, increased 5-HT availability can stimulate some splenic immune functions in an age and sex dependent manner. One of the objectives of the present study was to determine whether subchronic d-FEN would produce similar immunological effects on peripheral blood cells.

We [3] also have examined the effects of subchronic d-FEN on splenic 5-HT and norepinephrine (NE) metabolism, 5-HT binding and uptake sites, and beta-adrenergic binding in the F344 rat. We found that the splenic levels of 5-HT were 3-4 times greater than those of NE. For example, in the young male control animals the concentrations (mean $\pm$ sem ng/g) of 5-HT and 5-HIAA were 1340 $\pm$ 30 and 298 $\pm$ 17, whereas the concentrations of NE and its metabolite, MHPG, were 317 $\pm$ 7 and 248 $\pm$ 9, respectively. d-FEN treatment increased splenic NE turnover and decreased splenic 5-HT synthesis in the young and old rats of both sexes. The effects of d-FEN on ^{3}H-5-HT, ^{3}H-paroxetine, and ^{3}H-dihydroalprenolol binding, however, were age and sex dependent: increasing binding in the young male rats, and reducing binding in the old female rats. These age and sex dependent effects were not due to group differences in splenic drug or metabolite concentrations. These observations suggest that the effects of subchronic d-FEN on splenic immune functions are mediated by changes in 5-HT and NE receptor densities rather than alterations in monoamine levels *per se*. In view of these findings, in the present study we examined the distribution of 5-HT immunoreactive processes in the spleen.

2. Effects of d-FEN on Immune Functions in Peripheral Blood

Barrier reared F344 male and female rats were obtained from the NIA colony maintained by Harlan Sprague-Dawley Inc. (Indianapolis, Indiana). The rats were housed individually with food and water available *ad libitum*. The rats were weighed every 3 days and their 24-hour fluid intakes measured daily. Beginning 2 weeks after their arrival in the laboratory, the rats were given vehicle (deionized water) or d-FEN (young, 1.2 mg/kg/day; old, 0.6 mg/kg/day) in their drinking water for 30-44 days, then were sacrificed by decapitation directly after removal from their home cage. The blood obtained from 4 female rats/treatment and 3 male rats/treatment were pooled and 19-35 x 10^6 cells/sample were harvested for analysis by methods detailed previously [1,4].

At the time of sacrifice all of the young rats appeared healthy. In contrast, 27 of the 106 old rats which started the study were eliminated at sacrifice because they exhibited gross pathological conditions such as splenic hypertrophy and hypophyseal tumors. The young (6 mo; n=28) and old (23 mo; n=40) female animals weighed

176-236 g and 241-337 g, respectively. The young (n=33) and old (n=39) male rats weighed 363-460 g and 382-510 g, respectively. d-FEN had no effect on fluid intake or body weight.

As seen in table 1, significant age, sex, and drug effects were observed on blood %LGL and NK activity. Thus, the old rats exhibited a greater number of blood LGL than the young animals, the highest %LGL being observed in the old d-FEN-treated females. d-FEN also augmented %LGL in the young males. d-FEN treatment enhanced basal NK cytotoxicity by 100+% but only in the old females. The young rats exhibited significantly greater rIL-2 stimulated NK cytotoxicity than the old rats. However, d-FEN treatment led to rIL-2-stimulated NK cytotoxicity which was significantly greater in the old female rats than in any of the other old groups. Similar results were obtained using splenocytes [1,2].

No significant age, sex, or drug effects were found on Con A-induced T-cell proliferation (young male values were $58 \pm 1 \times 10^6$ cpm; background radioactivity was 1-2 x 10^6 cpm) or LPS-stimulated B-cell blastogenesis (young male values were $8.8 \pm 0.3 \times 10^6$ cpm). Further, using monoclonal antibodies to surface proteins on subsets of rat lymphocytes, as previously described [1], there were no d-FEN induced alterations in peripheral blood CD5 (Ox19), CD4 (W3/25), CD8 (Ox8), 3.2.3 (Fc τ receptor), or CD25 (IL-2 receptor) positive cells.

These results suggest that the effects of subchronic d-FEN on the immunological functions of peripheral blood cells are not only age and sex dependent but restricted to NK cytotoxicity.

3. 5-HT Immunoreactive Processes in the Spleen

Five young and three old rats of both sexes were deeply anesthetized with sodium pentobarbital, then perfused transcardially with heparinized phosphate buffer saline (PBS) followed by fixative (4.0% paraformaldelhyde and 0.1% glutaraldehyde in PBS). No other drugs were employed. Frozen sections (40 μm) through the spleen were obtained, incubated in rabbit-anti-5-HT (1;1000; INCSTAR, Stillwater, Minnesota) for 48 hours, then exposed to rhodamine labelled goat anti-rabbit IgG (1:100; Jackson ImmunoResearch, West Grove, Pennsylvania) for 1.0 hour prior to coverslipping. Control sections (primary antibody omitted) were processed in the same manner as the experimental tissue. Sections were examined and photographed using a Lietz Dialux 20 microscope outfitted with Ploemoptics.

TABLE 1. Effects of subchronic d-fenfluramine (d-FEN) treatment on blood (mean ± sem) percent large granular lymphocytes (%LGL), basal (LU) and rIL-2 stimulated (rIL-2/LU) natural killer cell cytotoxicity (expressed in lytic units) in young (6 mo) and old (23 mo) F344 rats.

Group (N)	%LGL	LU	rIL-2/LU
Young Male			
Vehicle (6)	2.8 ± 0.1	7.0 ± 0.9^D	15.2 ± 0.1^E
d-FEN (5)	4.5 ± 0.2	7.3 ± 0.2^D	15.4 ± 0.4^E
Young Female			
Vehicle (4)	3.0 ± 0.1	4.8 ± 0.1	16.3 ± 0.4^E
d-FEN (3)	3.3 ± 0.1	6.1 ± 0.3	15.8 ± 1.2^E
Old Male			
Vehicle (6)	11.2 ± 0.3^B	3.8 ± 0.2	8.2 ± 0.2
d-FEN (7)	8.5 ± 0.4^B	5.1 ± 0.2	9.0 ± 0.2
Old Female			
Vehicle (5)	15.0 ± 0.3^C	3.0 ± 0.3	8.9 ± 0.1
d-FEN (5)	17.7 ± 0.2^C	7.1 ± 0.1^D	13.4 ± 0.1^F

Statistical statements are based on a three-way analysis of variance followed by a Newman-Keuls' multiple range test.

A: Greater ($p<0.05$) than the other young groups.
B: Greater ($p<0.01$) than all of the young groups.
C: Greater ($p<0.01$) than the young groups and old male groups.
D: Greater ($p<0.05$) than both female vehicle groups and the old male groups.
E: Greater ($p<0.01$) than all of the old groups.
F: Greater ($p<0.01$) than the old male groups and the old female vehicle group.

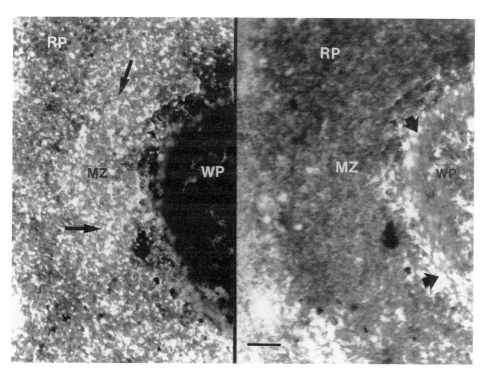

Figure 1. Punctate (2-4 μm) splenic 5-HT IR elements (long thin arrows in left panel) are found throughout the marginal zone (MZ) and red pulp (RP). Very little 5-HT IR was discerned in the white pulp (WP). The distribution of splenic autofluorescent processes as seen via the FITC filter (short thick arrows in right panel) is quite distinct from the distribution of the rhodamine labelled 5-HT IR cells seen in the left panel. Calibration bar = 25 μm.

As seen in figure 1, 5-HT immunoreactive (IR) processes (2-4 μm) are densely distributed throughout the splenic marginal zone and red pulp. Similar results were obtained in the young and old rats of both sexes. This is in marked contrast to the reported [5,6] localization of NE and peptidergic fibers in the white pulp. Since beaded fibers were not seen, it is uncertain as to whether the 5-HT IR elements visualized are platelets which were not washed out during the perfusion process, 5-HT containing splenocytes, or neuronal processes emanating from 5-HT cell bodies [7] in the celiac-superior mesenteric plexus. In view of the fact that d-FEN primarily affects NK activity, it is noteworthy that LGLs and NK cells [8] as well as

³H-5-HT recognition sites (G. Fillion, personal communication) and mRNA for the 5-HT transporter (B. Hoffman, personal communication) have been localized predominantly in the marginal zone and red pulp.

Steinbusch et al. [6] have reported that they were unable to differentiate specific 5-HT IR processes from the intense autofluorescence inherent to the spleen. These workers employed a fluoroscein (FITC) labelled secondary antibody whereas we used a rhodamine tagged secondary antibody. Autofluorescence is most intensely seen via a FITC filter combination. Since there was very little autofluorescent bleed-through via our rhodamine filter cube, specific 5-HT IR could be visualized.

4. Conclusions

Subchronic administration of d-FEN enhances some immune functions in an age and sex dependent manner. The effects of d-FEN were greater on splenocytes [1,2] than on peripheral blood cells. Thus, d-FEN increased splenic percent large granular lymphocytes, NK cytotoxicity, and B- (old females) and T-cell (young males) mitogenesis in old (21-23 mo) female and young (5-6 mo) male F344 rats. In contrast, subchronic d-FEN only increased blood NK activity in old female rats.

5-HT recognition sites, mRNA for the 5-HT transporter, and 5-HT immunoreactive processes are localized predominantly in the splenic marginal zone and red pulp, a region which contains large granular lymphocytes and NK cells. 5-HT binding sites also have been found on circulating lymphocytes (Fillion et al., this volume). The spleen contains high concentrations of 5-HT and its metabolite, 5-HIAA, as well as 5-HT reuptake sites. Thus, morphological and biochemical evidence provides support for the view that 5-HT can play an important role in the regulation of immunological functions.

d-FEN may enhance T- and B- cell function by altering beta-adrenergic receptor density in the white pulp and may augment NK activity by modifying 5-HT availability and receptor density in the marginal zone and red pulp. The age and sex related nature of these effects suggest that d-FEN enhances immune functions only if plasma testosterone levels are high (young males) or estrogen levels are low (old females).

5. References

1. Clancy Jr., J., Petrovic, L.M., Gordon, B.H., Handa, R.J., Campbell, D.B., and Lorens, S.A. (1991) 'Effects of subchronic d-fenfluramine on splenic immune functions in young and old male and female Fischer 344 rats', Int. J. Immunopharmac. 13, 1203-1212.
2. Petrovic, L.M., Lorens, S.A., George, M., Cabrera, T., Gordon, B.H., Handa, R.J., Campbell, D.B., and Clancy Jr., J. (1991) 'Subchronic d-fenfluramine treatment enhances the immunological competance of old female Fischer 344 rats', in J.R. Fozard and P.R. Saxena (eds.), Serotonin Molecular Biology,

Receptors and Functional Effects, Birkhäuser Verlag, Basel, pp. 299-397.
3. Lorens, S.A., George, M., Dersch, C., Hejna, G., Gordon, B.H., Campbell D.B., Clancy Jr., J., and Zaczek, R. (1992) 'Splenic monoamines: age and sex dependent effects of d-fenfluramine', submitted for publication.
4. Lorens, S.A., Hata, N. Handa, R.H., Van de Kar, L.D., Guschwan, M., Goral, J., Lee, J.M., Hamilton, M.E., Bethea, C.L., and Clancy Jr., J. (1990) 'Neurochemical, endocrine and immunological responses to stress in young and old Fischer 344 male rats', Neurobiol. Aging 11, 139-150.
5. Felten, S.Y. and Felten, D.L. (1991) 'Innervation of lymphoid tissue', in R. Ader, D.L. Felten and N. Cohen (eds.), Psychoneuroimmunology, 2nd Ed., Academic Press, Inc., New York, pp. 27-69.
6. Steinbusch, H.W.M., Van der Meer, E.G., Van Rooijen, N., and Eikelenboom, P. (1987) 'The presence and distribution of noradrenaline- and the absence of serotonin- immunofluorescent fibres in the rat spleen', in A. Nobin, C. Owman and B. Arneklo-Nobin (eds.), Neuronal Messengers in Vascular Function, Elsevier, Science Publishers, Amsterdam, pp. 371-383.
7. Ma, R.C., Horwitz, J., Kiraly, M., Perlman, R.L., and Dun, N.J. (1985) 'Immunohistochemical and biochemical detection of serotonin in the guinea pig celiac-superior mesenteric plexus', Neurosci. Lett. 56, 107-112.
8. Rolstad, B., Herberman, R.B., and Reynolds, C.W. (1986) 'Natural killer cell activity in the rat: V. The circulation patterns and tissue localization of peripheral blood large granular lymphocytes (LGL)', J. Immunol. 136, 2800-2808.

SEROTONERGIC REGULATION OF NATURAL KILLER CELLS: A MINIREVIEW

KRISTOFFER HELLSTRAND[1], CLAES DAHLGREN[2], and
SVANTE HERMODSSON[1]
[1]*Department of Virology and*
[2]*Institute of Medical Microbiology*
University of Göteborg
Sweden

ABSTRACT. Natural killer (NK) cells are a subset of lymphocytes that kill tumor cells and virus-infected target cells in a non-MHC-restricted fashion. Serotonin augments several functions of human NK cells, including anti-tumor cytotoxicity, proliferation, and lymphokine production by an indirect mechanism of action. Thus, activation of 5-HT$_{1A}$ type receptors abrogates a cell-contact-dependent suppressive signal delivered to NK cells by monocytes and granulocytes. The suppressive signal is closely related to the 'respiratory burst' activity of monocytes/granulocytes since catalase, a scavenger of reactive oxygen metabolites, abrogates the suppression as effectively as serotonin. Further, serotonin depresses the generation of oxygen metabolites by monocytes and granulocytes. The presented data are suggestive of a novel type of communication between subsets of leukocytes involved in nonadaptive immunity and its serotonergic regulation.

1. Review

In the beginning of the 1970s, it was observed that lymphocytes from humans and experimental animals are cytotoxic for a variety of target cells *in vitro* including cultured cell lines, virus-infected cells, immature cells hematopoietic origin, and some tumor cells (for review, [1]). In contrast to the cytotoxicity exerted by T cells, the killing activity required no previous sensitization and occurred independently of expression of major histocompatibility (MHC) products on the surface of target cells. This type of nonadaptive non-MHC-restricted cell-mediated cytotoxicity was defined as 'natural', and the effector cells were functionally defined as natural killer (NK) cells.

Human NK cells can be distinguished from other subsets of leukocytes by the expression of certain cell surface structures [2]. Thus, NK cells carry CD56 (Leu-19, NKH-1) which is identical to the neural cell adhesion molecule (N-CAM) expressed by most neuronal cells [3]. The functional role of CD56 is unknown, but recent studies suggest that NK cells can bind to and destroy tumor cells of neuroectodermal origin by homotypic binding to this molecule [4]. The majority of $CD56^+$ NK cells express a low-affinity cell surface receptor for IgG, CD16 (Leu-11, FcRIII) [1,2]. Via binding of IgG to FcR, NK cells can attach to and lyse IgG-coated target cells; this type of cytotoxicity is referred to as antibody-dependent cytotoxicity (ADCC). NK cells are functionally distinct from T cells: they do not require maturation in the thymus, lack immunological memory, and do not rearrange genes encoding T-cell receptors. Further, although NK cells share some surface molecules with T cells, they do not react with epitopes recognized by most conventional antibodies against the T-cell antigen CD3 [2].

With the exception of the ADCC reaction, the cytotoxicity is antibody-independent and the result of a cell-to-cell-mediated interaction between NK cells and target cells followed by transfer of a cytotoxic, pore-forming material to the target cell membrane [1]. One of the cytotoxic molecules detected in NK cell granules shows sequence homology with perforin. As regards the interaction between NK cells and targets, the recognition structure(s) remain unknown, although blocking of CD11a-ICAM-1 as well as CD2-ICAM-3 interactions partly reduce NK-cell conjugation with target cells (for review, [1]). Several cytokines including interferons (IFNs) and interleukin-2 (IL-2) effectively boost the cytolytic activity of NK cells *in vitro* and *in vivo* [5]. In addition, IL-2 induces proliferation of NK cells as well as production of IFN-gamma [1]. cAMP-inducing agents, including catecholamines (via $\beta 2$-type receptors on NK cells [6] and prostaglandins of the E series, effectively suppress NK cell function *in vitro*.

In peripheral tissues, the neurotransmitter serotonin (5-hydroxytryptamine, 5-HT) is stored in platelets, mast cells, and enterochromaffin cells of the gut mucosa. Relatively high concentrations of serotonin are present in immune and inflammatory reactions, mainly as the result of release from aggregated platelets [7]. Expression of serotonin receptors (5-HTR) of $5-HT_1$, $5-HT_2$, and $5-HT_3$ subtypes has been demonstrated on cells pivotal to immune reactivity [8,9], but the available information about immunoregulatory effects transduced via 5-HTR on leukocytes is surprisingly scarce. Sternberg and co-workers [10] have shown, that serotonin modulates phagocytosis and expression of MHC products in bone-marrow-derived murine macrophages. The effect was blocked by ketanserin and several other 5-HT_2R antagonists [10]. Further, 5-HT_2R on T cells reportedly play a role in the elicitation of delayed-type hypersensitivity reactions in mice [11]. Choquet and Korn [12] have demonstrated the existence of a potassium-gated voltage channel in pre-B-cells that is controlled by $5-HT_1$-like and $5-HT_3$ receptors.

Recent data suggest that serotonin may be an important modulator of NK-cell function. The original observation in this area was that presence of serotonin (at concentrations exceeding 10^{-7} M) markedly augments the cytotoxicity of human NK cells against susceptible tumor target cells *in vitro* [13]. This effect of serotonin is independent of formation of IFNs or IL-2 and shown to reflect a novel type of interaction between monocytes and NK cells. Thus, serotonin does not affect the cytotoxicity of enriched, monocyte-depleted NK cells; instead the enhancement of tumor cell killing is indirectly mediated by serotonin-induced abrogation of a cell-contact-dependent, suppressive signal delivered by monocytes [14,15]. The suppression induced by monocytes is completed with 30-45 minutes of incubation of monocytes with NK cells; thereafter, the baseline ('natural') cytotoxicity of NK cells is seemingly irreversibly suppressed for at the least 72 hours [16]. Further, NK cells suppressed by monocytes cannot respond to INFs [17] or IL-2 [16]. Serotonin completely prevents induction of the suppressive signal but does not affect the suppression when added later than 30-45 minutes after the onset of incubation of NK cells with monocytes [16].

Most known functions ascribed to human NK cells are sensitive to the serotonin-reversible suppression induced by monocytes. Thus, serotonin regulates, in a fashion similar to that observed for cytotoxicity against tumor cells, also the proliferation of NK cells and the production of IFN-gamma by NK cells (in response to IL-2) [16,18]. The inhibition of IFN-gamma production is of a pretranslational nature, since NK cells do not accumulate mRNA encoding IFN-gamma in the presence of monocytes unless serotonin is present to abrogate the suppressive signal [18]. Also, ADCC reactivity of NK-cells is suppressed by monocytes and restored in the presence of serotonin [16]. An unexpected finding was that incubation of enriched NK cells with monocytes induces the disappearance ('down-modulation') of cell surface structures of NK cells. Sixteen-to-twenty-four hours after a one-hour incubation of NK cells with monocytes, the CD16 and CD56 modules completely disappear from the surface of NK cells; monocytes thus induce a shift of NK-cell phenotype from $CD3^-/16^+/56^+$ to $CD3^-/16^-/56^-$ [16]. Other antigens expressed by NK cells such as CD2, CD11a, CD11b, and CD57 are subject to similar monocyte-induced down-modulation; in the case of CD11a and CD11b, these antigens do not disappear completely, but the density of expression is strongly reduced. Only $CD56^{dim}$ cells (i.e. the vast majority of NK cells) are sensitive to the monocyte-induced, down-modulatory signal, but a small subset of $CD56^{bright}$ cells (approximately 10% of all NK cells) are not affected. As is the case with NK-cell functional activities, the presence of serotonin completely abrogates the monocyte-induced, down-modulatory signal [16]. Thus, in addition to regulating functions ascribed to NK cells, serotonin also provides signals of importance for the maintenance of their phenotype markers.

Studies on which subset of serotonin receptors mediates effects on NK-cell function indicate a predominant role of $5\text{-}HT_{1A}R$. Cyproheptadine, a mixed $5\text{-}HT_1/5\text{-}HT_2R$ antagonist, and pindolol, a $5\text{-}HT_1R$ antagonist, block effects of serotonin and display agonist-like properties at high concentrations ($>10^{-5}$ M) [16]. Neither ketanserin (a $5\text{-}HT_2R$ antagonist), nor ondansetron (a $5\text{-}HT_3R$ antagonist) block the effect of serotonin. The prototypic $5\text{-}HT_{1A}R$ agonist 8-OH-DPAT mimics the effect; in most type of assays, 8-OH-DPAT is at least a partial agonist [13,16]. Two other $5\text{-}HT_{1A}R$ agonists, ALK-3 and BMY 738, are partial agonists in assays of NK-cell function, but the potency of these compounds is lower than that of 8-

OH-DPAT [16]. Both (+) and (-) forms of ALK-3 augment NK-cell cytotoxicity [16]; in contrast, neuronal cells can distinguish between enantiomers of ALK-3 [19]. 8-OH-DPAT, ALK-3, and BMY 738 show only agonist activity; neither of the compounds block the response to serotonin even at relatively high concentrations. Thus, although the receptor mediating NK-cell-regulatory effects of serotonin is similar to 5-HT$_{1A}$R, neuronal and the putative mononuclear cell 5-HT$_{1A}$R may not be identical. The cellular distribution of 5-HT$_{1A}$R on leukocytes is not known, but mRNA encoding 5-HT$_{1A}$R is expressed in human lymphoid tissues [8].

As is the case with monocytes, granulocytes suppress NK-cell function in a serotonin-reversible fashion. A difference between monocytes and granulocytes in this respect is that the suppression observed after incubation of monocytes with NK cells is almost completely reversed by the presence of serotonin, but part of the suppression exerted by granulocytes remains in spite of the presence of serotonin [20].

TABLE 1. Serotonergic regulation of NK-cell surface markers; role of monocytes and oxygen metabolism.

Treatment	Monocytes	NK cells (%)
Control	-	43
Serotonin	-	42
Catalase	-	44
Serotonin + Catalase	-	41
SOD	-	42
Control	+	12
Serotonin	+	43
Catalase	+	42
Serotonin + Catalase	+	44
SOD	+	10

The frequency of cells carrying NK-cell phenotype (CD3$^-$/56$^+$) was estimated by use of a fluorescence-activated cell sorter (FACScan) in NK-cell-enriched human lymphocytes and monocytes, separated by counter-current centrifugal elutriation. Serotonin: 10^{-5} M; catalase: 100 U/ml; SOD: 200/ml.

Recent data suggest that the effects induced by serotonin are related to the formation of reactive oxygen metabolites ('respiratory burst') by monocytes/granulocytes. Thus, catalase, a scavenger of oxygen metabolites, abrogates the serotonin-reversible monocyte/granulocyte-induced suppression of NK-cell function as well as the down-modulation of NK-cell phenotype markers with efficiency similar to serotonin. In contrast, superoxide dismutase (SOD) fails to affect the suppression (table 1). Further, serotonin markedly suppresses the fMLP-induced formation of oxygen metabolites (as measured by chemiluminescence) by monocytes/granulocytes (table 2). That catalase, but no SOD, can reverse the suppression strongly suggests a role for hydrogen peroxide as a mediator of suppressive effects. Thus, although the suppression of NK cells by monocytes/granulocytes is dependent of cellular proximity [15,16], the suppressive mediator may be a soluble product. In addition, it seems clear from these data that the 5-HT receptors mediating NK-cell-regulating effects of serotonin are located on monocytes/granulocytes rather than on NK cells.

TABLE 2. Serotonergic regulation oxygen metabolism in monocytes and granulocytes.

Treatment	Concentration	Cell Type	Chemo-luminescence
Control	-	MO	32 ± 2
Serotonin	10^{-5} M	MO	4 ± 0.5
Control	-	GR	233 ± 11
Serotonin	10^{-5} M	GR	89 ± 2

Monocytes (MO) and granulocytes (GR) were separated from peripheral blood by counter-current centrifugal elutriation. Chemoluminescence was estimated using a Luminoscan (Kebo, Stockholm, Sweden).

The *in vivo* role of these mechanisms remains to be established. That serotonin may positively regulate NK-cell function as *in vivo* is suggested by the recent finding that subchronic treatment of rats with d-fenfluramine (d-fen), a serotonin reuptake inhibitor and releaser, augments the baseline and IL-2-induced cytolytic activity of spleen cells against YAC-1 cells (a mouse lymphoma that is sensitive to rat NK cells). Also, the number of splenic lymphocytes with NK cells morphology (large granular lymphocytes) is higher in d-fen-treated animals. Interestingly, these effects of d-fen were observed only in young male and old female animals [21]. A similar sex- and age-dependence has not been observed in studies of human NK-cell function *in vitro* [20].

The reviewed data are suggestive of a role for serotonin in regulation of an earlier unrecognized interplay between NK cells and other cellular components of nonadaptive immunity. It seems reasonable to assume that serotonin exerts its putative immunoregulatory effects locally rather than systematically, since systemic serotonin concentrations rarely reach those required to affect NK-cell or monocyte/granulocyte function. This assumption is also compatible with the notion that the close cellular proximity required for the serotonin-regulated interaction between NK cells and monocytes/granulocytes probably only can be achieved outside of the circulation.

2. References

1. Trinchieri, G. (1989) 'Biology of natural killer cells', Adv. Immunol. 47, 187-376.
2. Ritz, J., Schmidt, R.E., Michon, J., Hercend, T., and Schlossman, S.F. (1988) 'Characterization of functional surface structures of human natural killer cells', Adv. Immunol. 42, 181-211.
3. Lanier, L.L., Testi, R., Bindl, J., Phillips, J.H. (1989) 'Identity of Leu-19 and (CD56) leukocyte differentiation antigen and neural cell adhesion molecule', J. Exp. Med. 169, 2233-2238.
4. Nitta, T., Yagita, H., Sato, K., and Okomura, K. (1989) 'Involvement of CD56 (NKH-1/Leu-19) antigen as an adhesion molecule in natural killer-target cell interactions', J. Exp. Med. 170, 1757-1761.
5. Rees, R.C. (1990) 'MHC restricted and non-restricted killer lymphocytes', Blood Rev. 4, 204-210.
6. Hellstrand, K., Hermodsson, S., and Strannegård, Ö. (1985) 'Evidence for a β-adrenoceptor-mediated regulation of human natural killer cells', J. Immunol. 134, 4095-4099.
7. Hellstrand, K. (1987) 'Biogenic amines in the regulation of human natural killer cell cytotoxicity', Thesis, University of Göteborg, pp.1-67.
8. Fargin, A., Raymond, J.R., Lohse, M.J., Kobilka, B.K., Caron, M.G., and Lefkowitz, R.J. (1988) 'The genomic clone G.21 which resembles a β-adrenergic receptor sequence encodes the 5-HT$_{1A}$ receptor', Nature (London) 335, 358-360.
9. Aune, T.M., Kelley, K.A., Ranges, G.E., and Bombara, M.P. (1990) 'Serotonin-activated signal transduction via serotonin receptors on Jurkat cells', J. Immunol. 145, 1826-1831.
10. Sternberg, E.M., Trial, J., and Parker, C.W. (1986) 'Effect of serotonin on murine macrophages: suppression of Ia expression by serotonin and its reversal by 5-HT$_2$ serotoninergic antagonists', J. Immunology. 137, 276-282.
11. Ameisen, J.C., Meade, R., Askenase, P.W. (1989) 'A new interpretation of the involvement of serotonin in delayed-type hypersensitivity. Serotonin-2 receptor antagonists inhibit contact sensitivity by and effect on T cells', J.

Immunol. 142, 3171-3176.
12. Choquet, D. and Korn, H. (1988) 'Dual effects of serotonin on a voltage-gated conductance channel in lymphocytes', Proc. Natl. Acad. Sci (U.S.A.) 85, 4557-4561.
13. Hellstrand, K. and Hermodsson, S. (1987) 'Role of serotonin in the regulation of human natural killer cell cytotoxicity', J. Immunol. 139, 869-875.
14. Hellstrand, K. and Hermodsson, S. (1990) 'Enhancement of human natural killer cell cytotoxicity by serotonin: role of non-T/CD16+ NK cells, monocytes, and 5-HT_{1A} receptors', Cell Immunol. 127, 199-214.
15. Hellstrand, K. and Hermodsson, S. (1990) 'Monocyte-mediated suppression of human natural killer cell cytotoxicity: regulation by serotonergic 5-HT_{1A} receptors', Scand. J. Immunol. 32, 123-145.
16. Hellstrand, K. and Hermodsson, S. (1992) "Serotonergic 5-HT_{1A} receptors regulate a cell-contact-mediated interaction between monocytes and natural killer cells', Scand. J. Immunol., in press.
17. Hellstrand, K. and Hermodsson, S. (1992) "Regulation of the NK cell response to IFN-alpha by biogenic amines', J. Interferon Res., in press.
18. Hellstrand, K., Czerkinsky, C., Kylefjord, H., Jansson, B., Ricksten, A., Asea, A., and Hermodsson, S. (1992) 'Role of serotonin in the regulation of IFN-gamma production by human natural killer cells', J. Interferon Res., in press.
19. Eriksson, E. and Humble, M. (1990) "Serotonin in psychiatric pathophysiology', Progr. Basic Clin. Pharmacol. 3, 66-119.
20. Hellstrand, K. and Hermodsson, S. (1992), unpublished.
21. Clancy Jr., J., Petrovic, L.M., Gordon, B.H., Handa, R.J., Campbell, B.D., and Lorens, S.A. (1991) 'Effects of subchronic d-fenfluramine on splenic immune functions in young and old male and female Fisher 344 rats', Int. J. Immunopharmacol. 13, 1203-1212.

INTERACTIONS BETWEEN SEROTONIN AND THE IMMUNE SYSTEM: AN OVERVIEW

J. CLANCY, JR.[1], G. FILLION[2], K. HELLSTRAND[3] AND S. A. LORENS[4]
Departments of [1]Cell Biology, Neurobiology and Anatomy and [4]Pharmacology
Loyola University Chicago Medical Center
2160 S. First Avenue
Maywood, Illinois 60153
USA,
[2]Unité de Pharmacologie Neuro-Immuno-Endocrinienne
Institut Pasteur
28 rue du Dr Roux
F75015 Paris
France, and
[3]Department of Virology
University of Göteborg
Guldhedsgatan 10 B
S-41346 Göteberg
Sweden

ABSTRACT. Experimental observations indicate that 5-HT receptors are located on a variety of immunocompetent cells and that 5-HT participates in controlling functional activity within the immune system, and in particular its natural killer (NK) arm. Reciprocally, changes in immune responsivity are able to alter CNS 5-HT activity. Thus, 5-HT appears to play a key role in the interactions between the CNS and the immune system.

1. Introduction

It is well established that 5-HT modulates the activity of other neurotransmitter systems in the CNS. The fact that the actions of 5-HT are mediated by several distinct receptor subtypes is consistent with the view that 5-HT is involved in fine tuning and optimizing CNS functions. 5-HT also appears to play a similar role in the regulation of the immune system and to serve as an important link in neuroimmune interactions [1,2]. On the one hand, 5-HT differentially regulates immune responses, especially natural killer (NK) cell activity, via distinct receptor subtypes. On the other hand, modifications in the immune response induce significant changes in the activity of 5-HT systems in the brain.

2. 5-HT Regulation of Immune Function

Fillion et al. (this volume) reviewed literature indicating that 5-HT had regulatory effects on the immune system. Overall, these data suggested that increased 5-HT activity or availability is associated with a decrease in the magnitude of the immune response, and, inversely, that decreased 5-HT levels or efficacy is correlated with an enhanced immune response. However, several reports, including those by Hellstrand et al. and Lorens et al. (this volume), provide exceptions to this rule, indicating that the mechanisms involved in these relationships are complex.

Some *in vitro* studies have attempted to identify the targets which mediate the activity of 5-HT on the immune system and have demonstrated the existence of binding sites to serotonergic ligands. These sites have yet to be fully characterized pharmacologically. Fillion et al., nevertheless, have demonstrated the presence of mRNAs specific for certain 5-HT receptor subtypes in murine thymocytes. Further, Fillion et al. have provided evidence for a functional link between the presence of 5-HT receptors in immunocompetent cells and the regulatory capacities of 5-HT on immune functions.

Hellstrand et al. (this volume) reviewed their studies documenting a role for 5-HT and the $5-HT_{1A}$ receptor in the regulation of human NK cells, a subset of lymphocytes that kill many types of target cells *in vitro*, including tumor cells, virus-infected cells, and some immature cells of haematopoietic origin. Thus, the presence of 5-HT abrogates a monocyte-derived signal that (i) suppresses several functions ascribed to NK-cells including cytotoxicity, proliferation, and cytokine production, and (ii) induces disappearance of NK-cell surface proteins such as CD16 and CD56 [3,4]. Recent data show that granulocytes exert a similar, 5-HT-reversible suppressive effect on NK-cell function and phenotype and that the monocyte/granulocyte-induced signal is inhibited by scavengers of oxidative metabolism. Further, 5-HT effectively blunts production of oxygen metabolites by monocytes and granulocytes. Thus, it seems reasonable to assume that the effects of 5-HT on human NK-cells reflect regulation of oxidative metabolism in monocytes/granulocytes [5]. The effects of 5-HT are mimicked by $5-HT_{1A}$ specific agonists, such as 8-OH-DPAT and (+)-ALK, and are blocked by cyproheptadine (a mixed $5-HT_1/5-HT_2$ antagonist) but not by ketanserin (a specific $5-HT_2$ antagonist) or ondansetron (a $5-HT_3$ antagonist). A preliminary conclusion is that the NK-cell regulatory effects of 5-HT are mediated by $5-HT_{1A}$ receptors. However, a definite conclusion on this point awaits the availability of $5-HT_{1A}$ specific antagonists.

Lorens et al. (this volume) reviewed their studies documenting the stimulatory effects of subchronic exposure to d-fenfluramine (d-FEN), a 5-HT releaser and reuptake inhibitor, on the NK arm of the immune system in young male and old female F344 rats. Similar age and sex related d-FEN-induced increases in NK activity were seen in both the splenic and blood compartments. Importantly, Hellstrand (unpublished data) has recently found that the administration of a single dose (10 mg/kg, IV, 6 hours prior to tumor inoculation) of d-FEN to mice significantly improved (63%) lung clearance of the NK-sensitive target cell YAC-1. Lorens et al. also observed that d-FEN significantly enhanced splenic but not blood Con-A stimulated T cell blastogenesis in young males and lipopolysaccharide (LPS)-induced B cell proliferation in old females. In addition, morphological and biochemical studies were presented which demonstrated high concentrations of 5-HT and its metabolite 5-HIAA, 5-HT immunoreactive processes, as well as 5-HT recognition and reuptake sites in the spleen. The 5-HT immunoreactive elements were found in the marginal zone and red pulp of the spleen, the site where most NK cells are localized.

Thus the studies presented in this symposium strongly suggest that 5-HT is a potential enhancer of the NK arm of the immune system. Since 5-HT can be synthesized or taken up by many different cell types in many different organs, as well as by circulatory platelets, the potential source and organ compartment effects may vary. However, during an immune response when there is elevation of interleukin 1 (IL-1) systemically, enhanced 5-HT activity in a number of sites (CNS as well as spleen) could help modulate any negative effects from the macrophage to maintain up regulation of the important NK arm. These interactions are illustrated in figure 1.

3. Impact of Immune Response on CNS 5-HT Activity

The interaction of the immune system with CNS 5-HT neurons represents part of the reciprocal activity of the immune system on the CNS and may account for the numerous changes in behavior observed after immunoreactions. Fillion's studies on the pharmacokinetics of 5-HT_{1A} receptors demonstrate that within 16 hours following sheep-red-blood-cell (SRBC) immunization of mice there was a significant reduction in the ability of 8-OH-DPAT to inhibit forskolin-induced adenyl cyclase activity in the hippocampus. Such modulation of functional 5-HT_{1A} receptors was reduced after 4 days. However, there was no effect of SRBC immunization on the number of $5\text{-HT}_{1A\text{-}E}$ or 5-HT_2 receptors in any area of the CNS.

Fillion et al. (this volume) also found that cyclosporine A, a specific inhibitor of interleukin 2 (IL-2) synthesis, markedly decreased the inhibitory activity of $5\text{-HT}_{1B/1D}$ receptors on acetylcholine release in hippocampal synaptosomes. These authors also showed that viral infection (rabies virus) in the rat induced a significant and specific reduction in cortical 5-HT binding. These results suggest that changes in immune activity may differentially affect 5-HT receptors in the brain and thereby induce diverse alterations in CNS function. The precise mechanism(s) which mediate these intereactions are not yet known. Interestingly, Gardier et al. [6] have reported that immunization of mice with SRBC induced a marked increase in hypothalamic 5-HT content 24 hours later, and a significant decrease (cyclosporine dependent) after 2-4 days. Moreover, Gemma et al. [7] have shown that systemic

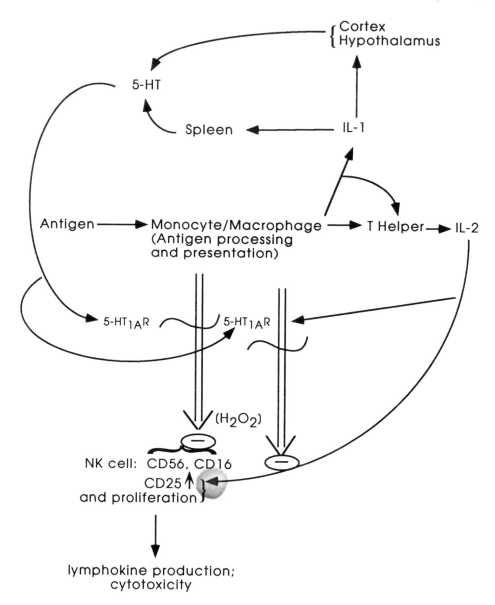

Figure 1. Potential interaction between 5-HT, CNS and various lymphoid cells, their cell surface antigens, and the cytokines produced during an immune response. IL-1 secretion during an immune response could cause 5-HT to be released in the CNS as well as the spleen. 5-HT (via its 5-HT$_{1A}$ receptor) inhibits the ability of monocytes (via H_2O_2) to decrease NK-cell surface antigen (CD 16, C56) expression, IL-2 induced CD25 expression, proliferation, lymphokine production, and activity.

recombinant interleukin 1 (rIL-1) produced increases in extracellular 5-HIAA in the anterior hypothalamus. Therefore, since IL-1 is released in substantial amounts during a SRBC response resulting in IL-2 secretion, it is reasonable to suppose that IL-1 will induce changes in 5-HT which will not only alter cerebral activity but will function to help modulate the immune response (figure 1).

It is hoped that future studies will further analyze the precise role, compartmental location (CNS, spleen, as well as possibly liver and lung), and molecular mechanisms involved in the interaction between 5-HT and the immune system.

4. References

1. Carlson, S.L., Felten, D.L., Livnat, S., and Felten, S.Y. (1987) 'Alterations of monoamines in specific central autonomic nuclei following immunization in mice', Brain Behav. Immunity 1, 52-63.
2. Gardier, A.M., Hachaner, S., Bohuon, C., Jacquot, C., and Pallardy, M. (1992) 'Time course of central variations of 5-HT and DA metabolism associated with a primary immune response to sheep red blood cells', Soc. Neurosci. Abstr. 18, 1012.
3. Gemma, C., Ghezzi, P., and De Simoni, M.G. (1991) 'Activation of the hypothalamic serotoninergic system by central interleukin-1', Eur. J. Pharmacol. 209, 139-140.
4. Hellstrand, K. and Hermodsson, S. (1987) 'Role of serotonin in the regulation of human natural killer cell cytotoxicity', J. Immunol. 139, 869-875.
5. Hellstrand, K. and Hermodsson, S. (1992) 'Serotonergic 5-HT_{1A}-receptors regulate a cell-contact-mediated interaction between natural killer cells and monocytes', Scand J. Immunol., in press.
6. Jackson, J.C., Cross, R.J., Walker, R.F., Markesbery, W.R., Brooks, W.H., and Roszman, T.L. (1985) 'Influence of serotonin on the immune response', Immunology 54, 505-512.
7. Silverman, D.H.S., Wu, H., and Karnovsky, M.L. (1985) 'Muramyl peptides and serotonin interact at specific binding sites on macrophages and enhance superoxide release', Biochem. Biophys. Res. Com. 131, 1160-1167.

THE ROLE OF SEROTONIN IN THE CELLULAR PHYSIOLOGICAL EFFECTS OF COCAINE

JOAN M. LAKOSKI and HUA ZHENG
Department of Human Biological Chemistry & Genetics
University of Texas Medical Branch
Galveston, Texas 77555-0498
USA

ABSTRACT. Serotonin has been widely recognized to mediate the psychopharmacology of numerous abused substances, including the stimulant cocaine. Cellular electrophysiological recording techniques are among the approaches used to facilitate the direct assessment of the interaction of cocaine with this neurotransmitter system. Utilizing both *in vivo* and *in vitro* preparations of the dorsal raphe nucleus we have investigated a series of potent phenyltropane cocaine analogs, RTI-Cocaine 31 and RTI-Cocaine 32, and characterized their inhibitory interactions with serotonergic neurons. Additional studies with a cocaine metabolite, cocaethylene, are also described to illustrate new approaches currently being utilized in elucidating the effects of cocaine on identified serotonergic receptors in selective brain regions. These results are briefly discussed in light of the potential usefulness of serotonergic compounds in the development of new therapeutic interventions for the treatment of stimulant abuse.

1. Introduction

The recognition of a role for serotonin in mediating the psychopharmacology of abused substances has gained new emphasis as investigators search for the mechanisms of cocaine's interactions with the nervous system. In addition to the critical interaction with dopamine-containing neuronal systems, behavioral and neurochemical actions of this psychostimulant have been demonstrated with serotonin (5-hydroxytryptamine;5-HT) neuronal systems [1,2]. With respect to the possible cellular physiological interactions of cocaine with serotonergic systems, both *in vivo* [3-6] and *in vitro* [7,8] recording studies have identified inhibitory effects of cocaine on serotonin-mediated cell function, including the 5-HT$_{1A}$ receptor of the somatodendritic autoreceptor recorded in the dorsal raphe nucleus (DRN).

1.1 COCAINE ANALOGS INTERACTION WITH SEROTONIN

Structure-activity studies of the cocaine molecule [9-11] have recently provided new tools designed to assess the mechanisms by which selective interactions with monoaminergic transporters occur. Using a strategy of structural modification to the phenyl and tropane rings, the effects of 3β-substitution on the tropane ring (figure 1) result in a marked increase in potency over cocaine to inhibit binding of [^{3}H]WIN 35,428 and dopamine binding with the analogs 3β-(4-methylphenyl)- and 3β-(4-chlorophenyl)-tropane-2β carboxylic acid methyl ester (RTI-32 and RTI-31, respectively). Recently, these analogs have been evaluated for cocaine-like discriminative stimulus properties in rats [12] and effects on locomotor activity [11] and demonstrated enhanced potency ranging from 6-30 fold over cocaine.

Our own electrophysiological studies have addressed the cellular physiological effects of cocaine on serotonergic neurons as this stimulant is a potent inhibitor of serotonin transport processes [3,4,7,13]. In agreement with the work of numerous investigators identifying WIN 35,428 as more potent on reuptake inhibition, binding and behavioral indices, we identified potent inhibitory effects of systemic administration of this cocaine analog on serotonin-containing neurons recorded in the rat dorsal raphe nucleus (DRN)[4]. These observations, coupled with the lack of cellular physiological data on the 3β-phenyltropane substituted compounds, led us to address the question of possible direct effects, and their potency, of RTI-31 and RTI-32 on serotonin cell function in the DRN.

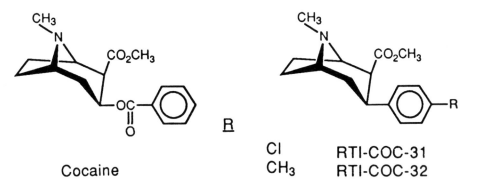

Figure 1. Chemical structures of cocaine and the *para*-substituted analogs used in the present studies.

2. Materials and Methods

Adult male Sprague-Dawley rats (225-275 gm; Harlan Sprague-Dawley, Houston, TX) were used in electrophysiological studies of serotonin-containing neurons recorded in the DRN. The effects of systemic administration of cocaine, RTI-31 and RTI-32 were investigated in the chloral hydrate (400 mg/kg, i.p.) anesthetized preparation as previously described [1,4]. These compounds were also evaluated using a 400 μm thick *in vitro* brain slice preparation containing the DRN [7]. Standard extracellular recording techniques for single-unit discrimination were utilized, which included both on- and off-line analysis of data. Only one cell was tested per animal using the *in vivo* preparation and only one cell was tested per brain slice. Both RTI-31 and RTI-32 were provided courtesy of Drs. John Boja and Michael J. Kuhar (NIDA Addiction Research Center, Baltimore) and compounds were prepared fresh or stored frozen at -20°C until use. Data were analyzed using an ANOVA with a $p < 0.05$ level of significance (Abstat, Bell-Anderson Labs). Where results are expressed relative to cocaine, both systemic drug administration and *in vitro* perfusion data were obtained for cocaine from this laboratory under identical conditions [3,4,7].

3. Results

Following systemic intravenous administration via a tail vein, both cocaine and the cocaine analogs RTI-31 and RTI-32 demonstrated the ability to completely inhibit the cell firing of spontaneously active serotonin neurons in the DRN in a reversible manner. As illustrated in figure 2, RTI-31 was consistently more potent than RTI-32 at producing a suppression of DRN cell firing. The intravenous administration of these analogs demonstrate an *in vivo* potency greater than the effects with cocaine (cocaine $ID_{50} = 0.6$ mg/kg, IV; Cunningham and Lakoski [4]).

Utilizing an *in vitro* slice preparation of the DRN, perfusion application of RTI-31 and RTI-32 demonstrated a rapid and reversible inhibition of 5-HT cell firing. As illustrated in figure 3, the perfusion of 5 μm RTI-32 completely inhibited spontaneous cell firing and a greater relative potency of these 3β-substituted phenyltropane analogs to suppress DRN activity over 5-HT perfusion alone.

4. Discussion

These studies have demonstrated a potent and direct effect of novel cocaine analogs, 3β-substituted phenyltropane derivatives RTI-31 and RTI-32, on serotonergic cellular physiology [14,15]. The marked potency of these analogs over cocaine at this somatodentric autoreceptor in the DRN confirm and extend previous neurochemical observations of a high-affinity binding for these compounds at monamine transporter sites, including transporters selective for serotonin located in the DRN. These data also further underscore the usefulness of cellular

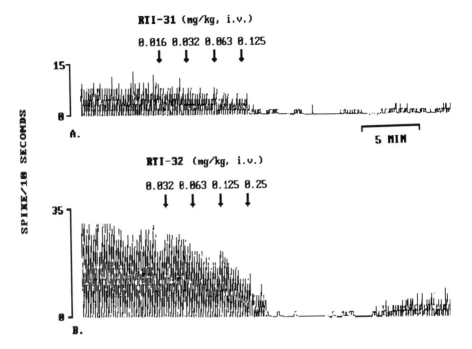

Figure 2. Systemic administration of RTI-31 and RTI-32 inhibition of spontaneous 5-HT cell firing recorded in the dorsal raphe nucleus. Intravenous administration of these cocaine analogs in a cumulative dose fashion inhibited unit activity recorded in the *in vivo* DRN preparation. Panel A. Administration of RTI-31 (0.016, 0.032, 0.063, and 0.125 mg/kg at 2 min intervals) potently and reversibly inhibited cell firing. Panel B. Similarly, administration of RTI-32 (0.032, 0.063, 0.125, 0.25 mg/kg) completely suppressed cell firing in a reversible manner.

physiological recording techniques in characterizing cocaine's interactions in a defined brain region. Clearly such "nuclei specific" observations can assist in identifying the neuronal circuits underlying the complex behavioral effects following acute and repeated administration of this stimulant.

We have recently begun to apply similar electrophysiological recording techniques to assess the effects of cocaethylene on serotonergic function in the DRN [16]. Cocaethylene, an ethyl ester of benzolyecgonine found in the urine of individuals using both cocaine and ethanol, has recently been demonstrated to potently interaction with dopamine and serotonin neuronal systems [17,18]. We have begun to investigate whether cocaethylene has direct effects on 5-HT DRN cell function and have observed a relative insensitivity of these neurons to either intravenous or bath perfusion application of this compound. These studies are being further elaborated to identify possible interactions between 5-HT or cocaine with this metabolite.

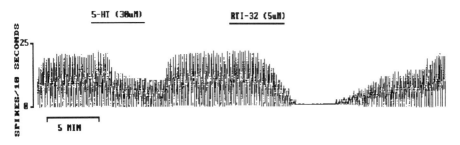

Figure 3. Comparison of the effects of 5-HT and RTI-32 applied by perfusion in an *in vitro* dorsal raphe nucleus preparation. Spontaneous cell firing was recorded in a DRN slice preparation (requiring constant perfusion with 2.5 μM phenylephrine and ACSF) demonstrated an acute and reversible inhibition of cell firing following 30 μM 5-HT or 5 μM RTI-32. Note the relative greater potency of RTI-32 to suppress 5-HT DRN cell firing over 5-HT, with a concomitant longer period of recovery of baseline activity.

In summary, these cellular electrophysiolgical studies have addressed the potent interactions of novel cocaine analogs on serotonergic cell function in the DRN. As the molecular interactions of these compounds become understood in more detail, including the mechanisms of interaction with 5-HT transporters, the selective actions of these compounds on specific 5-HT receptor subtypes in a region-specific manner will continue to merit close attention.

5. Acknowledgements

The laboratories of Dr. Michael J. Kuhar and Dr. Robert H. Roth have generously provided RTI-COC-31, RTI-COC-32 and cocaethylene, respectively. JML is a recipient of a Research Career Development Award from the National Institute on Aging.

6. References

1. Galloway, M.P. (1992) 'Neurochemical modulation of monoamines by

cocaine', in J.M. Lakoski, M.P. Galloway and F.J. White (eds.), Cocaine: Cocaine and Serotonin Cellular Physiology Pharmacology, Physiology and Clinical Strategies, CRC Press, Inc., Boca Raton, pp. 163-190.
2. Lakoski, J. M., Galloway, M.P., and White, F.J. (eds.)(1992) Cocaine: Pharmacology, Physiology and Clinical Strategies, CRC Press, Inc., Boca Raton.
3. Cunningham, K.A. and Lakoski, J.M. (1988) 'Electrophysiological effects of cocaine and procaine on dorsal raphe serotonin neurons, Eur. J.Pharmacol. 148, 457-462.
4. Cunningham, K.A. and Lakoski, J.M. (1990) 'The interaction of cocaine with serotonin: single-unit extracellular recording studies, Neuropsychopharmacology 3, 41-50.
5. Henry, D.J., Greene, M., and White, F.J. (1989) 'Electrophysiological effects of cocaine in the mesoaccumbens dopamine system: repeated administration, J. Pharmacol. Exp. Ther. 251, 833-839.
6. Pitts, D.K. and Marwah, J. (1987) 'Cocaine modulation of central monoaminergic neurotransmission', Pharmacol. Biochem. Behav. 26, 453-461.
7. Black, E.W. and Lakoski, J.M. (1990) 'In vitro electrophysiology of dorsal raphe serotonergic neurons in subchronic cocaine treated rats: development of tolerance to acute cocaine administration', Mol. Cell. Neurosci. 1, 84-91.
8. Pan, Z.Z. and Williams, J.T. (1989) 'Differential actions of cocaine and amphetamine on dorsal raphe neurons *in vitro*', J. Pharmacol. Exp. Ther. 251, 56-62.
9. Boja, J.W., Carroll, F.I., Rahman, M.A., Philip, A. Lewin, A.H., and Kuhar, M.J. (1990) 'New, potent cocaine analogs: ligand binding and transport studies in rat striatum', Eur. J. Pharmacol. 184, 329-332.
10. Carroll, F.I., Lewin A.H., Boja, J.W., and Kuhar, M.J. (1992) 'Cocaine receptor; Biochemical characterization and structure-activity relationships of cocaine analogs at the dopamine transporter, J. Med. Chem. 35, 969-980.
11. Cline, E.J., Scheffel. U., Boja, J.W., Carroll, F.I., Katz, J.L., and Kuhar, M.J.(1992) 'Behavioral effects of novel cocaine analogs; A comparison with *in vivo* receptor binding potency', J. Pharmacol. Exp. Ther. 260, 1174-1179.
12. Balster, R.L., Carroll, F.I., Graham, J.H., Mansbach, R.S., Rahman, M.A., Philip, A., Lewin, A., and Showalter, V.M. (1991) 'Potent substituted-3β-phenyltropane analogs of cocaine have cocaine-like discriminative stimulus properties', Drug Alcoh. Depend. 29, 145-151.
13. Lakoski, J.M., Black, E.W., and Moday, H.J. (1992) 'Electrophysiological effects of cocaine on serotonin neuronal systems', in J.M. Lakoski, M.P. Galloway and F.J. White (eds.), Cocaine: Pharmacology, Physiology and Clinical Strategies, CRC Press, Inc., Boca Raton, pp. 295-312.
14. Lakoski, J.M., Zheng, H, Boja, J.W., Carroll, F.I., and Kuhar, M.J. (1992) 'Potent phenyltropane cocaine analogs inhibit serotonin (5-HT) dorsal raphe cell firing', Soc. Neurosci. Abst. 18, Part 1, 772.

15. Lakoski, J.M., Zheng, H., Boja, J.W., Carroll, F.I., and Kuhar, M.J. 'Electrophysiological assessment of substituted-3β-phenyltropane cocaine analogs on serotonin neurons recorded in the dorsal raphe nucleus, in preparation.
16. Lakoski, J.M., Zheng, H., Bradberry, C.W., Jatlow, P., and Roth, R.H., 'Cellular electrophysiological effects of cocaethylene on serotonergic dorsal raphe neurons', in preparation.
17. Bradberry, C.W., Nobiletti, J.B., Elsworth, J.D., Murphy, B., Jatlow, P., and Roth, R.H., 'Cocaine and cocaethylene: Microdialysis comparison of brain drug levels and effects on dopamine and serotonin', J. Neurochem., in press.
18. Jatlow, P, Elsworth, J.D., Bradberry, C.W., Winger, G., Taylor, J.R., Russell, R., and Roth, R.H. (1991) "Cocaethylene: A neuropharmacologically active metabolite associated with concurrent cocaine-ethanol ingestion', Life Sci. 48, 1787-1794.

SEROTONIN MECHANISMS IN ETHANOL ABUSE

H. LAL, S.M. REZAZADEH, and C.J. WALLIS
Department of Pharmacology
Texas College of Osteopathic Medicine
3500 Camp Bowie Boulevard
Fort Worth, Texas 76107

ABSTRACT. Serotonergic (5-HT) receptor systems have been implicated in the control of many ethanol-related behaviors, such as voluntary ethanol consumption, anxiety-like behaviors during ethanol withdrawal, discrimination of ethanol as a stimulus, and modulation of ethanol intoxication. Several classes of serotonin receptors (5-HT$_{1A}$, 5-HT$_{1C}$, 5-HT$_2$ and 5-HT$_3$) may be important in mechanisms underlying the development and/or maintenance of ethanol tolerance and dependence by modulating emotions (anxiety), homeostasis, and adaptive processes. The impact of drugs selective for various types of serotonin receptors are discussed in relation to their effect on anxiety-like behaviors during ethanol withdrawal and voluntary ethanol consumption.

1. Potential Role for Specific Serotonin Receptor Systems in Ethanol Abuse

Neurons using serotonin as their neurotransmitter are located in the raphe nuclei of the mesencephalon and brainstem. These neurons innervate regions of the limbic system and forebrain that are involved in the control of slow wave sleep, temperature regulation, impulsivity, hormonal control, and appetitive behavior [1]. Several major classes of serotonin receptors have been described.

1.1   5-HT$_{1A}$ Receptors

5-HT$_{1A}$ receptors occur on serotonin cells of the raphe nuclei as autoreceptors and on other cells as postsynaptic receptors. 5-HT$_{1A}$ receptor stimulation results in hyperpolarization by increasing potassium flow and inhibits adenylate cyclase activity by a G-protein-linked mechanism. Stimulation of 5-HT$_{1A}$ receptors has been shown to reduce dopamine (DA) release. In correlative studies, it was shown that ethanol preferring rats (P) have more 1A receptors than nonpreferring rats [2]. Acute administration of ethanol increases dopamine release in the nucleus accumbens which is believed to contribute to the rewarding properties of ethanol [3,4]. In ethanol preferring rats (P), increased 5-HT$_{1A}$ receptor activity may be important in reducing baseline DA release, increasing the release of DA in response to ethanol [5], thus intensifying the rewarding properties of ethanol.

1.2   5-HT$_{1C/2}$ RECEPTORS

5-HT$_{1C/2}$ receptors are similar in terms of their pharmacological and binding profiles and their known second messenger effects, but differ in regional localization and receptor affinity. 5-HT$_{1C/2}$ receptors elicit phosphatidylinositol turnover by a G-protein-linked mechanism [6]. Ethanol preferring rats (P) have fewer 2 receptors than nonpreferring rats [7]. During ethanol withdrawal, Long-Evans male rats are less sensitive to DOI, a predominantly 5HT$_2$ agonist, as measured by the head shake response [8], while they are more sensitive to the anxiogenic properties of mCPP, a predominantly 5-HT$_{1C/1B}$ agonist [9]. These data suggest that low 2 receptor concentration may predispose animals to drink ethanol and that during ethanol consumption the 2 receptor system may be further down-regulated. In contrast, the 5-HT$_{1C}$ receptors may be up-regulated or unaffected by ethanol consumption contributing to the anxiety experienced during ethanol withdrawal.

1.3   5-HT$_3$ RECEPTORS

5-HT$_3$ receptors are the only known serotonin-gated ion channel receptor complexes [10]. These receptors are found in several mesolimbic regions including the nucleus accumbens [11]. Lovinger [12] has shown that acute ethanol treatment increases cation flow through 5-HT$_3$ channels. Stimulation of 5HT$_3$ receptors by a selective agonist increases dopamine release in the nucleus accumbens [13,14]. 5-HT$_3$ antagonists block or attenuate ethanol-induced dopamine release in the nucleus accumbens, indicating that 5-HT$_3$ receptor activity modulates dopamine release in the nucleus accumbens [15]. Since the dopamine and GABA systems of this region have been implicated in the rewarding properties of ethanol, 3 receptors may also modulate the rewarding properties of ethanol.

The following sections will include a short discussion of possible mechanisms by which serotonin may influence ethanol abuse. Briefly, these include serotonin action in: a) anxiety during ethanol withdrawal, b) modulation of the rewarding properties of ethanol, and c) homeostasis.

2.   Serotonin Drug Effects on Anxiety-like Behaviors During Ethanol Withdrawal

Anxiety is a symptom of ethanol withdrawal which occurs early, outlasts the acute signs of withdrawal, and persists for an extended period [16]. The appearance of this symptom may result in continued drinking or contribute to relapse after the acute signs of withdrawal have ended. Anxiety is a complex behavioral phenomenon accompanied by increased glucocorticoid and catecholamine release. Several neurotransmitter systems are involved in the display and perception of anxiety. Sedative drugs that increase GABA activity (benzodiazepines and barbiturates) are particularly effective in reducing anxiety in humans and anxiety-like behaviors in animals, as has been demonstrated by the use of drug discrimination paradigms [17,18]. Drugs acting at several types of serotonin receptors have proven useful for reducing anxiety-like behaviors in response to novel stimuli [19]. In the elevated plus-maze acute ethanol withdrawal results in an anxiogenic pattern characterized by a significant reduction in the number of entries to and time spent in the open arms. Table 1 is a short list of serotonin drugs that have been tested in this model and their respective effects on anxiety-like behavior in rodents.

TABLE 1. Serotonin drug effects on anxiety-like behaviors during ethanol withdrawal in rodents tested on the elevated plus-maze.

Drug	Action	Response
Buspirone [20]	5-HT_{1A} agonist	↓
Gepirone [21]	5-HT_{1A} agonist	↓
Mianserin [8]	$5\text{-HT}_{1C/2}$ antagonist	↓
mCPP [9]	5-HT_1 agonist	↑
Clozapine [22]	$5\text{-HT}_{1C/DA}$ antagonist	↓
MDL72222 [23]	5-HT_3 antagonist	↑

Reduction of postsynaptic $5\text{-HT}_{1C/2}$ receptor activity by direct inhibition or presynaptic inhibition alleviates some aspects of ethanol withdrawal anxiety. However, these drugs fail to block a pentylenetetrazol-like interoceptive stimulus present during ethanol withdrawal, suggesting that they are only partially effective in reducing anxiety [unpublished observations].

Repeated injections of 5HT_3 drugs like ondansetron reduce anxiety-like behavior as measured by tests for social interaction and light/dark box during mild ethanol withdrawal [24], but have been repeatedly shown to be ineffective or anxiogenic in single injection paradigms [25]. In fact, the 5HT_3 antagonist, MDL72222, reverses the anxiolytic activity of ethanol as measured on the plus-maze [23]. These data suggest that chronic blockade of 5HT_3 receptors probably has its anxiolytic effect by eliciting a change in receptor concentration or transduction in one or more receptor types. Since ethanol's properties as a weak anxiolytic agent may contribute to relapse in patients recovering from ethanol addiction, serotonin drugs which relieve anxiety may also reduce ethanol craving and/or intake.

3. Serotonin Drug Effects on Voluntary Ethanol Consumption

Voluntary ethanol consumption is influenced by several properties of ethanol including: a) reward, b) caloric, c) fluid, and d) psychogenic (anxiolytic, sedative) properties. Ethanol, given acutely, increases the release of dopamine in the nucleus accumbens [4]. Increased dopamine activity in the nucleus accumbens has been linked to the rewarding properties of ethanol [26]. A stimulatory action of serotonin on dopamine activity in the accumbens is suggested by the fact that destruction of the 5-HT terminals in the nucleus accumbens or lesions of the 5-HT cell bodies that project to the nucleus accumbens result in an increase in voluntary ethanol consumption [27]. Increasing levels of serotonin in the synaptic cleft by blocking reuptake with fluoxetine or clomipramine, reduces voluntary ethanol consumption [27]. In humans, the relatively selective serotonin uptake inhibitor citalopram also reduces voluntary ethanol consumption [28]. Treatment with drugs that have such a nonselective effect on serotonin systems may be influencing general appetitive behavior. For example, in many studies that report a decrease in ethanol consumption, there is also a decrease in food or water intake. Figure 1 demonstrates the effect of acute treatment with fluoxetine on voluntary ethanol consumption in a two-bottle-choice paradigm. The left panel demonstrates the increase in ethanol consumption (g/kg) as a function of increasing ethanol concentration, while ethanol preference is relatively constant. The right panel demonstrates the reduction of both ethanol and water consumption by fluoxetine (10 mg/kg), with no effect on ethanol preference.

Treatment with more selective serotonin drugs should help us refine our understanding of how serotonin influences ethanol consumption. Buspirone, a $5HT_{1A}$ partial agonist at presynaptic receptors, reduces both voluntary ethanol consumption [29] and anxiety [20]. Treatment with the $5HT_3$ antagonist, ondansetron, reduces voluntary ethanol consumption [30]. Such 3 antagonists have been reported to block the discrimination of ethanol's stimulus properties [31] and reduce ethanol-stimulated dopamine release in the nucleus accumbens [15]. Thus, $5-HT_3$ antagonists may reduce ethanol consumption by blocking the anxiolytic and rewarding properties of ethanol. Treatment with ritanserin, a $5-HT_{2/1C}$ antagonist, that has little or no anxiolytic effect after a single injection, also reduces ethanol consumption [32].

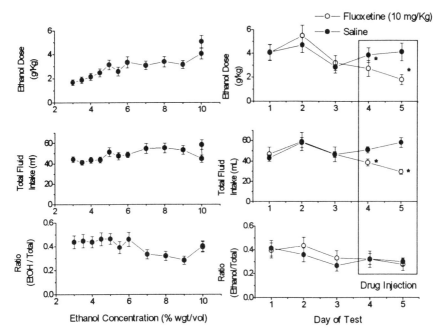

Figure 1. Male Long-Evans hooded rats were given a two-bottle choice between an ethanol solution (3-10%) or water. Ethanol was available on alternate days, with increasing concentration on each day until animals stabilized their intake at 10% ethanol. The left panel demonstrates the effect of increasing ethanol concentration on ethanol preference ratio, ethanol and fluid consumption. The right panel demonstrates the effect of acute administration of fluoxetine (10 mg/kg, ip), a serotonin reuptake blocker, on baseline consumption of 10% ethanol. * indicates $p < 0.05$ versus vehicle injected controls.

In conclusion, the findings from genetic, biochemical, and behavioral studies indicate that serotonergic receptor systems are involved in the predisposition to abuse ethanol, modulation of ethanol's rewarding and anxiolytic properties, the anxiety associated with ethanol withdrawal, and in modulating the appetitive aspects of ethanol consumption.

4. References

1. Soubrie, P. (1986) 'Reconciling the role of central serotonin neurons in human and animal behaviour', Behav. Brain Sci. 9, 319-364.
2. Wong, D.T., Threlkeld, P.G., Lumeng, L., and Li, T.-K. (1990) 'Higher density of serotonin-1A receptor in the hippocampus and cerebral cortex of

alcohol-preferring P rats', Life Sci. 46, 231-235.
3. Harris, R.A., Brodie, M.S., and Dunwiddie, T.V. (1992) 'Possible substrates of ethanol reinforcement: GABA and dopamine', Ann. N.Y. Acad. Sci. 654, 61-69.
4. Imperato, A. and DiChiara, G. (1986) 'Preferential stimulation of dopamine release in the nucleus accumbens of freely moving rats by ethanol', J. Pharmacol. Exp. Ther. 239, 219-228.
5. Fadda, F., Mosca, E., Colombo, G., and Gessa, G.L. (1990) 'Alcohol-preferring rats: Genetic sensitivity to alcohol induced stimulation of dopamine metabolism', Physiol. Behav. 47, 727-729.
6. Hoyer, D., (1988) 'Functional correlates of serotonin 1 recognition sites', J. Recept. Res. 8, 59-81.
7. Zhou, F.C., Bledsoe, S., Lumeng, L., and Li, T.-K. (1990) 'Serotonergic immunostained terminal fibers are decreased in selected brain areas of alcohol-preferring P rats', Alc: Clin. Exp. Res. 14, 355A.
8. Lal, H., Rezazadeh, S.M., and Prather, P.L. (1993) 'Potential role of 5-HT_{1C} and/or 5-HT2 receptors in mianserin-induced prevention of anxiogenic behaviors occurring during ethanol withdrawal', Alc: Clin. Exp. Res. , in press.
9. Rezazadeh, S.M., Prather, P.L., and Lal, H. (1993) 'Sensitization to 5-HT_{1C} receptor agonist in rats observed following withdrawal from chronic ethanol', Alcohol 10, in press.
10. DerKach, V., Suprenant, A., and North, R.A. (1989) '5-HT3 receptors are membrane ion channels', Nature 339, 706-709.
11. Kilpatrick, G.J., Jones, B.J., and Tyer, M.B. (1987) 'Identification and distribution of 5-HT3 receptors in rat brain using radioligand binding', Nature 330, 746-748.
12. Lovinger, D.M. (1991) 'Ethanol potentiates ion current mediated by 5-HT_3 receptors on neuroblastoma cells and isolated neurons', Alcohol and Alcoholism Suppl. 1, pp. 181-185.
13. Chen, J., van Praag, H.M., and Gardner, E.L. (1991) 'Activation of 5-HT3 receptors by 1-phenylbiguanide increases dopamine release in the rat nucleus accumbens', Brain Res. 543, 354-357.
14. Jiang, L.H., Ashby, C.R., Jr., Kasser, R.J., and Wang, R.Y. (1990) 'The effect of intraventricular administration of the 5-HT3 receptor agonist 2-methylserotonin on the release of dopamine in the nucleus accumbens: an in vivo chronocoulometric study', Brain Res. 513, 156-160.
15. Wozniak, K.M., Pert, A. and Linnoila, M. (1990) 'Antagonism of 5-HT_3 receptors attenuates the effects of ethanol on extracellular dopamine', Eur. J. Pharmacol. 187, 287-289.
16. Prather, P.L. and Lal, H. (1991) 'Protracted withdrawal: Sensitization of the anxiogenic response to cocaine in rats concurrently treated with ethanol', Neuropsychopharm. 6, 23-29.

17. Lal, H. and Emmett-Oglesby, M.W. (1983) 'Behavioral analogues of anxiety: animal models', Neuropharmacology 22, 1423-1441.
18. Lal, H., Harris, C.M., Benjamin, D., Springfield, A.C., Bhadra, S., and Emmett-Oglesby, M.W. (1988) 'Characterization of a pentylenetetrazol-like interoceptive stimulus produced by ethanol withdrawal', J. Pharmacol., Exp. Ther. 247, 508-518.
19. Pellow, S., Johnston, A.L., and File, S.E. (1987) 'Selective agonists and antagonists for 5-hydroxytryptamine receptor subtypes, and interactions with yohimbine and FG 7142 using the elevated plus-maze test in the rat', J. Pharm. Pharmacol. 39, 917-928.
20. Lal, H., Prather, P.L., and Rezazadeh, S.M. (1991) 'Anxiogenic behavior in rats during acute and protracted ethanol withdrawal: Reversal by buspirone', Alcohol 8, 467-471.
21. Benjamin, D., Saiff, E.I., Coupet, J., and Lal, H. (1990) 'Down regulation of brain 5-HT2 receptors underlies anxiolytic affect produced by sustained treatment with gepirone', Soc. for Neurosci. Abs. 16, 1323.
22. Lal, H., Rezazadeh, S.M., and Prather, P.L. (1991) 'Clozapine reverses anxiogenic behaviors observed during ethanol withdrawal', Pharmacologist 33, 165.
23. Wallis, C.J., Rezazadeh, S.M., and Lal, H. (1992) 'Behavioral evidence for down-regulation of 5HT3 receptors produced by chronic ethanol', Soc. Neurosci. Abst. 18, 1431.
24. Jones, B.J., Costall, B., Domeney, A.M., Kelly, M.E., Naylor, R.J., Oakley, N.R., and Tyers, M.B. (1988) 'The potential anxiolytic activity of GR38032F, a 5-HT3 receptor antagonist', Br. J., Pharmacol. 93, 985-993.
25. File, S.E. and Johnston, A.L. (1989) 'Lack of effects of 5HT3 receptor antagonists in the social interaction and elevated plus-maze tests of anxiety in the rat', Psychopharmacol. 99, 248-251.
26. Kornetsky, C., Bain, G.T., Unterwald, E.M., and Lewis, M.J. (1988) 'Brain stimulation reward: Effects of ethanol' Alc: Clin. Exp. Res. 12, 609-616.
27. McBride, W.J., Murphy, J.M., Lumeng, L., and Li, T.-K. (1989) 'Serotonin and ethanol preference', in M. Galanter (ed.), Recent developments in alcoholism vol. 7, pp. 187-209.
28. Naranjo, C., Sellers, F.M., Sullivan, J.T., Woodley, D.V., Kadlec, K., and Sykora, K. (1987) 'The serotonin uptake inhibitor citalopram attenuates ethanol intake', Clin. Pharmacol. Ther. 41, 266-274.
29. Privette, T.H. Hornsby, R.L., Myers, R.D. (1988) 'Buspirone alters alcohol drinking induced in rats by tetrahydropapaveroline injected into brain monoaminergic pathways', Alcohol 5, 147-152.
30. Sellers, E.M., Kaplan, H.L., Lawrin, M.D., Somer, C.A., Naranjo, C.A., and Frecker, R.C. (1988) 'The 5-HT$_3$ antagonist GR38032F decreases alcohol consumption in rats', Soc. Neurosci. Abst. 14, 41.
31. Meert, T.F. and Janssen, P.A.J. (1991) 'Ritanserin, a new therapeutic

approach for drug abuse. Part 1: Effects on alcohol', Drug Dev. Res. 24, 235-249.
32. Grant, K.A. (1990) '5-HT$_3$ Antagonists block behavioral effects of ethanol', Alc: Clin. Exp. Res. 14, 294.

SEROTONIN REGULATION OF ALCOHOL DRINKING

W.J. McBRIDE, J.M. MURPHY, L. LUMENG, and T.-K. LI
Departments of Psychiatry, Medicine and Biochemistry
Indiana University School of Medicine and VA Medical Center, and
Department of Psychology
Purdue School of Science
Indianapolis, Indiana 46202
USA

ABSTRACT. Rat lines selectively bred for their disparate alcohol drinking behavior exhibit differences in their CNS serotonin (5-HT) systems. The alcohol-preferring P line has lower contents of 5-HT and fewer immunostained 5-HT fibers than the alcohol-nonpreferring NP line in several CNS regions. In addition, the densities of 5-HT$_{1A}$ receptors are higher while values for 5-HT$_{1B}$ and 5-HT$_2$ sites are lower in the P compared to the NP line. Alcohol administration (i.p. and local) increases the synaptic levels of 5-HT and dopamine (DA) in the nucleus accumbens (ACB). Moreover, the alcohol-stimulated release of DA appears to be mediated by 5-HT$_3$ receptors. In addition, manipulation of the dorsal raphe nucleus 5-HT system produces a corresponding change in ACB DA release, suggesting that 5-HT is regulating the mesolimbic DA system. Systemic administration of agents that increase synaptic levels of 5-HT reduce the alcohol intake of P rats. Overall, the data suggest that CNS 5-HT systems are involved in regulating the actions of alcohol and that a deficiency in 5-HT may be an important factor contributing to the high alcohol drinking characteristics of the P rats.

1. Introduction

Serotonin (5-HT) has long been implicated as having a role in regulating alcohol drinking behavior [1]. However, the overall findings have been inconsistent and have failed to establish a firm link between alterations of CNS 5-HT and changes in alcohol drinking. Myers and Melchior [1] pointed out that the manner of ethanol administration and differences in the experimental animals can confound the results. One experimental approach toward establishing an involvement of 5-HT in regulating alcohol drinking would be with the use of animal models that have been carefully characterized and satisfy an established set of criteria.

There is convincing evidence that genetic factors can contribute to alcoholism and alcohol abuse. Selective breeding studies have established rat lines which exhibit disparate alcohol-drinking behavior [2,3]. Such selectively bred lines offer more homogeneous populations of rats with clearly distinct alcohol drinking preferences. Therefore, by determining neurobiological factors which are different between lines with high and low alcohol intakes, it might be possible to gain a better understanding of the CNS neuronal systems involved in regulating alcohol consumption.

Among the rat lines with high alcohol drinking behavior, only the P rat has been systematically characterized as satisfying the criteria, described by Lester and Freed [4] and Cicero [5], for an animal model of alcoholism. The P rat orally consumes, under free-choice conditions with food and water available, greater than 5 g ethanol/kg body wt./day [2]. This amount would be equivalent to a 70 kg man drinking approximately a fifth of whiskey a day. The oral self-administration of ethanol by P rats produces pharmacologically intoxicating (50-200 mg%) blood alcohol levels [6].

The P line of rats find alcohol reinforcing since they will work under operant-responding conditions for oral ethanol self-administration [7,8]. Moreover, alcohol appears to have post-ingestive rewarding properties in the P rats since these animals will self-administer ethanol directly into the stomach [9]. In addition, the volitional oral consumption of ethanol by P rats leads to the development of tolerance [10] and dependence [11]. Therefore, the P line of rats appears to be a good animal model for studying some of the neurobiological mechanisms involved in excessive alcohol drinking.

2. Results

In the results section, data will be reviewed indicating that (a) there are innate differences in limbic 5-HT systems between P and NP rats; (b) agents which increase synaptic concentrations of 5-HT can attenuate alcohol intake in the P rats; (c) ethanol itself can increase synaptic levels of 5-HT; (d) 5-HT can regulate the dopamine (DA) pathway to the nucleus accumbens (ACB); and (e) 5-HT_3 receptors are involved in the actions of ethanol on the DA system in the ACB.

2.1. INNATE DIFFERENCES IN LIMBIC 5-HT SYSTEMS

The contents of 5-HT and/or its major metabolite 5-hydroxyindoleacetic acid (5-HIAA) were generally 10-20% lower in many CNS regions of the P compared to the NP line [12,13]. This 10-20% difference was readily evident in limbic structures such as the ACB, olfactory tubercle (OTU), frontal cortex, and hypothalamus. The lower contents of 5-HT in the CNS of the P line may be due to lower 5-HT innervation since Zhou et al. [14] reported fewer immunostained 5-HT fibers in several CNS regions of the P relative to the NP line. These results provide support for the hypothesis that a deficiency in certain CNS systems may be associated with high alcohol drinking behavior.

Using standard membrane binding techniques, Wong et al. [15] reported higher densities of 5-HT$_{1A}$ receptors in the hippocampus and cerebral cortex of P than of NP rats. Autoradiographic analysis confirmed these differences in the densities of 5-HT$_{1A}$ [16,17] and further indicated that significantly higher values were found in various layers and subregions of the cerebral cortex, including the medial prefrontal cortex (MPF). On the other hand, 20-25% lower values for 5-HT$_{1B}$ and/or 5-HT$_2$ binding sites have been found in the ACB, OTU, and MPF of the P compared to the NP rats [18]. These findings provide additional support for 5-HT abnormalities in the CNS of alcohol-preferring P rats.

2.2. PHARMACOLOGICAL INCREASES IN CNS 5-HT AND ALCOHOL INTAKE

If decreased innervation and/or functioning of certain CNS 5-HT systems are associated with high alcohol drinking of the P line of rats, then increasing the physiologically active pool of 5-HT might decrease alcohol consumption. This approach was undertaken using i.p. administration of fluoxetine (a 5-HT re-uptake inhibitor), d-fenfluramine (a 5-HT releaser), and D, L-5-hydroxytryptophan (the immediate precursor of 5-HT) into P rats that were given scheduled access (one 4-hour period daily) to a 10% ethanol solution [19]. All three agents produced a significant dose-dependent decrease in alcohol drinking by the P rats. These pharmacological findings provide additional evidence for the involvement of 5-HT in regulating alcohol drinking.

2.3. EFFECTS OF ALCOHOL ON SYNAPTIC LEVELS OF 5-HT

If 5-HT is involved in regulating alcohol drinking, then ethanol might be expected to increase the release of 5-HT. Indirect neurochemical data suggest that systemic administration of ethanol can increase the activity of the 5-HT pathway innervating the ACB of P rats [20]. In order to obtain more direct, definitive data that ethanol can increase synaptic levels of 5-HT, a microdialysis study was undertaken [21]. The i.p. administration of ethanol (2.0 g/kg) elevated the extracellular levels of 5-HT in the ACB of Wistar rats to 175% of control values within the first 30 minutes [21]. Furthermore, local perfusion of the ACB with 100 mM ethanol markedly elevated the extracellular concentrations of 5-HT to 200-225% of control values [21]. The results of these experiments suggest that alcohol can stimulate the release of 5-HT from dorsal raphe nucleus (DRN) projections to the ACB. Since the ACB is thought to play an important role in mediating the actions of drugs of abuse, including ethanol [22], these data are consistent with the notion that 5-HT is involved in regulating alcohol drinking.

2.4. REGULATION OF MESOLIMBIC DA SYSTEM BY 5-HT

One possible mechanism by which 5-HT might affect alcohol drinking may be through its regulation of the activity of the mesolimbic DA system. The DA projection from the ventral tegmental area (VTA) to the ACB has been implicated in being an important part of the brain reward system [22]. Neuroanatomical studies have established the presence of 5-HT fibers from the raphe nuclei to the VTA [23].

If the DRN 5-HT system is regulating the activity of the VTA DA projection to the ACB, then stimulating or inhibiting the DRN should alter the activity of this DA system. One experimental approach to examine this possibility would be with the use of a microdialysis probe inserted into the ACB to monitor DA and 5-HT release while microinjecting excitatory or inhibitory agents directly into the DRN. Using this approach, Yoshimoto and McBride [24] established that microinjecting 0.4-0.7 ug L-glutamate (an excitatory neurotransmitter) into the DRN increased the release of both 5-HT and DA in the ACB to approximately 150% of control levels ($P < 0.05$). Conversely, local injection of 8-OH-DPAT, a 5-HT_{1A} agonist which reduces 5-HT neuronal firing via activation of cell body autoreceptors [25], produced a 25-50% decrease ($P < 0.05$) in the release of both 5-HT and DA in the ACB [24]. The results of this study indicate that 5-HT neurons in the DRN are regulating VTA DA projections to the ACB. However, the data do not establish whether 5-HT is acting directly on DA neurons (at the cell bodies and/or axon terminals) and/or if 5-HT is indirectly affecting DA neurons via interactions with other transmitter systems.

2.5. 5-HT_3 RECEPTOR INVOLVEMENT IN ALCOHOL-STIMULATED DA RELEASE

Systematic administration of a 5-HT_3 antagonist has been shown to reduce the alcohol-stimulated release of DA in the ACB following i.p. ethanol administration [26]. Studies from our laboratory demonstrated that local application of 100 uM ICS 205-930, a 5-HT_3 antagonist, completely blocked the elevated extracellular levels of DA produced by local perfusion of the ACB with 100 mM ethanol [21]. The above studies are consistent with the involvement of 5-HT_3 receptors in mediating the actions of alcohol on the ACB DA system.

3. Conclusions

Evidence has been presented that there is lower innervation and/or functioning of 5-HT pathways in several limbic regions of the P rats than were found in the NP line. Since the alcohol intake of the NP line is similar to that of unselected Wistar rats, while the intake of the P line is much higher, the neurochemical content data and the neuroanatomical findings indicate that there is a deficiency in certain 5-HT systems. This innate 5-HT deficiency might then alter the normal functioning of other CNS sites receiving these 5-HT inputs.

The MPF, ACB, and OTU are three limbic structures thought to play a role in mediating the rewarding actions of alcohol [22]. All three regions appear to have a 5-HT abnormality in the P rat. In addition, there may also be lower 5-HT innervation to the VTA which could result in altered neuronal functioning of this site, which is considered to be an important part of the brain reward system [22].

It is postulated that in the CNS of the P rat, because of the 5-HT deficiencies is certain limbic structures, there is altered regulation of neuronal circuits that are affected by alcohol, thereby producing an abnormal response to ethanol. This abnormal ethanol response could lead to alcohol acting as a strong reinforcer in the P line, or, alternatively, the 5-HT system could be regulating the aversive properties of alcohol and the loss of 5-HT inputs might raise the aversive threshold. In either case, high alcohol drinking would result. Treatment with agents that increase the synaptic levels of 5-HT would restore some of the 5-HT regulation and result in decreased alcohol drinking. This is in fact what happens following administration of a 5-HT uptake inhibitor, a 5-HT releaser, or the immediate precursor of 5-HT [19].

There is evidence that alcohol may be acting as a strong reinforcer in the VTA system of the P line of rats. For one, the P line, but not the NP rats, will self-administer 50-150 mg% ethanol directly into the VTA [17,18]. In addition, the oral self-administration of ethanol increased the extracellular levels of DA significantly more in the ACB of P rats than in stock Wistar rats [27]. However, conditioned taste aversion studies suggest that P rats have an innately higher threshold to the aversive properties of ethanol than do NP rats [28]. Therefore, a combination of enhanced sensitivity to the rewarding properties of ethanol and a higher threshold to the aversive properties of ethanol are contributing to the high alcohol preference of the P line of rats. These innate characteristics may be a result of CNS abnormalities produced by a deficient 5-HT system.

4. Acknowledgements

The research done in our laboratories and described in this article was supported by grants from the National Institute on Alcohol Abuse and Alcoholism (AA 07611, 07462, 08533, and 09090).

5. References

1. Myers, R.D. and Melchior, C.L. (1977) 'Alcohol and alcoholism, role of serotonin', in W.B. Essman (ed.), Serotonin in Health and Disease, Vol. 2, Physiological Regulation and Pharmacological Action, Spectrum, New York, pp. 373-430.
2. Lumeng, L., Hawkins, T.D., and Li, T.-K. (1977) 'New strains of rats with

alcohol preference and nonpreference', in R.G. Thurman, J.R. Williamson, H. Drott and B. Chance (eds.), Alcohol and Aldehyde Metabolizing Systems, Vol. 3, Academic Press, New York, pp. 537-544.

3. Li, T.-K., Lumeng, L., Doolittle, D.P., McBride, W.J., Murphy, J.M., Froehlich, J.C., and Morzorati, S. (1988) 'Behavioral and neurochemical associations of alcohol-seeking behavior' in K. Kuriyama, A. Takada and H. Ishii (eds.), Biomedical and Social Aspects of Alcohol and Alcoholism, Elsevier, Amsterdam, pp. 435-438.

4. Lester, D. and Freed, E.X. (1973) 'Criteria for an animal model of alcoholism', Pharmacol. Biochem. Behav. 1, 103-107.

5. Cicero, T.J. (1980) 'Animal models of alcoholism?' in J.D. Sinclair and K. Kiianma (eds.), Animal Models in Alcohol Research, Academic Press, New York, pp. 99-117.

6. Murphy, J.M., Gatto, G.J., Waller, M.B., McBride, W.J., Lumeng, L., and Li, T.-K. (1986) 'Effects of scheduled access on ethanol intake by the alcohol-preferring (P) line of rats', Alcohol 3, 331-336.

7. Penn, P.E., McBride, W.J., Lumeng, L., Gaff, T.M., and Li, T.-K. (1978) 'Neurochemical and operant behavioral studies of a strain of alcohol-preferring rats', Pharmacol. Biochem. Behav. 8, 475-481.

8. Murphy, J.M., Gatto, G.J., McBride, W.J., Lumeng, L., and Li, T.-K. (1989) 'Operant responding for oral ethanol in the alcohol-preferring P and alcohol-nonpreferring NP lines of rats', Alcohol 6, 127-131.

9. Waller, M.B., McBride, W.J., Gatto, G.J., Lumeng, L., and Li, T.-K. (1984) 'Intragastric self-infusion of ethanol by ethanol-preferring and -nonpreferring lines of rats', Science 225, 78-80.

10. Gatto, G.J., Murphy, J.M., Waller, M.B., McBride, W.J., Lumeng, L., and Li, T.-K. (1987) 'Chronic ethanol tolerance through free-choice drinking in the P line of alcohol-preferring rats', Pharmacol. Biochem. Behav. 28, 111-115.

11. Waller, M.B., McBride, W.J., Lumeng, L., and Li, T.-K. (1982) 'Induction of dependence on ethanol by free-choice drinking in alcohol-preferring rats', Pharmacol. Biochem. Behav. 16, 501-507.

12. Murphy, J.M., McBride, W.J., Lumeng, L., and Li, T.-K. (1982) 'Regional brain levels of monoamines in alcohol-preferring and -nonpreferring lines of rats', Pharmacol. Biochem. Behav. 16, 145-149.

13. Murphy, J.M., McBride, W.J., Lumeng, L., and Li, T.-K. (1987) 'Contents of monoamines in forebrain regions of alcohol-preferring (P) and -nonpreferring (NP) lines of rats', Pharmacol. Biochem. Behav. 26, 389-392.

14. Zhou, F.C., Bledsoe, S., Lumeng, L., and Li, T.-K. (1991) 'Immunostained serotonergic fibers are decreased in selected brain regions of alcohol-preferring rats', Alcohol 8, 425-431.

15. Wong, D.T., Threlkeld, P.G., Lumeng, L., and Li, T.-K. (1990) 'Higher density of serotonin-1A receptors in the hippocampus and cerebral cortex of alcohol-preferring P rats', Life Sci. 46, 231-235.

16. McBride, W.J., Murphy, J.M., Lumeng, L., and Li, T.-K. (1990) 'Serotonin, dopamine and GABA involvement in alcohol drinking of selectively bred rats', Alcohol 7, 199-205.
17. McBride, W.J., Murphy, J.M., Gatto, G.J., Levy, A.D., Lumeng, L., and Li T.-K. (1991) 'Serotonin and dopamine systems regulating alcohol intake. Alcohol Alcoholism', Suppl. 1, 411-416.
18. McBride, W.J., Murphy, J.M., Gatto, G.J., Levy, A.D., Yoshimoto, K., Lumeng, L., and Li, T.-K. (1992) 'CNS mechanisms of alcohol self-administration', Alcohol Alcoholism, in press.
19. McBride, W.J., Murphy, J.M., Lumeng, L., and Li, T.-K. (1992) 'Serotonin and alcohol consumption', in C.A. Naranjo and E.M. Seller (eds.), Novel Pharmacological Interventions for Alcoholism, Springer-Verlag, New York, pp. 59-67.
20. Murphy, J.M., McBride, W.J., Gatto, G.J., Lumeng, L., and Li, T.-K. (1988) 'Effects of acute ethanol administration on monoamine and metabolite content in forebrain regions of ethanol-tolerant and -nontolerant alcohol-preferring (P) rats', Pharmacol. Biochem. Behav. 29, 169-174.
21. Yoshimoto, K., McBride, W.J., Lumeng, L., and Li, T.-K. (1991) 'Alcohol stimulates the release of dopamine and serotonin in the nucleus accumbens', Alcohol 9, 17-22.
22. Koob, G.F. and Bloom, F.E. (1988) 'Cellular and molecular mechanisms of drug dependence', Science 242, 715-723.
23. Moore, R.Y., Halaris, A.E., and Jones, B.E. (1978) 'Serotonin neurons of the midbrain raphe, ascending projections', J. Comp. Neurol. 180, 417-437.
24. Yoshimoto, K. and McBride, W.J. (1992) 'Regulation of nucleus accumbens dopamine release by the dorsal raphe nucleus in the rat', Neurochem. Res. 17, 401-407.
25. Hutson, P.H., Sarna, G.S., O'Connell, M.T., and Curzon, G. (1989) 'Hippocampal 5-HT synthesis and release in vivo is decreased by infusion of 8-OH DPAT into the nucleus raphe dorsalis', Neurosci. Lett. 100, 276-280.
26. Carboni, E., Acquas, E., Frau, R., and DiChiara, G. (1989) 'Differential inhibitory effects of a $5\text{-}HT_3$ antagonist on drug-induced stimulation of dopamine release', Eur. J. Pharmacol. 164, 515-519.
27. Weiss, F., Hurd, Y.L., Ungerstedt, U., Markou, A., Plotsky, P.M., and Koob, G.F. (1992) 'Neurochemical correlates of cocaine and ethanol self-administration', in P.A. Kalivas and H.H. Samson (eds.), The Neurobiology of Drug and Alcohol Addiction, New York Academy of Sciences, New York, pp. 220-241.
28. Froehlich, J.C., Harts, J., Lumeng, L., and Li, T.-K. (1988) 'Differences in response to the aversive properties of ethanol in rats selectively bred for oral ethanol preference', Pharmacol. Biochem. Behav. 31, 215-222.

HYPOTHALAMIC SEROTONIN IN RELATION TO APPETITE FOR MACRONUTRIENTS AND EATING DISORDERS

SARAH R. LEIBOWITZ
The Rockefeller University
1230 York Avenue
New York, New York 10021
USA

ABSTRACT. Brain serotonin (5-HT) is known to have an inhibitory effect on feeding behavior. Recent studies have characterized more precisely the nature of this effect, identified the brain sites and receptors involved, and defined a possible physiological role of endogenous 5-HT in controlling natural patterns of eating and nutrient selection. The evidence reviewed here suggests that 5-HT in the hypothalamus modulates the circadian pattern of feeding, as well as the proportion of carbohydrate in the diet. The neural substrates for this modulation include serotonergic projections to the paraventricular and ventromedial nuclei of the medial hypothalamus, where there exist neurons that control satiety in relation to the body's energy stores, and also to the suprachiasmatic nucleus, a primary circadian pacemaker that controls temporal patterns of feeding. Rather than tonically inhibiting eating behavior, endogenous 5-HT appears to act phasically according to a distinct circadian rhythm, with a peak at the beginning of the active feeding cycle. At this time, freely-feeding rats consume meals that are rich in carbohydrate. It is proposed that gradually over the course of these meals, medial hypothalamic 5-HT increases its activity and then, through stimulation of 5-HT_{1B} receptors, evokes a state of satiety that is specific to carbohydrate. Biochemical evidence in humans suggests that disturbances in brain 5-HT may mediate certain symptoms of eating disorders which may be ameliorated by pharmacological agents that potentiate serotonergic activity.

1. Hypothalamic Serotonin Reduces Food Intake and Increases Energy Metabolism

Studies indicate that 5-HT has a suppressive effect on food intake and body weight [1-3]. This effect of serotonergic stimulation has been demonstrated with both peripheral and central injections of serotonergic agonists. Moreover, the opposite effect, an enhancement of food consumption, has been observed with receptor antagonists and other drugs which reduce 5-HT activity. This report will focus on brain mechanisms involved in this effect of 5-HT. Hypothalamic microinjection of serotonergic agents into chronically brain-cannulated rats produces a potent and selective effect on feeding patterns and food choice, at much smaller doses than those used in intraventricular and peripheral injection studies [3]. This phenomenon is localized specifically to nuclei in the medial hypothalamus, most particularly the paraventricular (PVN), ventromedial (VMN), and suprachiasmatic (SCN) nuclei, which are believed to be essential for normal control of nutrient intake, as well as for normal responsiveness to peripherally injected serotonergic drugs [3-10]. The suppressive effect on food intake can be seen with exogenous 5-HT, as well as with agents that enhance synaptic availability of endogenous 5-HT. These include d-norfenfluramine and fluoxetine, which enhance the release or block synaptic uptake of endogenous 5-HT [3,5-8]. In addition to reducing food intake, serotonergic stimulation of the paraventricular nucleus has been shown to stimulate energy metabolism [11].

A specific role for medial hypothalamic serotonergic receptors in the control of feeding is supported by the finding that hypothalamic administration of general 5-HT antagonists effectively and dose-dependently blocks the feeding inhibitory effect of 5-HT in the PVN [3,5]. It has been proposed that $5-HT_1$, as opposed to $5-HT_2$ or $5-HT_3$, receptor subtypes mediate the feeding-suppressive action of serotonergic stimulation in the medial hypothalamus [12]. This is supported by the evidence that peripheral or hypothalamic injections of serotonergic or ß-adrenergic antagonists with relatively high affinity for $5-HT_1$ receptors, but not selective antagonists of $5-HT_2$ or $5-HT_3$ receptors, significantly attenuate the feeding-suppressive action of serotonergic agonists injected peripherally or into the PVN [5,13-16]. These serotonergic receptor antagonists on their own may additionally potentiate feeding behavior [16], presumably by blocking the action of the endogenous 5-HT at its postsynaptic $5-HT_1$ receptors. Further differentiation of the $5-HT_1$ receptor indicates that, in the rat, the $5-HT_{1B}$ and possibly the $5-HT_{1C}$ subtypes are specifically involved in 5-HT-induced hypophagia, in contrast to the $5-HT_{1A}$ receptor which may mediate the opposite response, hyperphagia [5,17-20].

In the rat, receptor binding assays in micropunched tissue [21] and autoradiographic analyses of brain sections [22] have identified a dense concentration of $5-HT_{1B}$ and $5-HT_{1C}$ receptor sites in the hypothalamus, in contrast to a relatively low concentration of $5-HT_{1A}$ receptors. The evidence indicates, further, that the highest density of $5-HT_{1B}$ receptor sites can be found in medial hypothalamic nuclei, in particular, the PVN, VMN, and SCN [21], precisely where endogenous 5-HT is believed to act in modulating nutrient intake in the rat [4,5,7,8]. While studies in humans, similar to rats, have revealed a suppressive effect of serotonergic agonists on eating behavior [23-25], the failure to detect $5-HT_{1B}$ receptors in human brain [26] suggests that $5-HT_{1C}$ or possibly $5-HT_{1D}$ receptors, which exhibit similar characteristics to $5-HT_{1B}$ receptors in rodents, may be involved in mediating the modulatory effects of serotonergic stimulation.

2. Characterization and Mechanism of 5-HT's Impact on Eating Patterns and Macronutrient Intake

Studies of meal patterns, using computer-automated procedures, demonstrate that medial hypothalamic 5-HT has a specific role in controlling the temporal aspects of feeding [1]. Hypothalamic as well as peripheral administration of serotonergic agonists affects feeding patterns by producing a significant decrease in the size and duration of individual meals, in association with a reduced rate of eating [1,2,3,6]. Since the latency to meal onset and the frequency of meals taken are not affected, it is proposed that endogenous 5-HT may influence primarily the termination rather than the initiation of eating.

Through hypothalamic administration of 5-HT, as well as studies employing systemic injection of serotonergic agents, evidence has accumulated to indicate a role for 5-HT in the modulation of the animals' appetite for specific foods [1,2,6-8,27,28]. This role appears to involve the control of carbohydrate and protein intake or the ratio of these two macronutrients, with serotonergic stimulation reducing the proportion of carbohydrate in the diet. This phenomenon was initially demonstrated in a two-diet self-selection paradigm, first with peripherally administered fenfluramine and fluoxetine [28] and then with PVN injection of 5-HT or norfenfluramine [4,6,7]. The opposite pattern has been detected with a reduction in brain 5-HT after intraventricular injection of a serotonergic neurotoxin [27].

Subsequent studies have examined this phenomenon in freely-feeding animals offered separate sources of pure macronutrient diets [3,7,8]. These studies reveal that injections of 5-HT, or the serotonergic agonists d-norfenfluramine and fluoxetine, into the medial hypothalamus selectively and dose-dependently suppress carbohydrate consumption, while having no effect on or possibly enhancing ingestion of protein or fat. A recent meal pattern analysis of this selective change in nutrient intake demonstrates that 5-HT in the PVN is active in terminating specifically carbohydrate-rich meals, while reducing the rate of carbohydrate eating and enhancing the satiating value of this nutrient [29]. This is consistent with other evidence obtained with electrolytic brain lesions, distinguishing these two nuclei as having an essential role in controlling satiety for carbohydrate [9,10].

When peripherally administered, the serotonergic agents d-fenfluramine, fluoxetine, and quipazine also suppress carbohydrate intake; however, depending upon the drug dose and nutritional state of the animal, they are found sometimes to be less selective in their effect on macronutrient choice, producing a reduction in fat as well as carbohydrate ingestion, while having lesser impact on or sometimes enhancing protein consumption [2,3,5-8]. With peripheral injection of 5-HT antagonists, the converse pattern is obtained, namely, a strong enhancement in feeding attributed primarily to a robust increase in carbohydrate and fat as opposed to protein ingestion [3,6,16]. A recent meal pattern study with peripherally administered metergoline demonstrates a specific effect of this antagonist in potentiating the carbohydrate concentration of a meal, specifically by reducing the satiating impact of this macronutrient [29].

Experiments indicate that a rhythm of serotonergic activity, in relation to feeding, exists across the light/dark cycle, as reflected by temporal shifts in responsiveness to medial hypothalamic 5-HT$_1$ receptor stimulation and in the release and utilization of endogenous 5-HT. The time of strongest serotonergic activity exists at the beginning of the active feeding period, just at the transition between light and dark [3,6-8]. At the onset of the dark cycle, the active period for the freely feeding rat, serotonergic stimulation of the PVN, VMN, or SCN is most effective in suppressing food intake and, in particular, carbohydrate ingestion. Protein and fat ingestion are unaffected or even enhanced by 5-HT at this time, and at other points in the dark cycle, the serotonergic agonists cause no change in either total food intake or macronutrient choice.

The proposal, that hypothalamic serotonergic control of feeding is expressed phasically primarily at the beginning of the natural feeding cycle, is supported by various biochemical studies which have revealed that endogenous 5-HT activity in the medial hypothalamus rises or peaks specifically at this time. This change at dark onset has been revealed in the PVN, VMN, and SCN with measurements of either synaptic 5-HT content, synthesis, metabolism, or uptake [30-35]. At this time, carbohydrate-rich meals are found to predominate under natural feeding conditions [36,37]. Moreover, a single hypothalamic injection of 5-HT reduces the size and carbohydrate content of only the first 3 meals of this nocturnal feeding cycle [29], and a similar effect can be observed with the serotonergic compounds, fluoxetine and d-norfenfluramine, which through microdialysis studies are found to enhance the synaptic concentration of endogenous 5-HT in the PVN (Leibowitz and Paez, unpublished data). Thus, it is proposed that 5-HT in the medial hypothalamus plays a specific role in terminating these initial carbohydrate meals, possibly by stimulating PVN or VMN "satiety" neurons known to control intake of this macronutrient [9,10] and SCN neurons that determine the circadian rhythms of physiological systems [31,33,34].

3. Clinical Studies: Possible Disturbances in Brain 5-HT and Therapeutic Value of Serotonergic Agonists in Eating Disorders

These findings in animals may have implications for understanding human brain mechanisms of appetite control, which exhibit some similarities in their responsiveness to serotonergic stimulation. As in animal studies, administration of serotonergic agents in humans suppresses eating behavior, possibly by reducing the rate of eating, as well as the size of large meals, through intensification of satiety [23-25]. In some [24], but not all [25] studies, they cause a preferential reduction in the consumption of carbohydrate, as well as a decrease in the "craving" for snacks rich in this macronutrient. Administration of agents that antagonize serotonergic activity are generally found to increase calorie intake and hunger ratings in human subjects and possibly enhance intake of "sweet high carbohydrate" foods [23].

Moreover, clinical studies have revealed disturbances in the activity of the serotonergic system in patients with eating disorders [38,39]. Of particular interest are bulimic patients, who appear to have a disorder in the production of satiety. These patients are found to overconsume specifically carbohydrate in all but their largest meals [40]. They also exhibit a deficiency of brain 5-HT activity, decreased postsynaptic serotonergic receptor responsiveness, and a tendency toward smaller post-meal increases in the plasma ratio of tryptophan/large neutral amino acids, which may determine brain 5-HT synthesis [38,39]. Disturbances in brain 5-HT have also been seen in patients with anorexia nervosa. In particular, after long-term weight restoration, anorexic subjects exhibit a significant enhancement of brain 5-HT activity, a pattern opposite to that detected in bulimic patients [38].

Some studies with serotonergic compounds, such as, fenfluramine and fluoxetine, have reported positive results in the treatment of patients with eating disorders [38,39]. These agents which enhance endogenous serotonergic activity substantially decrease binge eating in bulimic patients, and this effect appears to occur independently of changes in symptoms of depression. Furthermore, a possible effectiveness of antidepressant agents that block 5-HT uptake, in maintaining a healthy body weight in anorexics, has also been described [38].

Eating disorders occur predominately in females and generally develop during the adolescent years [41]. Moreover, disturbances in their carbohydrate preferences, as well as the distribution of their carbohydrate diet across the 24-hour cycle, are common [42]. This is of interest in light of recent studies showing gender differences in the eating patterns of animals. For example, female rats exhibit a stronger preference for carbohydrate than male rats; they consume more calories of carbohydrate relative to protein or fat, and they show a diurnal shift compared to males, with a larger portion of their carbohydrate consumed during the inactive period [43]. This preference pattern, similarly seen in humans [44], is evident sometime between weaning and puberty, with peak carbohydrate intake in females occurring at the onset of puberty. The possibility that this gender difference in appetite for carbohydrate may be linked to brain 5-HT [42] is supported by the evidence that female rats exhibit a higher density of medial hypothalamic 5-HT$_{1B}$ as well as 5-HT$_{1A}$ receptors compared to males, as well as a greater responsiveness to serotonergic agents [45], and they also show faster 5-HT turnover rate and responsiveness to agents that modulate brain 5-HT synthesis [46]. Moreover, clinical evidence indicates that dieting alters the responsiveness of women, but not men, to the 5-HT precursor tryptophan [47]. Investigation of the relationship of this neurochemical profile to gender differences in natural meal patterns and macronutrient preferences should help us to understand how the brain serotonergic system might contribute to the development of disturbances in appetite and eating patterns in humans.

4. Acknowledgements

This research was supported by U.S. Public Health Service grant MH 43422.

5. References

1. Blundell, J.E. (1984) 'Serotonin and appetite', Neuropharmacology 23, 1537-1551.
2. Blundell, J.E. (1986) 'Serotonin manipulations and the structure of feeding behaviour', Appetite 7, 39-56.
3. Leibowitz, S.F., Weiss, G.F., and Shor-Posner, G. (1988) 'Hypothalamic serotonin: Pharmacological, biochemical and behavioral analyses of its feeding-suppressive action', Clin. Neuropharmacol. 11, S51-S71.
4. Leibowitz, S.F., Weiss, G.F., and Suh, J.S. (1990) 'Medial hypothalamic nuclei mediate serotonin's inhibitory effect on feeding behavior', Pharmacol. Biochem. Behav. 37, 735-742.
5. Weiss, G.F., Papadakos, P., Knudson, K., and Leibowitz, S.F. (1986) 'Medial hypothalamic serotonin: Effects on deprivation and norepinephrine-induced eating', Pharmacol. Biochem. Behav. 25, 1223-1230.
6. Shor-Posner, G., Grinker, J.A., Marinescu, C., Brown, O., and Leibowitz, S.F. (1986) 'Hypothalamic serotonin in the control of meal patterns and macronutrient selection', Brain Res. Bull. 17, 663-671.
7. Weiss, G.F., Rogacki, N., Fueg, A., Buchen, D., and Leibowitz, S.F. (1990) 'Impact of hypothalamic d-norfenfluramine and peripheral d-fenfluramine injection on macronutrient intake in the rat', Brain Res. Bull. 25, 849-859.
8. Weiss, G.F., Rogacki, N., Fueg, A., Buchen, D., Suh, J.S., Wong, D.T., and Leibowitz, S.F. (1991) 'Effect of hypothalamic and peripheral fluoxetine injection on natural patterns of macronutrient intake in the rat', Psychopharmacology 105, 467-476.
9. Shor-Posner, G., Azar, A.P., Insinga, S., and Leibowitz, S.F. (1985) 'Deficits in the control of food intake after hypothalamic paraventricular nucleus lesions', Physiol. Behav. 35, 883-890.
10. Sclafani, A. and Aravich, P.F. (1983) 'Macronutrient self-selection in three forms of hypothalamic obesity', Am. J. Physiol. 244, R686-R694.
11. Sakaguchi, T. and Bray, G.A. (1989) 'Effect of norepinephrine, serotonin and tryptophan on the firing rate of sympathetic nerves', Brain Res. 492, 271-280.
12. Curzon, G. (1990) 'Serotonin and appetite', Ann. N.Y. Acad. Sci. 600, 521-531.
13. Massi, M. and Marini, S. (1987) 'Effect of the $5HT_2$ antagonist ritanserin on food intake and on 5HT-induced anorexia in the rat', Pharmacol. Biochem. Behav. 26, 333-340.
14. Kennett, G.A., Dourish, C.T., and Curzon, G. (1987) '$5\text{-}HT_{1B}$ agonists induce

14. Kennett, G.A., Dourish, C.T., and Curzon, G. (1987) '5-HT_{1B} agonists induce anorexia at a postsynaptic site', Eur. J. Pharmacol. 141, 429-435.
15. Neill, J.C. and Cooper, S.J. (1989) 'Evidence that d-fenfluramine anorexia is mediated by 5-HT_1 receptors', Psychopharmacology 97, 213-218.
16. Dourish, C.T., Clark, M.L., Fletcher, A., and Iversen, S.D. (1989) 'Evidence that blockade of post-synaptic 5-HT_1 receptors elicits feeding in satiated rats', Psychopharmacology 97, 54-58.
17. Hutson, P.H., Donohoe, T.P., and Curzon, G. (1988) 'Infusion of the 5-hydroxytryptamine agonists RU24969 and TFMPP into the paraventricular nucleus of the hypothalamus causes hypophagia', Psychopharmacology 95, 550-552.
18. Dourish, C.T., Hutson, P.H., Kennett, G.A., and Curzon, G. (1986) '8-OH-DPAT-induced hyperphagia: Its neural basis and possible therapeutic relevance', Appetite 7, 127-140.
19. Kennett, G.A. and Curzon, G. (1988) 'Evidence that mCPP may have behavioural effects mediated by central 5-HT_{1C} receptors', Br. J. Pharmacol. 94, 137-147.
20. Kennett, G.A. and Curzon, G. (1988) 'Evidence that hypophagia induced by mCPP and TFMPP requires 5-HT_{1C} and 5-HT_{1B} receptors; hypophagia induced by RU 24969 only requires 5-HT_{1B} receptors', Psychopharmacology 96, 93-100.
21. Leibowitz, S.F. and Jhanwar-Uniyal, M. (1989) '5-HT_{1A} and 5-HT_{1B} receptor binding sites in discrete hypothalamic nuclei: Relation to feeding', Soc. Neurosci. Abstr. 15, 655.
22. Pazos, A. and Palacios, J.M. (1985) 'Quantitative autoradiographic mapping of serotonin receptors in the rat brain. I. Serotonin-1 receptors', Brain Res. 346,, 205-230.
23. Silverstone, T. and Goodall, E. (1986) 'Serotoninergic mechanisms in human feeding: The pharmacological evidence', Appetite 7, 85-97.
24. Wurtman, J.J. and Wurtman, R.J. (1984) 'd-Fenfluramine selectively decreases carbohydrate but not protein intake in obese subjects', Int. J. Obesity 8, 79-84.
25. Hill, A.J. and Blundell, J.E. (1986) 'Model system for investigating the actions of anorectic drugs: Effect of d-fenfluramine on food intake, nutrient selection, food preferences, meal patterns, hunger and satiety in healthy human subjects', in E. Ferrari and F. Brambilla (eds.), Advances in the Biosciences, vol. 60, Pergamon Press, New York, pp. 377-389.
26. Peroutka, S.J. (1988) '5-Hydroxytryptamine receptor subtypes', Ann. Rev. Neurosci. 11, 45-60.
27. Li, E.T.S. and Anderson, G.H. (1984) '5-Hydroxytryptamine: A modulator of food composition but not quantity?', Life Sci. 34, 2453-2460.
28. Wurtman, J.J. and Wurtman, R.J. (1977) 'Fenfluramine and fluoxetine spare protein consumption while suppressing caloric intake', Science 198, 1178-

29. Leibowitz, S.F., Alexander, J.T., Cheung, W.K., and Weiss, G.F. (1993) 'Effects of serotonin and the serotonin blocker metergoline on meal patterns and macronutrient selection', Pharmacol. Biochem. Behav., in press.
30. Martin, K.F. and Marsden, C.A. (1985) 'In vivo diurnal variations of 5HT release in hypothalamic nuclei', in P.H. Redfern, I.C. Campbell, J. A. Davies and K.F. Martin (eds.), Circadian Rhythms in the Central Nervous System, The Macmillan Press, London, pp. 81-92.
31. Meyer, D.C. and Quay, W.B. (1976) 'Hypothalamic and suprachiasmatic uptake of serotonin in vitro: Twenty-four-hour changes in male and proestrous female rats', Endocrinology 98, 1160-1165.
32. Stanley, B.G., Schwartz, D.H., Hernandez, L., Leibowitz, S.F., and Hoebel, B.G. (1989) 'Patterns of extracellular 5-hydroxyindoleacetic acid (5-HIAA) in the paraventricular hypothalamus (PVN): Relation to circadian rhythm and deprivation-induced eating behavior', Pharmacol. Biochem. Behav. 33, 257-260.
33. Hery, M., Faudon, M., Dusticier, G., and Hery, F. (1982) 'Daily variations in serotonin metabolism in the suprachiasmatic nucleus of the rat: Influence of oestradiol impregnation', J. Endocrinol. 94, 157-166.
34. Faradji, H., Despuglio, R., and Jouvet, M. (1983) 'Voltammetric measurements of 5-hydroxyindole compounds in the suprachiasmatic nuclei: Circadian fluctuations', Brain Res. 279, 111-119.
35. Mason, R. (1986) 'Circadian variation in sensitivity of suprachiasmatic and lateral geniculate neurones to 5-hydroxytryptamine in the rat', J. Physiol. 377, 1-13.
36. Tempel, D.L., Shor-Posner, G., Dwyer, D., and Leibowitz, S.F. (1989) 'Nocturnal patterns of macronutrient intake in freely feeding and food-deprived rats', Am. J. Physiol. 256, R541-548.
37. Shor-Posner, G., Ian, C., Brennan, G., Cohn, T., Moy, H., Ning, A., and Leibowitz, S.F. (1991) 'Self-selecting albino rats exhibit differential preferences for pure macronutrient diets: Characterization of three subpopulations', Physiol. Behav. 50, 1187-1195.
38. Kaye, W.H. and Weltzin, T.E. (1991) 'Serotonin activity in anorexia and bulimia nervosa: relationship to the modulation of feeding and mood', J. Clin. Psychiatry 52, 41-48.
39. Jimerson, D.C., Lesem, M.D., Kaye, W.H., Hegg, A.P., and Brewerton, T.D. (1990) 'Eating disorders and depression: is there a serotonin connection?', Biol. Psychiatry 28, 443-454.
40. Kaye, W.H., Weltzin, T.E., McKee, M., McConaha, C., Hansen, D., and Hsu, L.K.G., (1992) 'Laboratory assessment of feeding behavior in bulimia nervosa and healthy women: methods for developing a human-feeding laboratory', Am. J. Clin. Nutri. 55, 372-380.
41. Silverman, J.A. (1983) 'Anorexia nervosa clinical and metabolic observations', Int. J. Eating Disorders 2, 159-166.

42. Wurtman, R.J. and Wurtman, J.J. (1989) 'Carbohydrate and depression', Sci. Am. 260, 68-75.
43. Leibowitz, S.F., Lucas, D.J., Leibowitz, K.L., and Jhanwar, Y.S. (1991) 'Developmental patterns of macronutrient intake in female and male rats from weaning to maturity', Physiol. Behav. 50, 1167-1174.
44. Drewnowski, A., Kurth, C., Holden-Wiltse, J., and Saari, J. (1992) 'Food preferences in human obesity: Carbohydrate versus fats', Appetite 18, 207-221.
45. Leibowitz, S.F. and Jhanwar-Uniyal, M. (1990) '5-HT_{1A} and 5-HT_{1B} receptor binding sites in discrete hypothalamic nuclei: Relation to circadian rhythm and gender', Soc. Neurosci. Abs. 16, 294.
46. Haleem, D.J., Kennett, G.A., and Curzon, G. (1990) 'Hippocampal 5-HT synthesis is greater in female rats than in males and more decreased by 5-HT_{1A} agonist 8-OH-DPAT', J. Transm. Gen. Sec. 79, 93-101.
47. Goodwin, G.M., Fairburn, C.G., and Cowen, P.J. (1987) 'Dieting changes serotonergic function in women, not men: Implications for the aetiology of anorexia nervosa?', Psychol. Med. 17, 839-842.

SEROTONERGIC PHARMACOLOGY OF APPETITE

G. CURZON, E.L. GIBSON, A.J. KENNEDY, and A.O. OLUYOMI
Institute of Neurology
1, Wakefield Street
London WC1N, 1PJ
UK

ABSTRACT. Feeding is influenced by 5-HTergic drugs. Thus, 5-HT_{1A} agonists cause hyperphagia in freely feeding rats by activating raphe presynaptic 5-HT_{1A} receptors so that extracellular 5-HT at terminals is decreased. 5-HT_{1B} and 5-HT_{1C} agonists cause hypophagia in freely feeding or food-deprived rats by acting on hypothalamic postsynaptic receptors. D-fenfluramine (d-F) increases brain extracellular 5-HT but its hypophagic effect and that of its metabolite d-norfenfluramine (d-NF) survive 5-HT depletion. In the case of d-NF, results suggest a direct action at 5-HT_{1C} sites. As appetite disorders are more frequent in females, sex differences in these mechanisms are considered. Results in general have implications for the understanding and control of disorders of appetite.

1. Introduction

Although 5-HT is one of numerous known determinants of appetite [1], particular attention has been paid to this transmitter in research on feeding partly because of indications that it mediates the action of the commonly used appetite suppressant fenfluramine [2]. There has also been interest in whether effects of dietary intake, composition, or restriction on brain 5-HT are involved in normal pathological appetite control [3,4] and hypotheses can be constructed for the precipitation of both anorexic and bulimic states by disturbances of 5-HTergic function. More recently, much research has concerned the roles of particular 5-HT receptor subtypes in feeding [3]. This chapter will mainly deal with the latter topic.

2. 5-HT_{1A} Receptors

8-OH-DPAT and other 5-HT_{1A} agonists [5,6] increase food intake by freely feeding (but not food-deprived) rats by activating presynaptic 5-HT_{1A} receptors on 5-HT cell bodies in the raphe so that release of 5-HT at terminals is decreased [7,8,9]. The hyperphagic effect of BMY 7378 [10], an agonist at raphe 5-HT_{1A} receptors with mainly antagonistic properties at postsynaptic 5-HT_{1A} receptors, and the blockade of 8-OH-DPAT hyperphagia by WAY 100135 [11], an antagonist at both sites, agree with this mechanism.

Do 5-HT_{1A} agonists cause hyperphagia directly or via general behavioral activation? Although injecting 8-OH-DPAT into the raphe has locomotor effects [12,13], the hyperphagia does not merely result from these. Thus, systemic [5] or raphe [7] injection at low dosage only increased feeding and a specific mechanism was indicated by a behavioral competition study [14]. Also, the hyperphagia does not merely reflect DA-dependent gnawing [5,15,16]. However, a DAergic link is indicated as 8-OH-DPAT hyperphagia was inhibited by DA antagonists [17]. Buspirone may be a special case; though a 5-HT_{1A} agonist, its hyperphagic effect was not prevented by inhibiting 5-HT synthesis [18].

8-OH-DPAT hyperphagia is the only available behavioral model of presynaptic 5-HT_{1A} receptor response. 8-OH-DPAT-induced hypothermia in the rat was previously thought to reflect a presynaptic response, but it has now been shown to involve postsynaptic 5-HT_{1A} receptors [10,19]. The hyperphagia is a useful index of presynaptic activation in young adult male rats, but higher doses of 8-OH-DPAT elicit the 5-HT syndrome which can interfere [5].

The hyperphagic effects of 5-HT_{1A} agonists raise the question of their utility in anorexia nervosa but there is little information on this. Interestingly, the anxiolytic 5-HT_{1A} agonist SM-3997 is reported to alleviate bulimia nervosa in some subjects [20].

3. 5-HT_{1B} and 5-HT_{1C} Receptors

Hyperphagia due to decreased 5-HT at postsynaptic sites is consistent with hypophagia when they are activated. 5-HT_{1B} and 5-HT_{1C} sites are implicated in the rat. Thus the 5-HT_{1B} agonists RU 24969 [21], the structurally similar compounds CP-93129 [22], CP-94253 [23], and the 5-HT_{1C} agonists mCPP, TFMPP [21], and MK 212 [24] decrease food intake by freely feeding and food-deprived rats. Involvement of 5-HT_{1B} receptors in RU 24969 hypophagia is indicated primarily by effects of antagonists [21,25]. 5-HT_{1B} mediation of the effects of CP-93129 and CP-94253 is suggested mainly by their selectivity for 5-HT_{1B} sites [22,23]. Mediation of mCPP, TFMPP, and MK212 hypophagias by 5-HT_{1C} sites is pointed to by antagonist experiments [24,25]. Evidence for mCPP is particularly strong as ID_{50}s of antagonists for inhibition of its hypophagic effect correlated specifically with their 5-HT_{1C} affinities [26]. RU 24969 [21] and mCPP [27] hypophagias are not prevented by inhibiting 5-HT synthesis; this points to a direct postsynaptic mechanism. The 5-HT_{1B} receptors appear to be downstream from the 5-HT_{1C} sites as mianserin (high 5-HT_{1C} affinity, low 5-HT_{1B} affinity) blocked mCPP but not RU 24969 hypophagia while (±) cyanopindolol (low 5-HT_{1C} affinity, high 5-HT_{1B} affinity) blocked both hypophagias [25].

5-HT_{1C} antagonists increased intake by freely feeding rats [25,28]; one of these drugs, cyproheptadine is claimed to be beneficial in anorexia [29]. None are very selective; specific 5-HT_{1C} antagonists could be clinically valuable. Selective drugs for 5-HT_{1B} receptors may not be relevant to the human brain as 5-HT_{1B} sites are not detectable therein [30]. However, the rat 5-HT_{1B} receptor and the human 5-$HT_{1D\beta}$ receptor are highly homologous [31]. If they are functionally equivalent, 5-HT_{1D} ligands able to penetrate to the brain could influence human appetite.

The hypophagic properties of 5-HT_{1B} and 5-HT_{1C} agonists are not due to their other effects. Thus, haloperidol and (±) pindolol [32,33] blocked the hyperlocomotor but not the hypophagic effect of RU 24969 [21] and infusing it into the hypothalamus only caused hypophagia [34]. Similarly, though mCPP (and TFMPP) cause hypophagia, hypolocomotion, and anxiety, chronic chlordiazepoxide prevented only the anxiety [35], cyanopindolol prevented only the hypophagia [33], and injecting of TFMPP into the hypothalamus caused hypophagia but not hypoactivity [34]. The hypophagias seemed not to be due to drug-induced malaise as trimethobenzamide (an antiemetic) prevented the hypophagic response of rats to acetyl salicylate (an emetic in man and dogs) but did not prevent RU 24969, mCPP, and TFMPP hypophagias [36]. However, interpretation is somewhat unclear as RU 24969 (but not TFMPP) caused emesis in cats that was not blocked by trimethobenzamide [37].

Infusions of RU 24969 or TFMPP aimed at the paraventricular nucleus (PVN) of the hypothalamus reduced food intake by food-deprived rats [34]. Similar results were obtained with CGS-93129 [22] and 5-HT [38]. However, in a recent study [39], RU 24969 and TFMPP given by this route had no effect. As injection volumes were smaller than those used previously [34,22] (0.4 μl vs. 1.0 μl) the authors suggest the earlier results may have been due to diffusion to other sites.

3.1 FENFLURAMINE

How d-fenfluramine (d-F) and its metabolite d-norfenfluramine (d-NF) inhibit feeding is unclear; both central and peripheral mediation [2,40] and different mechanisms for the two drugs [41] have been proposed. Although both increase hypothalamic extracellular 5-HT [42,43], its role is in doubt as the hypophagias survive inhibition of 5-HT synthesis [27,41]. d-NF (but not d-F) has considerable affinity for 5-HT_{1C} receptors [44] and their involvement is suggested by effects of antagonists on d-NF but not d-F [45] hypophagia. The increased hypophagic response to d-NF after inhibition of 5-HT synthesis also points to 5-HT_{1C} sites in view of their selective sensitization on 5-HT depletion [46]. Dependence on 5-HT_{1C} receptors is further indicated by the attenuation of d-NF hypophagia by chronic pretreatment with the 5-HT_{1C} agonist mCPP [27,45].

4. Sex Differences

Female rats were less responsive than males to the hyperphagic action of 8-OH-DPAT [47]. Similarly, estrogen decreased the inhibition of firing of dorsal raphe cells by 8-OH-DPAT [48] and also decreased its hyperphagic effect [49]. These findings suggest that weaker 5-HT dependent presynaptic appetite enhancement in females could play a part in anorexic disorders. A more active 5-HTergic postsynaptic suppression of feeding in female rats than in males is consistent with a report of higher densities in females of 5-HT_{1B} receptors in some hypothalamic nuclei [50] and with the finding in our laboratory that the 5-HT_{1B} agonist RU 24969 was more hypophagic in females. A similar sex difference with mCPP is however largely explicable by drug kinetics [51] and thus tends to cast doubt on the RU 24969 data. Sex differences in central drug effects are often of merely pharmacokinetic origin [52]. Although d-F shows a sex difference in its pharmacokinetics in the rat, both sexes have very similar ED_{50} values for its acute hypophagic effect. Suppression of tail-pinch induced feeding is an interesting exception; we find that females become relatively more resistant to d-F on maturation.

These are rather fragmentary findings on sex differences in the 5-HTergic control of feeding but the greater vulnerability of women to disorders of feeding, the possibility that these could involve 5-HT abnormalities [3], and many reports of sex differences of 5-HTergic function [53,54,55] encourage further study.

5. References

1. Blundell, J. (1991) 'Pharmacological approaches to appetite suppression', Trends Pharmacol. Sci. 12, 147-157.
2. Campbell, D.B. (1991) 'Dexfenfluramine: an overview of its mechanisms of action', Rev. Contemp. Pharmacother. 2, 93-113.
3. Curzon, G. (1992) 'Serotonin and eating disorders: pharmacological relationships', in S.Z. Langer, N. Brunello, G. Racagni and J. Mendlewicz (eds.) Serotonin Receptor Subtypes: Pharmacological Significance and Clinical Implications, Karger, Basel, pp. 112-128.
4. Curzon, G. (1992) 'Tryptophan as precursor of brain 5-hydroxytryptamine: an update', in I. Ishiguro, R. Kido, T. Nagatsu and Y. Nagamura (eds.) Advances in Tryptophan Research, Fujita Health University, Toyoake, in press.
5. Dourish, C.T., Hutson, P.H., Kennett, G.A., and Curzon, G. (1986) '8-OH-DPAT induced hyperphagia: its neural basis and possible therapeutic relevance', Appetite 7 (Suppl. 1), 127-140.
6. Hutson, P.H., Donohoe, T.P., and Curzon, G. (1987) 'Neurochemical and behavioral evidence for an agonist action of 1-[2-(4-aminophenyl)-4-(3-trifluoromethylphenyl) piperazine at central 5-HT receptors', Eur. J. Pharmacol. 138, 215-223.
7. Hutson, P.H., Dourish, C.T., and Curzon, G. (1986) 'Neurochemical and behavioral evidence for mediation of the hyperphagic action of 8-OH-DPAT by 5-HT cell body autoreceptors', Eur. J. Pharmacol. 129, 347-352.
8. Hutson, P.H., Dourish, C.T., and Curzon, G. (1988) 'Evidence that the hyperphagic response to 8-OH-DPAT is mediated by 5-HT_{1A} receptors', Eur. J. Pharmacol. 150, 361-366.
9. Hutson, P.H., Sarna, G.S., O'Connell, M.T., and Curzon, G. (1989) 'Hippocampal 5-HT synthesis and release in vivo is decreased by infusion of 8-OH-DPAT into the nucleus raphe dorsalis', Neurosci. Lett. 100, 276-280.
10. O'Connell, M.T., Sarna, G.S., and Curzon, G. (1992) 'Evidence for postsynaptic mediation of the hypothermic effect of 5-HT_{1A} receptor activation', Br. J. Pharmacol. 106, 603-609.
11. Hartley, J.E. and Fletcher, A. (1992) 'The 5-HT_{1A} receptor antagonist WAY 100135 attenuates 8-OH-DPAT-induced feeding in rats', J. Psychopharmacol. Abst. BAP/EBPS Meeting 329.

12. Hillegart, V. (1990) 'Effects of local application of 5-HT and 8-OH-DPAT into the dorsal and median raphe nuclei on motor activity in the rat', Physiol. Behav. 48, 143-148.
13. Higgins, G.A. and Elliot, P.J. (1991) 'Differential behavioral activation following intra-raphe infusion of 5-HT$_{1A}$ receptor agonists', Eur. J. Pharmacol. 193, 351-356.
14. Shepherd, J.K. and Rodgers, R.J. (1990) '8-OH-DPAT specifically enhances feeding behavior in mice: evidence from behavioral competition', Psychopharmacology 101, 408-413.
15. Dourish, C.T., Clark, M.L., and Iversen, S.D. (1988) '8-OH-DPAT elicits feeding and not chewing: evidence from liquid diet studies and a diet choice test', Psychopharmacology 95, 185-188.
16. Fletcher, P.J., Zack M.H., and Coscina, D.V. (1991) 'Influence of taste and food texture on the feeding responses induced by 8-OH-DPAT and gepirone', Psychopharmacology 10, 302-306.
17. Fletcher, P.J. and Davies, M. (1990) 'A pharmacological analysis of the eating response induced by 8-OH-DPAT injected into the dorsal raphe nucleus reveals the involvement of a dopaminergic mechanism', Psychopharmacology 100, 188-194.
18. Fletcher, P.J. and Davies, M. (1990) 'The involvement of 5-hydroxytryptaminergic and dopaminergic mechanisms in the eating induced by buspirone, gepirone and ipsapirone', Br. J. Pharmacol. 99, 519-525.
19. Bill, D.J., Knight, M., Forster, E.A., and Fletcher, A. (1991) 'Direct evidence for an important species difference in the mechanism of 8-OH-DPAT-induced hypothermia', Br. J. Pharmacol. 103, 1857-1864.
20. Tamai, H., Komaki, G., Kubota, S. et al. (1990) 'The clinical efficacy of a 5-HT agonist, SM-3997 in the treatment of bulimia', Int. J. Obes. 14, 289-292.
21. Kennett, G.A., Dourish, C.T., and Curzon, G. (1987) '5-HT$_{1B}$ agonists induce anorexia at a postsynaptic site', Eur. J. Pharmacol. 141, 429-435.
22. Macor, J.E., Burkhart, C.A., Heym, J.H. et al. (1990) '3-(1,2,5,6-tetrahydropyrid-4-yl) pyrrolo [3,2b] pyrid-5-one: a potent and selective serotonin (5-HT$_{1B}$) agonist and rotationally restricted phenolic analogue of 5-methoxy-3-(1,2,5,6-tetrahydpyrid-4-yl)indole', J. Med. Chem. 33, 2087-2093.
23. Koe, B.K., Nielsen, J.A., Macor, J.E., and Heym, J. (1992) 'Biochemical and behavioral studies of the 5-HT$_{1B}$ receptor agonist, CP-94253', Drug. Dev. Res. 26, 241-250.
24. Snyder, S.M., Reid, L.R., Buelke-Sam, J.L., and Wong, D.T. (1990) 'Suppression of food intake by MK 212 and other arylpiperazines in food-deprived rats', Soc. Neurosc. Abst. 376.10.
25. Kennett, G.A. and Curzon, G. (1988) 'Evidence that hypophagia induced by mCPP and TFMPP requires 5-HT$_{1C}$ and 5-HT$_{1B}$ receptors; hypophagia induced by RU 24969 only requires 5-HT$_{1B}$ receptors, Psychopharmacology 96, 93-100.
26. Kennett, G.A. and Curzon, G. (1991) 'Potencies of antagonists indicate that 5-HT$_{1C}$ receptors mediate mCPP-induced hypophagia', Br. J. Pharmacol. 103, 2016-2020.
27. Gibson, E.L., Kennedy, A.J., and Curzon, G. (1992) 'D-Fenfluramine and D-

norfenfluramine hypophagia: involvement of postsynaptic 5-HT$_{1C}$ receptors', J. Psychopharmacol. Abst. BAP/EBPS Meeting 330.

28. Dourish C.T., Clark, M.L., Fletcher, A., and Iversen, S.D. (1989) 'Evidence that blockade of post-synaptic 5-HT$_1$ receptors elicits feeding in satiated rats', Psychopharmacology 97, 54-58.
29. Halmi, K.A., Eckert, E., La Du, T.J., and Cohen, J. (1986) 'Anorexia nervosa. Treatment efficacy of cyproheptadine and amitryptyline', Arch. Gen. Psychiat. 43, 177-181.
30. Martial, J., Lal, S., Dalpe, M. et al. (1989) 'Apparent absence of serotonin$_{1B}$ receptors in biopsied and post-mortem human brain', Synapse 4, 203-209.
31. Adham, N., Romanienko, P., Hartig, P. et al. (1992) 'The rat 5-hydroxytryptamine$_{1B}$ receptor is the species homologue of the human 5-hydroxytryptamine$_{1D}$ receptor', Mol. Pharmacol. 41, 1-7.
32. Tricklebank, M.D., Middlemiss, D.M., and Neil, J. (1986) 'Pharmacological analysis of the behavioral and thermoregulatory effects of the putative 5-HT$_1$ receptor agonist RU 24969 in the rat, Neuropharmacology 25, 877-886.
33. Kennett, G.A. and Curzon, G. (1988) 'Evidence that mCPP may have behavioral effects mediated by 5-HT$_{1C}$ receptors', Br. J. Pharmacol. 94, 137-147.
34. Hutson, P.H., Donohoe, T.P., and Curzon, G. (1988) 'Infusion of the 5-hydroxytryptamine agonists RU 24969 and TFMPP into the paraventricular nucleus of the hypothalamus causes hypophagia', Psychopharmacology 95, 550-552.
35. Kennett, G.A., Whitton, P., Shah, K., and Curzon, G. (1989) 'Anxiogenic-like effects of mCPP and TFMPP in animal models are opposed by 5-HT$_{1C}$ receptor antagonists', Eur. J. Pharmacol. 164, 445-454.
36. Kennett, G.A. and Curzon, G. (1988) 'The antiemetic drug trimethobenzamide prevents hypophagia due to acetyl salicylate but not to 5-HT$_{1B}$ or 5-HT$_{1C}$ agonists', Psychopharmacology 96, 101-103.
37. Lucot, J.B. (1990) 'RU 24969-induced emesis in the rat: 5-HT$_1$ sites other than 5-HT$_{1A}$, 5-HT$_{1B}$ or 5-HT$_{1C}$ implicated', Eur. J. Pharmacol. 180, 193-199.
38. Weiss, G.F., Papadakos, P, Knudson, K., and Leibowitz, S.F. (1986) 'Medial hypothalamic serotonin: effects of deprivation and nor-epinephrine-induced eating', Pharmacol. Biochem. Behav. 25, 1223-1230.
39. Fletcher, P.J., Ming, Z.H., Zack, M.H., and Coscina, D.V. (1992) 'A comparison of the effects of the 5-HT agonists TFMPP and RU 24969 on feeding following peripheral or medial hypothalamic injection', Brain Res. 580, 265-272.
40. Rowland, N.E. and Carter, J. (1986) 'Neurobiology of an anorectic drug fenfluramine', Prog. Neurobiol. 27, 13-62.
41. Borsini, F., Bendotti, C., Aleotti, A. et al. (1982) 'D-fenfluramine and D-norfenfluramine reduce food intake by acting on different serotonin mechanisms in the rat brain', Pharmacol. Res. Comm. 14, 671-678.
42. Schwartz, D., Hernandez, L., and Hoebel, B.G. (1989) 'Fenfluramine administered systemically or locally increases extracellular serotonin in the lateral hypothalamus

as measured by microdialysis', Brain Res. 482, 261-270.
43. Rogacki, N., Weiss, G.R., Fueg, A. et al. (1989) 'Impact of hypothalamic serotonin on macronutrient intake', Ann. N.Y. Acad. Sci., 575, 619-622.
44. Mennini, T., Bizzi, A., Caccia, S., Codegoni, A. et al. (1991) 'Comparative studies on the anorectic activity of d-fenfluramine in mice, rats, and guinea pigs', Naunyn-Schmiedeberg's Arch. Pharmacol. 343, 483-490.
45. Gibson, E.L., Kennedy, A.J., and Curzon, G. 'An investigation of how systemically administered d-norfenfluramine suppresses feeding', in preparation.
46. Berendsen, H.H.G., Broekkamp, C.L.E., and Van Delft, A.M.L. (1991) 'Depletion of brain serotonin differently affects behaviors induced by 5-HT_{1A}, 5-HT_{1C}, and 5-HT_2 receptor activation in rats', Behav. Neural Biol. 55, 214-226.
47. Uphouse, L., Salamanca, S., and Caldarola-Pastuszka, M. (1991) 'Gender and estrous cycle differences in the response to 5-HT_{1A} agonist 8-OH-DPAT', Pharmacol. Biochem. Behav. 40, 901-906.
48. Lakoski, J.L. (1988) 'Estrogen-induced modulation of serotonin 5-HT_{1A} mediated responses in the dorsal raphe nucleus (DRN)', The Pharmacologist 30, 94.6.
49. Salamanca, S. and Uphouse, L. (1992) 'Estradiol modulation of the hyperphagia induced by the 5-HT_{1A} agonist, 8-OH-DPAT', Pharmacol. Biochem. Behav. 43, in press.
50. Leibowitz, S.F. and Jhanwar-Uniyal, M. (1990) '5-HT_{1A} and 5-HT_{1B} receptor binding sites in discrete hypothalamic nuclei: relation to circadian rhythm and gender', Soc. Neurosci. Abstracts 131-5.
51. Barnfield, A.M., Clifton, P.G., and Curzon, G. 'Effects of food deprivation and mCPP treatment on the microstructure of ingestive behavior of male and female rats', in preparation.
52. Yonkers, K.A., Kando, J.C., Cole, J.O., and Blumenthal, S. (1992) 'Gender differences in pharmacokinetics of psychotropic medication', Am. J. Psychiat. 149, 587-595.
53. Haleem, D.J., Kennett, G.A., Whitton, P.S., and Curzon, G. (1989) '8-OH-DPAT increased corticosterone but not other 5-HT_{1A} receptor-dependent responses more in females', Eur. J. Pharmacol. 164, 435-443.
54. Carlsson, M. and Carlsson, A. (1988) 'In vivo evidence for a greater brain tryptophan hydroxylase capacity in female than in male rats', Naunyn-Schmiedeberg's Arch. Pharmacol. 338, 345-349.
55. Haleem, D.J., Kennett, G.A., and Curzon, G. (1990) 'Hippocampal 5-HT synthesis is greater in female rats than in males and more decreased by the 5-HT_{1A} agonist 8-OH-DPAT', J. Neural Transm. Gen. Sec. 79, 93-101.

SEROTONIN (5-HT)$_{1A}$ RECEPTOR AGONISTS AND NEUROPROTECTION

J. TRABER, K.M. BODE-GREUEL, and E. HORVÁTH
Bayer-Tropon Institute for Neurobiology
Troponwerke GmbH & Co. KG
Berliner Str. 156
D-5000 Cologne 80
Federal Republic Germany

ABSTRACT. The activation of serotonin (5-HT)$_{1A}$ receptors exerts an inhibitory influence on neuronal activity in a way similar to the activation of adenosine A1 receptors. Adenosine A$_1$ receptor agonists have been shown to exert neuroprotective effects in animal models of cerebral ischemia. Therefore, we hypothesized that 5-HT$_{1A}$ receptor agonists might also show neuroprotective activity. We tested the partial agonists ipsapirone and gepirone and the full agonists Bay R 1531 and 8-OH-DPAT in the model of transient global ischemia in the mongolian gerbil. At a dose of 3 mg/kg i.p., ipsapirone protected 53% of the pyramidal neurons in the CA1 area of the hippocampus from ischemic damage. While gepirone was ineffective, Bay R 1531 and 8-OH-DPAT showed a pronounced neuroprotective activity with 75% and 93.8% preservation of neurons, respectively. The protective activity correlated with the affinity of the tested compounds to 5-HT$_{1A}$ receptors. These results suggest that 5-HT$_{1A}$ agonists might be used for the therapy of cerebral ischemia.

1. Introduction

Transient cerebral ischemia may lead to a selective pattern of neuronal death that typically develops with a delay of about 2 days [1]. Although the precise mechanisms underlying this phenomenon are unclear, evidence is accumulating that excessive neuronal activity correlating with an increased release of glutamate is one of the main pathophysiologic factors [2,3]. It is assumed that this delayed degeneration could result from a gradually developing neuronal Ca^{2+} overload induced by the excitotoxic action of glutamate [4]. One of the receptors for glutamate, the N-methyl-D-aspartate receptor, is coupled to an ion channel that shows a marked Ca^{2+} conductance when the membrane is strongly depolarized [5]. At strong membrane depolarization, voltage-regulated Ca^{2+} channels, in particular L-channels, are also active and contribute to the neuronal Ca^{2+} accumulation [6].

Intracellular Ca^{2+} increase and membrane depolarization constitute a positive feedback system that might finally lead to pathologic Ca^{2+} overload. Any intervention that interrupts this feedback loop should therefore have a neuroprotective effect. In the present study, we investigated the neuroprotective effect of 5-hydroxytryptamine $(5-HT)_{1A}$ receptor agonists, because this class of compounds mimics the hyperpolarizing action of 5-HT on the resting membrane potential and reduces the input resistance of neurons in the hippocampus [7,8], cortex [9], and raphe nuclei [10]. Hippocampal CA1 cells discharge fewer action potentials during exposure to $5-HT_{1A}$ receptor agonists. Based on these findings, we tested the neuroprotective activity of $5-HT_{1A}$ receptor agonists in the mongolian gerbil as a model for transient global forebrain ischemia.

2. Materials and Methods

Mongolian gerbils (Winkelmann Versuchstierzucht, Borken, F.R.G.) of either sex weighing 50-80 grams were used in this study. Five minutes prior to the operation, the gerbils obtained a subcutaneous injection of atropine (1 μg/0.1 ml). Anesthesia was induced with 3% halothane in room air, which was reduced to 1% during surgery. For the operation, the gerbils were fixed on a heating pad that was adjusted to 37°C, and body temperature was monitored with a rectal probe. Great care was taken to avoid a drop of body temperature. The carotid arteries were exposed through a midline cervical incision and simultaneously occluded with Biemer clips (Bartels and Reiger, Köln, F.R.G.) for 5 minutes. After ischemia, the gerbils were allowed to survive for 7 days.

For histology, the animals were perfused and embedded in paraffin according to standard procedures. We cut coronal sections through the hippocampus at 4 predetermined levels (-1.4 to -2 mm to bregma according to Loskota et al. [11]) and stained with cresyl fast violet. These sections were evaluated with the aid of camera lucida projection. Surviving pyramidal cells showing a nucleolus were counted in the CA1 area within two standardized frames. All counted cells of one animal were summed. The significance of the differences of means was assessed by analysis of variance according to Scheffé. The level of significance was $p < 0.05$.

Drugs were applied intraperitoneally in 300 μl of vehicle (saline) 15 minutes before ischemia and twice daily on the following 3 days. We tested the partial $5-HT_{1A}$ receptor agonists ipsapirone (0.5-10 mg/kg) and gepirone (1-15 mg/kg) and the full agonists 8-OH-DPAT (0.1-5 mg/kg) and Bay R 1531 (0.5-4 mg/kg). The compounds were synthesized by the Chemistry Department of Bayer AG, Wuppertal, F.R.G.

3. Results

A complete forebrain ischemia of 5 minutes resulted in a neuronal degeneration of the pyramidal cells in the CA1 area in 86% of the vehicle-treated gerbils. In comparison to sham-operated controls, only about 13% of the pyramidal cells survived. During the operation, the body temperature varied in the range of ± 1.7°C. There was no consistent correlation of these variations with the doses of the applied drugs or with neuroprotective effects.

The neuroprotective activity of the 5-HT_{1A} receptor agonists tested in this study was variable. The partial 5-HT_{1A} receptor agonist ipsapirone, for example (figure 1A), showed a significant neuroprotective effect only at a dose of 3 mg/kg. Higher or lower doses were inefficacious. At 3 mg/kg, the cell counts reached 53% in comparison to the sham-operated controls. Gepirone being also a partial agonist was without neuroprotective effect over a dose range of 1 to 15 mg/kg (figure 1B). In contrast, the full agonist Bay R 1531 (figure 1C) was dose-dependently active. A significant protection was already achieved with 2 mg/kg. At a dose of 3 mg/kg, the neuronal survival reached a maximum with 76% protection of neurons. The full 5-HT_{1A} receptor agonist 8-OH-DPAT also showed pronounced neuroprotective effects at 3 and 5 mg/kg (84% and 94%, respectively).

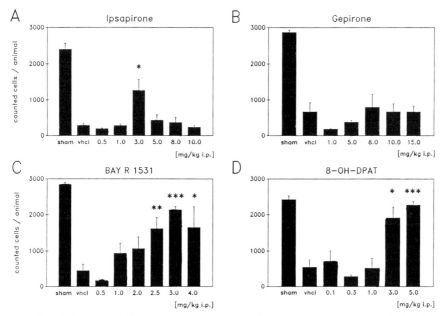

Figure 1. Effects of 5-HT_{1A} receptor agonists on neuronal survival of CA1 pyramidal cells in gerbil hippocampus after transient forebrain ischemia. Panel A: Ipsapirone showe a neuroprotective effect at 3 mg/kg with 53% preservation of neurons. Panel B: Gepirone was ineffective. Panel C: Bay R 1531 was strongly effective, leading to 76% protection at a dose od 3 mg/kg. Panel D: 8-OH-DPAT also showed pronounced effects at 3 and 5 mg/kg (84% and 94% protection, respectively.
Values are means +SEM. $^*p < 0.05$; $^{**}p < 0.01$; $^{***}p < 0.001$ (Scheffé test); n=5-12

4. Discussion

In the gerbil model of transient global ischemia, the neuroprotective effects of 5-HT_{1A} receptor agonists were variable. Bay R 1531 and 8-OH-DPAT showed highly significant effects. Bielenberg and Burkhardt [12] investigated the activity of several 5-HT_{1A} receptor agonists in a stroke model (permanent occlusion of the middle cerebral artery in rats and mice) and also observed neuroprotective activity. In the gerbil model of transient cerebral ischemia, neuroprotection was only obtained at doses of Bay R 1531 and 8-OH-DPAT that were higher than those usually applied to mice and rats [12,13,14]. This may be explained by our observation (de Vry and Bode-Greuel, unpublished) that gerbils develop the serotonin syndrome only at higher doses of 8-OH-DPAT than needed in rats and mice suggesting a lower sensitivity of gerbils to 5-HT_{1A} receptor agonists. Binding studies using tritated ipsapirone as ligand indeed revealed a lower affinity to 5-HT_{1A} receptors in the hippocampus and cortex of gerbils compared to tissue taken from calf and rat [15].

Compounds characterized as full 5-HT_{1A} receptor agonists at the presynaptic level may show antagonistic properties in postsynaptic areas such as the hippocampus [16,17]. These differences could be of relevance especially for the gerbil model because neuroprotection is assessed in the hippocampus. The antagonists activity may vary among the compounds used in this study. These variations might at least in part explain the variable neuroprotective effects of the tested compounds in the gerbil and the variable doses that are required.

In summary, the results of this study show that 5-HT_{1A} receptor agonists might be effective tools for the treatment of disorders of the cerebral circulation.

5. Acknowledgements

We would like to thank Mrs. S. Lederer, Mrs. C. Osten, and Mrs. W. Scheip for excellent technical assistance.

6. References

1. Wieloch, T. (1985) 'Neurochemical correlates to selective neuronal vulnerability', in K. Kogure, K.A. Hossmann, B.K. Siesjö and F.A. Welsh (eds.), Progress in Brain Research, Elsevier, New York, vol. 63, pp. 69-85.
2. Wieloch, T., Lindvall, O., Blomqvist, P., and Gage, F.H. (1985) 'Evidence for amelioration of ischaemic neuronal damage in the hippocampal formation by lesions of the perforant path', Neurol. Res., 24-26.
3. Benveniste, H., Jorgensen, M.B., Sandberg, M., Christensen, T., Hagberg, H., and Diemer, N.H. (1989) 'Ischemic damage in hippocampal CA1 is dependent on glutamate release and intact innervation from CA3', J. Cereb. Blood Flow Metab. 9, 629-639.
4. Meyer, F.B. (1989) 'Calcium, neuronal hyperexcitability and ischemic injury',

Brain Research, Rev. 14, 227-243.
5. MacDermott, A.B., Mayer, M.L., Westbrook, G.L., Smith, S.J., and Barker, J.L. (1986) 'NMDA-receptor activation increases cytoplasmic calcium concentration in cultured spinal cord neurones', Nature 321, 519-522.
6. Nowycky, M.C., Fox, A.P., and Tsien, R.W. (1985) 'Three types of neuronal calcium channel with different calcium agonist sensitivity', Nature 316, 440-443.
7. Andrade, R. and Nicoll, R.A. (1987) 'Novel anxiolytics discriminate between postsynaptic serotonin receptors mediating different physiological responses on single neurons of the rat hippocampus', Naunyn-Schmiedeberg's, Arch. Pharmacol. 336, 5-10.
8. Colino, A. and Halliwell, J.V. (1987) 'Differential modulation of three separate K-conductances in hippocampal CA1 neurons by serotonin', Nature 328, 73-77.
9. Davies, M.F., Deisz, R.A., Prince, D.A., Peroutka, S.J. (1987) 'Two distinct effects of 5-hydroxytryptamine on single cortical neurons', Brain Res. 423, 347-352.
10. Sprouse, J.S. and Aghajanian, G.K. (1987) 'Electrophysiological responses of serotoninergic dorsal raphe neurons to $5-HT_{1A}$ and $5-HT_{1B}$ agonists', Synapse 1, 3-9.
11. Loskota, W.J., Lomax, P., and Verity, M.A. (eds.) (1973) A Sterotactic Atlas of the Mongolian Gerbil Brain, Arbor Science Publishers, Ann Arbor, Michigan.
12. Bielenberg, G.W. and Burkhardt, M. (1990) '5-Hydroxytryptamine$_{1A}$ agonists: A new therapeutic principle for stroke treatment', Stroke 21 (suppl. IV), IV-161-IV-163.
13. De Vry, J., Glaser, T., Schuurman, T., Schreiber, R., and Traber, J. (1991) '$5-HT_{1A}$ receptors in anxiety', in M. Briley and S. E. File (eds.), New concepts in anxiety, MacMillan Press, London.
14. Glaser, T., Dompert, W.U., Schuurman, T., Spencer, D.G., and Traber, J. (1987) 'Differential pharmacology of the novel $5-HT_{1A}$ receptor ligands 8-OH-DPAT, Bay R 1531 and ipsapirone', in C. T. Dourish, S. Ahlenius and P. H. Hudson (eds.), Brain $5-HT_{1A}$ receptors, Ellis Horwood Ltd., Chichester, UK.
15. Bode-Greuel, K.M., Horváth, E., de Jonge, M., Glaser, T., and Traber J. (1989) '$5-HT_{1A}$ receptors in the brain of the mongolian gerbil' (abstract), Soc. Neurosci. Abstr. 15, 222.
16. Colino, A. and Halliwell, J.V. (1986) '8-OH-DPAT is a strong antagonist of 5-HT action in rat hippocampus', Eur. J. Pharmacol. 130, 151-152.
17. Segal, M., Azmitia, E.C., and Whitaker-Azmitia, P. (1989) 'Physiological effects of selective $5-HT_{1A}$ and $5-HT_{1B}$ ligands in rat hippocampus: Comparison to 5-HT', Brain Res. 502, 67-74.

NEW ANTIDEMENTIA MOLECULES WHICH SELECTIVELY INFLUENCE SEROTONIN RECEPTOR SUBTYPES

B. COSTALL, A.M. DOMENEY, M.E. KELLY, and R.J. NAYLOR
Postgraduate Studies in Pharmacology
The School of Pharmacy
University of Bradford
Bradford BD7 1DP
UK

ABSTRACT. A battery of animal tests is described which have allowed the detection of novel compounds which can inhibit memory impairments. The compounds include selective 5-HT$_3$ receptor antagonists such as ondansetron, zacopride, and tropisetron, and certain new ligands for the 5-HT$_{1A}$ receptor, including lesopitron. The preclinical tests which have been employed have included a mouse habituation model, rat T-maze and water maze tests, and a primate object discrimination and reversal task. Impairments have been induced by scopolamine and by lesions of the nucleus basalis magnocellularis, and the natural impairments associated with old age have also been assessed. The abilities of the 5-HT$_3$ receptor antagonists to inhibit memory impairments may relate to an ability to inhibit a reduced acetylcholine release, as indicated by biochemical experiments, and this would form a correlate with the cholinergic hypothesis describing memory disorders. Initial clinical trials have investigated the potential of ondansetron to improve Age Associated Memory Impairments. The data has been sufficiently encouraging to provide a validity for the animal studies.

1. Introduction

For decades there has been a search for compounds which will improve an impaired memory, but both a lack of suitable preclinical tests and a lack of novel compounds has meant few advances. The memory disorders have long been associated with cholinergic deficits, particularly in cortical and mesocortical systems [1], but the cholinergic agents have been precluded from use by the severity of their autonomic side effects. In animal tests the cholinomimetics are effective, but again the side-effect potential has proven prohibitive. Nevertheless, the cholinomimetic agents have been used to show that certain molecules which selectively influence serotonin receptor subtypes are able to inhibit memory impairments. In particular efficacy has been demonstrated for the 5-HT$_3$ receptor antagonists, ondansetron, zacopride, tropisetron, granisetron, and, more recently, for a select group of agents which selectively influence the 5-HT$_{1A}$ receptor subtype, for example, lesopitron [1, unpublished].

The test procedures have used mouse, rat, and primate, and have included procedures such as habituation, a T-maze task, and a water maze task in rodents, and an object reversal discrimination task using marmosets. Natural behaviors have been used or impairments have been caused by old-age, scopolamine, or lesions of the nucleus basalis magnocellularis. All tests are responsive to the cholinomimetic agents but not to the so-called nootropic agents which include cerebral vasodilators, dilantin and the calcium antagonists, the piracetam-type drugs, and hydergine.

2. Use of the Mouse Habituation Test

The studies used male albino BKW mice (27-35 g, 6-8 weeks old, i.e. 'young adult' and 35-42 g, 8-10 months old, i.e. 'aged') which were normally housed in groups of 10 and given free access to food and water. The mice were kept on a 12-hour light/dark cycle with lights off at 07.00 am.

Three types of impairment in habituation behavior were measured: firstly, that associated with old age; secondly, that induced in young adult mice using scopolamine (0.25 mg/kg sc bd); and, thirdly, that induced by lesion of the nucleus basalis of young adult mice. In order to induce such lesions mice were anesthetized with chloral hydrate (150 mg/kg ip) and placed in a Kopf stereotaxic frame. Electrolesions of the nucleus basalis magnocellularis were induced at Ant. 2.3 mm (relative to the zero of the Kopf frame), Vert. 4.5 mm (below skull surface), and Lat. ± 2.1 mm from the midline. The electrode used was constructed of 0.3 mm external diameter stainless steel which was insulated except at the tip to which a current of 1 mA was delivered for 10 seconds. The correct placement of the lesions was confirmed on termination of the experiments both histologically and biochemically (determination of ChAT levels in septum, frontal cortex, hippocampus, and striatum: lesions reduced levels of ChAT by 34-57% selectively in the frontal cortex).

Habituation testing was carried out daily between 08.30 and 12.30 hours. Mice were taken from a dark home environment in a dark container to the experimental room maintained in low red lighting, and placed in the center of the white section of a white and black test box. The box (45 x 27 x 27 cm high) was divided. Forty percent of the area was painted bland and illuminated under a red light (1 x 60 W, 0 lux) and the other painted white and brightly illuminated with white light (1 x 60 W, 400 lux) located 17 cm above the box. Access between the two areas was enabled by a 7.5 x 7.5 cm opening located at floor level in the center of the partition. Behavior was assessed via remote video-recording and the latency to move from the white to the black section was measured. The brightly-lit area of the black and white test box has aversive properties, mice normally distributing their behavior preferentially in the black compartment. However, on repeated daily testing young adult mice habituate to the test system with a reduced latency in movement from the white to the black section. Thus, while latency is delayed for some 10-15 seconds on the first three days of testing, this latency is reduced to some 2-3 seconds on days four to six of test [2].

In contrast to these habituation patterns consistently exhibited by young adult mice, old mice failed to show any habituation over a 6-day period, and habituation also failed to occur when young mice were treated repeatedly with scopolamine or had been subject to nucleus basalis lesions (figure 1).

However, when the aged mice, scopolamine-treated mice, or nucleus basalis lesioned mice were treated throughout the 6-day test period with ondansetron, zacopride, or tropisetron, habituation patterns were not only restored but were improved above the basal level observed for young adult mice (figure 1). Thus, aged mice given a 5-HT_3 receptor antagonist, scopolamine-treated mice given a 5-HT_3 receptor antagonist, and nucleus basalis lesioned mice given a 5-HT_3 receptor antagonist all showed maximal habituation by the third day of test, with latencies as low as 1 second. Such marked improvements were maintained throughout the remainder of the 6-day test period (figure 1).

While 5-HT_{1A} ligands such as buspirone, ipsapirone, and gepirone are without effect on mouse habituation or impairments in habituation patterns, agents from a new class of 5-HT_{1A} ligands [3], the prototype being lesopitron, can be shown to be effective (figure 2).

3. Use of a Rat T-maze Food Reinforced Alternation Task

Male Lister Hooded rats (250-300 g, 11-15 weeks old, i.e. 'young adults') were trained on a food reinforced alternation task. Food was withdrawn 2 days prior to testing and animals were deprived of food for 23 hours per day. Water was available ad lib and body weight was maintained at 85% of normal. Animals were taken from the holding room to the dimly lit test room 30 minutes before testing. Experiments were carried out between 08.00 and 15.00 hours using a T-maze elevated 30 cm above the ground. The start arm measured 80 x 10 cm and the side arms were 60 x 10 cm with food wells 3 cm deep at each end.

On day 1 each rat was allowed 10 minutes habituation to the maze. Both food wells were baited with banana flavored pellets and pellets were also scattered along the approach arm. The rats were then subjected to a period of reinforced alternation training, days 2-5 being designated 'pretraining' days with days 6-9 'training' days. All reinforced alternation training consisted of paired trials (each pair consisting of a 'run'). The first trials were the 'forced' trials in that one arm was blocked while the other arm was baited. The second trial of the pair was a 'choice' trial in which reward pellets were placed in the arm opposite to that reinforced in the first trial of the pair. A correct choice was when the rat entered the arm and passed a point 20 cm along the arm containing the food in the choice trial. In addition to correct/incorrect choice, latency to reward was recorded for both forced and choice trials.

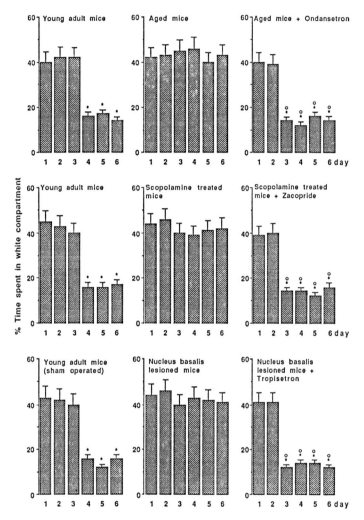

Figure 1. Habituation profiles of young adult mice and mice with memory impairments caused by old age, scopolamine treatment, and nucleus basalis lesions, and effect on these impairments of daily treatment with ondansetron, zacopride, and tropisetron (10 ng/kg ip bd). n=5. *P < 0.001 (improved performance as compared with day 1). ∘P < 0.001 (inhibition of deficit) (one-way ANOVA followed by Dunnett's t test). The measurement from the black:white box taken for presentation is the % of time spent in the white compartment (sem calculated from original data).

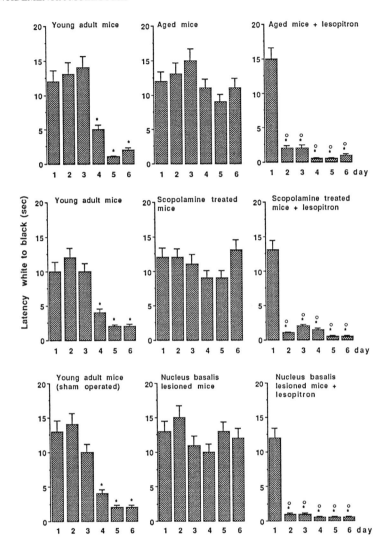

Figure 2. Habituation profiles of young adult mice and mice with memory impairments caused by old age, scopolamine treatment, and nucleus basalis lesions, and effect on these impairments of daily treatment with lesopitron (1 ng/kg ip bd). n=5. *P < 0.001 (improved performance as compared with day 1). ₀P < 0.001 (inhibition of deficit) (one-way ANOVA followed by Dunnett's t test). The measurement from the black:white box taken for presentation is latency of the initial move from the white to the black compartment.

Four runs per day were carried out on pretraining days (intertrial interval 0 seconds, interrun interval 30 seconds) and 6 runs per day during training (intertrial interval 30 seconds; interrun interval 60 seconds). The number of lefts and rights was pseudorandom (following the Gellerman schedule) and was balanced across the test groups. Behavioral results were analyzed using two-way analysis of variance (repeated measure analysis) followed by Dunnett's test.

To effect destruction of the nucleus basalis magnocellularis, rats were anesthetized with sodium pentobarbitone and, using standard stereotaxic techniques, 5 µg of ibotenic acid (or vehicle) was infused in 5 µl at Ant. 5.8, Lat. 2.6, Vert. -8.2 mm (Atlas of Paxinos and Watson, [4]). All rats surviving for 14 days and appearing to be in good health were used in subsequent experiments. A more general destruction of forebrain cholinergic function was achieved by chronic infusion of hemicholinium-3 (HC-3) 2.5 µg/day into the ventricular system throughout the 9-day test period. The correct placement of lesions was confirmed histologically and biochemically (determination of ChAT levels) [2].

A 40-60% reduction in the percentage correct responses was observed in aged animals, and in young adult rats receiving scopolamine or in those that had received an ibotenic acid lesion of the nucleus basalis magnocellularis. As in the mouse test, it could be shown that these impairments in performance could be inhibited by 5-HT_3 receptor antagonists, ondansetron, zacopride, tropisetron, or the new 5-HT_{1A} ligand, lesopitron (see figure 3 for examples).

It is interesting that impairments caused by hemicholinium-3 or by diazepam could not be inhibited by the test compounds influencing serotonin.

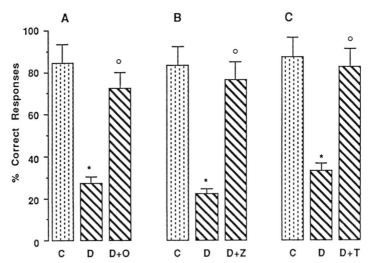

Figure 3. Abilities of ondansetron (O), zacopride (Z), and tropisetron (T) to inhibit deficits (D) in performance in a T-maze reinforced alternation task. Deficits were induced by A. old age, B. scopolamine treatment, and C. lesions of the nucleus

basalis. C indicated the response of control animals. *P < 0.001 (impairments). ◦P < 0.001 (inhibition of impairments) (one-way ANOVA followed by Dunnett's t test). All 5-HT$_3$ receptor antagonists were given at a dose of 100 ng/kg ip bd. n=6.

4. Use of a Rat Morris Water Maze Task

Male Lister Hooded rats 250-300 g, housed in groups of 5, were given free access to food and water. Rats were placed in a square (120 x 120 cm) pool of water rendered opaque by the addition of an emulsion. A white painted platform was located 2 cm below the surface of the water. The rats were trained to locate and escape onto the island using spatial strategies: a 2-day test protocol was utilized. On day 1 each rat was placed on the island for 30 seconds immediately before testing commenced. For each rat the island was kept in constant position (although the position was randomized and balanced across groups), each rat beginning each trial at a different corner in the pool. Using a timer and tracking device the time was recorded for the animal to escape from the water onto the platform. The rat remained on the platform for 10 seconds before being placed in the pool for trial 2. On each trial the rat was allowed 100 seconds to find the island and the latency, swim speed, and percentage of time spent in the island quadrant were measured. (If the rat failed to find the island in 100 seconds it was placed on it for 10 seconds before being removed.) Each rat received 6 trials on day 1. On day 2 the same procedure was carried out as on day 1, the basis of the test being that rats had formed a strategy to find the island on day 1 which could be modified by drug treatments. In addition, a seventh trial with a black visible island was also carried out to ensure that no visual/locomotor effects were influencing performance [5].

On the first day of testing scopolamine (0.25 mg/kg ip bd) tended to increase escape latency but this failed to achieve significance (results not shown). However, on the second day, the continued treatment with scopolamine significantly impaired the escape latency by some 60% and the percentage of time spent in the island quadrant by some 50%. The administration of ondansetron, zacopride, or tropisetron, all given at a dose of 1 μg/kg ip, significantly antagonized the scopolamine-induced deficit such that the performance to escape and the percentage of time in the island quadrant were indistinguishable from values determined for vehicle-treated control animals (figure 4). The lack of effect of the drug treatments on swim speed and ability to locate the black visible island indicated an absence of effect on visual and locomotor performance.

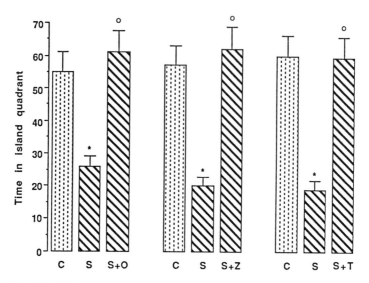

Figure 4. Abilities of ondansetron (O), zacopride (Z), and tropisetron (T) to inhibit scopolamine-induced deficits (S) in Morris water maze. All 5-HT$_3$ receptor antagonists were given at a dose of 1 μg/kg ip. n=6 *$P < 0.001$ (impairments). ∘$P < 0.001$ (inhibition of impairments) (one-way ANOVA followed by Dunnett's t test). % time spent in the correct island quadrant was determined, and sem calculated from original data.

5. Use of a Marmoset Object Discrimination and Reversal Learning Task

Object discrimination and reversal learning tasks were assessed using a Wisconsin General Test apparatus. Behavioral testing was carried out between 10.00 and 15.30 hours in a room where temperature and lighting conditions were identical to those of the holding rooms. Following the initial object discrimination training using plastic junk objects to 90% correct performance, the task set for the marmosets was to select between two different objects covering two food wells, one of which contained a food reward. The food reward object was presented to the animal on either the left- or right-hand side, according to the pseudorandom schedule of Gellerman. On completing six consecutive correct responses on the first food rewarded object (initial discrimination task) the reward paradigm was changed so that the marmoset was required to select the second, initially unrewarded object, to the same criterion (reversal task). Objects remained constant throughout the 5-day test periods: the last rewarded object of one day always became the first rewarded object of the following day. Marmosets received ondansetron or vehicle twice daily (morning and evening) of a 5-day test period. A blind, randomized cross-over design was employed for drug or vehicle administration. The mean ($\pm$se) differences between drug and vehicle controls for the number of trials to criterion for all marmosets within a dose group on all days were analyzed using a paired t test [6].

Treatment with 10 ng/kg sc bd ondansetron (a maximally effective dose) and zacopride or tropisetron, again at 10 ng/kg sc bd throughout the 5-day test period, significantly decreased the number of trials to criteria in both the object discrimination and reversal learning task. The object reversal task was the more difficult test for the marmosets to perform and more trials were required before reaching criterion. Generally, the improvement in performance caused by the 5-HT$_3$ receptor antagonists was greater in the reversal task. Within 2 days following cessation of drug treatment the performance of marmosets returned to predrug levels for both discrimination and reversal learning.

6. Biochemical and Clinical Correlates

The mode of action of the novel 5-HT$_{1A}$ ligands is still under investigation, but studies on the release of tritiated acetylcholine have indicated that the 5-HT$_3$ receptor antagonists are able to inhibit a reduction in acetylcholine release. The experimental set-up was described by Barnes and colleagues [7] and the observations fit well with the preferred requirements that any new agent which may inhibit memory impairment should in some way enhance acetylcholine function.

Initial clinical studies [8] have used Age Associated Memory Impairments (AAMI). Ondansetron was given according to careful clinical trial procedures and was shown to improve performance on a number of primary outcome measures, e.g. acquisition of name-face association, delayed recall on name-face association, and on the number recognized before the first error in a name-facial recognition test. The data obtained using 0.01, 0.25 and 1 mg ondansetron were encouraging and, overall, the authors concluded that ondansetron was of considerable interest and merits further study as a putative treatment for AAMI and perhaps other adult-onset cognitive disorders.

7. Conclusions

There are now new animal tests available which will allow the measurement of impairments in memory. There are biochemical and some initial clinical correlates which would confirm the validity of our approaches. New chemical compounds have been developed and it is proposed that the 5-HT$_3$ receptor antagonists and selected 5-HT$_{1A}$ ligands may herald unique approaches to the treatment of memory impairments, and this would include those associated with aging.

8. References

1. Bartus, R.T., Dean, R.L., Beer, B., and Lippa, A.S. (1982) 'The cholinergic

hypothesis of geriatric memory dysfunction', Science 217, 408-417.
2. Barnes, J.M., Costall, B., Coughlan, J., Domeney, A.M., Gerrard, P.A., Kelly, M.E., Naylor, R.J., Onaivi, E.S., Tomkins, E.M., and Tyers, M.B. (1990) 'The effects of ondansetron, a 5-HT_3 receptor antagonist, on cognition in rodents and primates', Pharmacol. Biochem. Behav. 35, 955-962.
3. Costall, B., Domeney, A.M., Farre, A.J., Kelly, M.E., Martinez, L., and Naylor, R.J. (1992) 'Profile of action of a novel 5-hydroxytryptamine receptor ligand E-4424 to inhibit aversive behavior in the mouse, rat and marmoset', J. Pharmacol. Exp. Ther. 262, 90-98.
4. Paxinos, G. and Watson, C. (1982) 'The rat brain in stereotaxic co-ordinates', Academic Press, New York.
5. Costall, B., Coughlan, J., Horovitz, Z.P., Kelly, M.E., Naylor, R.J., and Tomkins, D.M. (1989) 'The effects of ACE inhibitors captropril and SQ29852 in rodent tests of cognition', Pharmacol. Biochem. Behav. 33, 573-579.
6. Domeney, A.M., Costall, B., Gerrard, P.A., Jones, D.N.C., Naylor, R.J., and Tyers, M.B. (1991) 'The effect of ondansetron on cognitive performance in the marmoset', Pharmacol. Biochem. Behav. 38, 169-175.
7. Barnes, J.M., Barnes, N.M., Costall, B., Naylor, R.J., and Tyers, M.B. (1989) '5-HT_3 receptors mediate inhibition of acetylcholine release in cortical tissue', Nature 338, 762-763.
8. Crook, T. and Lakin, M. (1991) 'Effects of ondansetron in age-associated memory impairments', in 'The role of ondansetron, a novel 5-HT_3 antagonist, in the treatment of psychiatric disorders' Satellite symposium, 5th World Congress of Biological Psychiatry, Florence, 9 June 1991, pp. 21-23.

PHARMACOLOGICAL MODULATION OF THE SEROTONERGIC SYSTEM: AN OVERVIEW OF THE EFFECTS ON NORMAL AND PATHOLOGICAL SLEEP

C. DUGOVIC and G.H.C. CLINCKE
Janssen Research Foundation
Turnhoutseweg 30
B-2340 Beerse
Belgium

ABSTRACT. Serotonergic sleep regulation has been studied through the use of a number of relatively specific agonists and antagonists at different serotonergic receptors and subtypes. An overview of the currently available data indicates that 5-HT_3 receptors have probably no major influence on sleep regulation. Both human and animal data on agonists of the 5-HT_{1A} receptor subtype point to a possible role in rapid eye movement (REM) sleep regulation. A functional role of 5-HT_{1A} receptors in arousal mechanisms is suggested by animal data but was not confirmed in man. 5-HT_2 and/or 5-HT_{1C} receptors are clearly mediating the inhibitory control of serotonin on deep slow wave sleep (SWS). Consequently, consistent sleep promoting effects have been observed in normal and pathological conditions after treatment with 5-$HT_{2/1C}$ antagonists such as ritanserin. The potential of this drug to alter the preference for different drugs of abuse might be related to the fact that it could normalize disturbed sleep during withdrawal from chronic cocaine treatment.

1. Introduction

Sleep disturbances are found in a variety of mental disorders which are also associated with alterations in brain serotonergic systems. Serotonin is well recognized to play a major role in the mechanisms of sleep regulation as was demonstrated in extensive electrophysiological, pharmacological, and brain lesion studies. However, in early pharmacological investigations some contradictory results were obtained. This was mainly due to the relatively nonspecificity of the available pharmacological, tools with respect to the serotonergic system.

In recent years, multiple serotonin receptors (5-HT_1, 5-HT_2, 5-HT_3, and 5-HT_4) and subtypes have been identified. Consequently, an increasing number of specific agonists and antagonists became available. This has stimulated new research in the field of serotonergic sleep regulation. The purpose of this paper is to give an overview of the investigations on normal and disturbed sleep with these new compounds, and to discuss the functional role of the different 5-HT receptor subtypes in the modulation of vigilance states.

Therapeutical implications are also considered in the light of recent clinical data obtained in different neurobiological disorders associated with impairments of sleep-wakefulness patterns.

2. Involvement of 5-HT$_1$ receptors

For 5-HT$_1$ receptors only data on receptor agonists are currently reported. With regard to the different subtypes, the 5-HT$_{1A}$ receptor has been most frequently studied because of the availability of relative specific compounds for that subtype.

2.1 EFFECTS OF 5-HT$_{1A}$ RECEPTORS AGONISTS

In rats the selective 5-HT$_{1A}$ agonist 8-OH-DPAT and the selective partial agonists buspirone and ipsapirone have a similar activity profile. In general these compounds increase the duration of wakefulness (W) and induce an initial decrease of sleep especially in rapid eye movement (REM) sleep, followed by a secondary rebound [1-3]. These effects seem to be mediated by post- rather than by presynaptic receptors since lesions of the nucleus raphe dorsalis did not abolish the effects of ipsapirone [4]. However, local injections of ipsapirone or 8-OH-DPAT, directly into the dorsal raphe nucleus resulted in similar sleep-wakefulness alterations [5]. Hence, the involvement of presynaptic receptors cannot be excluded.

In patients with either insomnia [6,7] or anxiety [8], buspirone (as in rats) induced an initial decrease in REM sleep which was compensated by a rebound at the end of the night.

2.2 EFFECTS OF MIXED 5-HT$_1$ AGONISTS

The mixed 5-HT$_{1B/1A}$ agonist RU 24969 mainly elicited a dose-related increase of W and a concomittant sleep reduction in the rat [9]. The mixed 5-HT$_{1C/1B}$ agonist mCPP similarly increased W and decreased both slow wave sleep (SWS) and REM sleep in the rat [10]. In healthy volunteers mCPP reduced total sleep time and sleep efficiency [11].

2.3 FUNCTIONAL ROLE OF 5-HT$_1$ RECEPTORS AND CLINICAL IMPLICATIONS

Consistent effects are seen with 5-HT$_{1A}$ agonists on REM sleep. An initial decrease and a compensatory rebound were observed in both animals and humans. 5-HT$_{1A}$ receptors may also be involved in arousal mechanisms in animals. Clinical studies indicate that therapeutic doses of 5-HT$_{1A}$ agonists have minimal action on sleep structure, but their effects in patients with objective sleep impairments have not been explored yet.

With respect to the other subtypes, no definitive conclusions can be drawn. Firstly, the primary effects of mixed 5-HT_1 agonists are behavioral alterations which probably interfere with sleep by postponing sleep onset. Secondly, no specific agonists at 5-HT_{1B} and 5-HT_{1C} receptors are available. Furthermore, the proper evaluation of direct subtype mediation through pharmacological agonism/antagonism studies is currently lacking in sleep research.

3. Involvement of 5-HT_2 Receptors

For 5-HT_2 receptors data exist on the proper effects of agonists and antagonists. Interactions between agonists and antagonists have been studied as well, including interactions with drugs acting on other 5-HT receptor sites.

With the exception of the antagonist cinanserin both agonists and antagonists are only relatively selective to 5-HT_2 receptors since they also have high affinity for 5-HT_{1C} receptors. This is not surprising given the close relationship between 5-HT_2 and 5-HT_{1C} receptors with regard to their binding, pharmacology, second-messenger coupling, and amino acid sequences [12]. Hence, responses attributed to interactions at the 5-HT_2 site might also be controlled by 5-HT_{1C} receptors.

3.1 ANIMAL STUDIES

3.1.1 *Antagonists.* In the rat, the mixed 5-HT_2/5-HT_{1C} agonist methoxphenethylamines DOM [13], DOB [14], and DOI [15] all induced the same sleep-wakefulness disturbances. W was dose-relatedly increased, while SWS and REM sleep were reduced.

3.1.2 *Antagonists.* Concerning the mixed 5-HT_2/5-HT_{1C} antagonists studied in rats, ritanserin increased the duration of deep SWS mainly at the expense of REM sleep [16]. Similarly, ICI 170,809 modified both SWS and REM sleep [17] whereas RP62203 only increased SWS [18] and ICI 169,369 only decreased REM sleep [17]. The selective 5-HT_2 antagonist cinanserin revealed effects on sleep comparable to those of ritanserin.

3.1.3 *Interaction studies.* At doses which by themselves had no effect on sleep, ritanserin, cinanserin, and RP62203 dose-dependently reversed the disruptive effects on W and deep SWS of the agonists DOM [13,16], DOB [14], and DOI [15,18]. REM sleep was only restored in immature rats [5]. In contrast, sleep-wakefulness alterations induced by 8-OHDPAT, RU 24969 and mCPP were not modified by pretreatment with ritanserin, cinanserin, or RP62203 [10,13,18].

3.2 HUMAN STUDIES

3.2.1 *Healthy volunteers.*
Ritanserin, seganserin, and ICI 169,369 all dose-relatedly increased the duration of SWS stages 3 and 4 at the expense of stages 1 and 2 whereas REM sleep was either unaffected or slightly reduced [19-21]. SWS was persistently increased during two weeks of chronic ritanserin treatment [22].

3.2.2 *Clinical studies.*
Ritanserin has been investigated in different clinical indications. Chronic treatment (two to four weeks) of poor sleepers [23], insomniacs [24], narcoleptic [25], and dysthymic patients [26] permanently increases SWS without alteration of REM sleep. A similar profile of activity was observed after acute treatment in major depression [27] and generalized anxiety disorder [28]. Sleep efficiency was improved in schizophrenics [29] and chronic alcoholics [30] treated for four weeks with only minor changes in SWS.

3.3 FUNCTIONAL ROLE OF 5-HT_2 RECEPTORS AND CLINICAL IMPLICATIONS

Both animal and human data suggest that serotonin exerts an inhibitory control on deep SWS through 5-HT_2 and/or 5-HT_{1C} receptors. It could be speculated that 5-HT_2 receptors are primarily involved in this modulation since the selective 5-HT_2 antagonist cinanserin acts in a similar way as the mixed 5-$HT_{2/1C}$ antagonists in animals. On the other hand there is clearly no functional interaction between 5-HT_2 and 5-HT_{1A} receptors in the modulation of sleep and wakefulness.

In most clinical studies, the specific SWS-increasing properties of the 5-HT_2/5-HT_{1C} antagonist ritanserin have been confirmed. A general sleep-promoting effect, i.e. increase of sleep efficiency and sleep maintenance, appears to be the more relevant activity of ritanserin, especially in patients with virtual absence of SWS.

4. Involvement of 5-HT_3 Receptors

A number of 5-HT_3 antagonists with different chemical structures have been tested for their effects on sleep-wakefulness. In general, MDL 72222, ICS 205-930, zacopride, and ondansetron do not induce major modifications of sleep-wakefulness patterns in both rats [3,31] and humans [32,33]. Combined treatment with MDL 72222 and ipsapirone in the rat [3] further indicates that stimulation of 5-HT_{1A} receptors is not modulated by blockade of 5-HT_3 receptors. Therefore, it seems that 5-HT_3 receptors are not primarily implicated in sleep-wakefulness regulation.

5. Sleep and Serotonin: New Developments in Addiction

Depressive mood, insomnia, and anxiety are persistent symptoms in abstinent alcoholic patients. The concomittant occurrence of these symptoms during withdrawal offers a rationale for treating these conditions with compounds that modulate the serotonergic system. In patients with alcohol dependence it was recently reported that the 5-$HT_{2/1C}$ antagonist ritanserin not only improved sleep continuity but also reduced the compulsion to drink [30]. Detailed pharmacological research in animals substantiated these initial clinical observations. In the rat, ritanserin not only reduced alcohol preference and intake but had similar effects on fentanyl and cocaine intake without inducing aversion or interfering with consumatory physiological processes [34].

In humans, cocaine induces acute sleep disturbances, and sleep pathology is also seen after abrupt cessation of cocaine abuse [35]. We therefore studied the effects on sleep-wakefulness of acute cocaine administration and withdrawal after chronic cocaine treatment in the rat. In a series of experiments we were able to demonstrate that ritanserin did not attenuate sleep disturbances after acute cocaine injections, but could successfully antagonize the sleep disturbances seen after discontinuation of chronic cocaine treatment from the first treatment day on [36].

These data suggest that 5-HT_2 antagonists might become useful tools in the treatment of relapse prevention during abstinence. The fact that preference for pharmacologically different drugs of abuse such as alcohol, fentanyl, and cocaine could be reduced indicates that fundamental and general mechanisms of abuse are altered by ritanserin. Seeking relief from withdrawal effects is a powerful controlling factor in relapse and drug seeking behavior in addicts. Therefore, attenuation of abstinence symptoms such as decreasing withdrawal anxiety and normalizing disrupted sleep might be important factors in the effects of serotonergic drugs in abuse. However, the exact contribution of receptor subtypes in these effects requires further experimental evaluation.

6. References

1. Depoortere, H. (1988) 'Effects of 5-HT1A agonists on sleep-wakefulness cycle in the rat', in W.P. Koella (ed.), Sleep '86, FisherVerlag, Stuttgart, pp. 346-348.
2. Lerman, J.A., Kaitin, K.I., Dement, W.C., and Peroutka, J.S. (1986) 'The efects of buspirone on sleep in the rat', Neurosci. Lett. 72, 64-68.
3. Tissier, M.H., Franc, B., Hamon, M., and Adrien J. (1990) 'Effects of 5-HT1A and 5-HT3 receptor ligands on sleep in the rat', in J. Horn (ed.), Sleep '90, Pontenagel Press, Bochum, pp. 126-128.
4. Adrien, J., Tissier, M.H., Fattaccini, C.M., and Hamon, M. (1991) 'Post-synaptic action of ipsapirone on sleep-wakefulness regulation', Biol. Psychiatry 29, 306S.
5. Lainey, E., Tissier, M.H., Hamon, M., and Adrien J. (1991) 'The action of ipsapirone on sleep regulation involves primarily the stimulation post-synaptic 5-HT1A receptors', Sleep Res. 20A, 153.
6. Manfredi, R.C., Kale, A., Vgontzas, A.N., Bixler, E.O., Isaac, M.A., and

Falcone, M. (1991) 'Buspirone: sedative or stimulant effect', Am. J. Psychiatry 148, 1213-1217.
7. Seidel, W.F., Cohen, S.A., Bliwise, N.G., and Dement, W.C. (1985), 'Buspirone: an anxiolytic without sedative effect', Psychopharmacology 87, 371-373.
8. De Roeck, J., Cluydts, R., Schotte, C., Rouckhout, D., and Cosyns, P. (1989), 'Explorative single-blind study on the sedative and hypnotic effects of buspirone in anxiety patients', Acta Psychiatr. Scand. 79, 129-135.
9. Dzoljic, M.R., Saxena, P.R., and Ukponmwan, O.E. (1987) '5-HT1 receptor agonists enhance alertness', Sleep Res. 16, 88.
10. Dugovic, C. and Van den Broeck W. (1991) 'Sleep-wakefulness alterations produced by the putative serotonin 5-HT1C agonist m-chlorophenylpiperazine in the rat', Sleep Res. 20A, 129.
11. Lawlor, B.A., Newhouse, P.A., Balkin, T.J., Molchan, S.E., Mellow, A.M., Murphy, D.L., and Sunderland, T. (1991) 'A preliminary study of the effect of nighttime administration of the serotonin agonist, m-CPP, on sleep architecture and behavior in healthy volunteers', Biol. Psychiatry 29, 281-286.
12. Hoyer, D. (1988) 'Molecular pharmacology and biology of 5-HT1C receptors', Trends Pharmacol. Sci. 9, 89-94.
13. Dugovic, C., Wauquier, A., Leysen, J.E., Marrannes, R., and Janssen, P.A.J. (1989) 'Functional role of 5-HT2 receptors in the regulation of sleep and wakefulness in the rat', Psychopharmacology 97, 436-442.
14. Davenne, D., Dugovic, C., Franc, B., and Adrien, J. (1989) 'Ontogeny of slow wave sleep', in A. Wauquier, C. Dugovic and M. Radulovacki (eds.), Slow Wave Sleep: Physiological, Pathophysiological and Functional Aspects, Raven Press, New York, pp. 21-30.
15. Monti, J.M., Piñeyro, G., Orellana, C., Boussard, M.J., Jantos, H., Labraga, P., Olivera, S., and Alvarino, F. (1990) '5-HT receptor agonists 1-(2,5-dimethoxy-4-iodophenyl)-2-aminopropane (DOI) and 8-OH-DPAT increase wakefulness in the rat', Biog. Amines 7, 145-151.
16. Dugovic, C. and Wauquier, A. (1987) '5-HT2 receptors could be primarily involved in the regulation of slow-wave sleep in the rat', Eur. J. Pharmacol. 137, 145-146.
17. Tortella, F.C., Echevarria, E., Pastel, R.H., Cox, B., and Blackburn, T.P. (1989) 'Suppressant effects of selective 5-HT2 antagonists on rapid eye movement sleep in rats', Brain Res. 485, 292-300.
18. Stutzmann, J.M., Eon, B., Lucas, M., Blanchard, J.C., and Laduron, P.M. (1992) 'RP 62203, a 5-Hydroxytryptamine2 antagonist, enhances deep NREM sleep in rats', Sleep 15, 119-124.
19. Dijk, D.J., Beersma, D.G.M., Daan, S., and van den Hoofdakker, R.H. (1989) 'Effects of seganserin, a 5-HT2 antagonist, and temazepam on human sleep stages and EEG power spectra', Eur. J. Pharmacol. 171, 207-218.
20. Idzikowski, C., Mills, F.J., and Glennard, R. (1986) '5-Hydroxytryptamine-2

antagonist increases human slow wave sleep', Brain Res., 378, 164-168.
21. Sharpley, A.L., Solomon, R.A., Fernando, A.I., da Roza Davis, J.M., and Cowen, P.J. (1990) 'Dose-related effects of selective 5-HT2 receptor antagonists on slow wave sleep in humans', Psychopharmacology 101, 568-569.
22. Idzikowski, C., Cowen, P.J., Nutt, D., and Mills, F.J. (1987) 'The effects of chronic ritanserin treatment on sleep and the neuroendocrine response to l-tryptophan', Psychopharmacology 93, 416-420.
23. Adam, K. and Oswald, I. (1989) 'Effects of repeated ritanserin on middle-aged poor sleepers', Psychopharmacology 99, 219-221.
24. Ruiz-Primo, E., Haro, R., and Valencia, M. (1989) 'Polysomnographic effects of ritanserin in insomniacs: a crossed double-blind controlled study', Sleep Res. 18, 72.
25. Lammers, G.J., Arends, J., Declerck, A.C., Kamphuisen, H.A.C., Schouwink, G., and Troost, J. (1991) 'Ritanserin, a 5-HT2 receptor blocker, as add-on treatment in narcolepsy', Sleep 14, 130-132.
26. Paiva, T., Arriaga, F., Wauquier, A., Lara, E., Largo, R., and Leitao, J.N. (1988) 'Effects of ritanserin on sleep disturbances of dysthymic patients', Psychopharmacology 96, 395-399.
27. Staner, L., Kempenaers, C., Simonnet, M.P., Fransolet, L., and Mendlewicz, J. (1992) '5-HT2 receptor antagonism and slow-wave sleep in major depression', Acta Psychiatr. Scand. 86, 133-137.
28. da Roza Davis, J.M., Sharpley, A.L., and Cowen, P.J. (1992) "Slow wave sleep and 5-HT2 receptor sensitivity in generalised anxiety disorder: a pilot study with ritanserin', Psychopharmacology 108, 387-389.
29. Rieman, D., Berger, M., and Olbrich, R. (1991) 'The effect of ritanserin on negative symptoms and sleep in schizophrenic patients', Biol. Psychiatry 20, 401S.
30. Monti, J.M. and Alterwain, P. (1991) 'Ritanserin decreases alcohol intake in chronic alcoholics', The Lancet 337, 60.
31. Adrien, J., Tissier, M.H., Lanfumey, L., Hay-Dahmane, S., Jolas, T., Franc, B., and Hamon, M. (1992) 'Central action of 5-HT3 receptor ligands in the regulation of sleep-wakefulness and raphe neuronal activity in the rat', Neuropharmacology 31, 519-529.
32. Guldner, J., Rothe, B., Lauer, C., Pollmächer, T., Steiger, A., Spiegel, R., and Holsboer, F. (1991) 'Effects of a 5-HT-3 receptor antagonist on sleep-EEG and sleep-associated secretion of cortisol and growth hormone', Biol. Psychiatry 29, 326S.
33. Sharpley, A.L., Ellis, P.M., and Cowen, P.J. (1991) 'The effect of the 5-HT3 antagonist ondansetron on sleep in humans', Biol. Psychiatry 29, 597S
34. Meert, T.F., Awouters, F., Niemegeers, C.J.E., Schellekens, K.H.L., and Janssen, P.A.J. (1991) 'Ritanserin reduces abuse of alcohol, cocaine and fentanyl in rats' Pharmacopsychiatry 24, 159-163.

35. Gawin, F.H. and Kleber, H.D. (1986) 'Abstinence symptomathology and psychiatric diagnosis in cocaine abusers.', Arch. Gen. Psychiatry 43, 107-113.
36. Dugovic, C., Meert, T.F., Ashton, D., and Clincke G.H.C. (1992) 'Effects of ritanserin and chlordiazepoxide on sleep-wakefulness alterations in rats following chronic cocaine treatment', Psychopharmacology 108, 263-270.

Index

α_1-adrenergic receptors 52
α1-adrenoceptors 151
α-adrenoceptors 297
acetylcholine 297, 299, 300, 302, 321, 332, 355
acetylcholine bromide 298
acetylcholine release 407, 415
actinomycin D 166
acute cisplatin-induced emesis 190
acute dystonic reactions 187
acute emesis 188
acute myocardial infarction 307
adenyl cyclase activity 355
adenylate cyclase 73, 85, 367
adenylyl cyclase 56, 60, 107, 331
adrenal hormones 240
adrenocorticotropic hormone 273
adriamycin 160, 166
afferent nerve fibers 166
affinity chromatography 66
Age Associated Memory Impairments (AAMI) 407, 415
aggressiveness 255
AIDs 192
albino BKW mice 408
alcohol 255
alcohol dependence 421
alcohol drinking 375
alcohol-preferring P line 375
alcoholism 201, 255
alizapride 187, 191
ALK-3 347
amino acid sequences 61
aminotetralins 51
amygdala 69
angina 297
anorexia 188, 394
anorexia nervosa 394
anorexic and bulimic states 393
anorexic disorders 395
ANOVA 173, 410, 411, 413, 414
anterior hypothalamus 357
anterior raphe nuclei 65
anti-5-HT$_{1A}$ receptor antibodies 66, 67, 70
anti-emetics 99
anti-tumor cytotoxicity 345

anticholine acetyltransferase antibodies 70
anticipatory emesis 188
anticipatory vomiting 160
antidepressants 13, 200
antidopaminergic toxicity 162
antiemetic 101
antiemetic activity 161
antiemetic effects 165
antiemetic efficacy 165
antiemetic therapy 188
antihypertensive drug 151
antimigraine drugs 127
antineoplastic drugs 171
antipsychotic 271
antipsychotic drugs 271, 289
anxiety 60, 205, 367, 368, 369, 395, 418
anxiety disorders 201, 223, 249, 420
anxiolytic 370
anxiolytic agents 188
apomorphine 166
aporphines 51
appetitive behavior 370
arachidonic acid 16
architecture residues 124
area postrema 165, 166, 167, 183
aromatic box 119, 125
arousal 159
arousal mechanisms 417
arterial blood pressure 141
arteriovenous anastomoses 128
aryloxyalkylamines 52
arylpiperazines 50, 52
aspiration pneumonia 191
astrocytes 69
atherosclerosis 321
atypical antipsychotic 289
autoreceptors 367
aversive properties 408
azipirones 223
β-adrenergic antagonism 52
β-adrenergic blockers 206

B cell proliferation 355
B_{max} 60
Bay R 1531 401

behaving animals 231
benzamides 53, 110
benzimidazolones 110
benzoates 53
blood pressure 151
BMY 42568 56
BMY 7378 56, 60, 393
BMY 738 347
bone-marrow-derived murine macrophages 346
brainstem 367
BRL 24924 (renzapride) 60
BRL 46470A 292
bulimia nervosa 394
buspirone 369, 370, 394, 409, 418
buspirone analogue 56
butyrophenones 160

c-fos protein 166
C-fos-like immunoreactivity 33
CA_1 331
CA_3 331
CA_3 pyramidal neurons 56
Ca^{2+}-activated Cl^- calcium antagonists 408
cancer 171
cannabinoids 188
carbamates 53
carbazoles 53
carbohydrate 385
carboxylic acid derivatives 53
cardiovascular disturbances 191
carinii pneumonia (PCP) 192
cat 166
catalase 345, 349
CD16 354
CD25 356
CD56 345, 354
cDNA 66
cell surface antigens 356
central gray 69
central hypotensive activity 151
central nervous system (CNS) 65, 329, 333, 353, 355, 356, 357
cerebellum 69
cerebral cortex 59, 69
cerebral ischemia 401
cerebral vasodilators 408
cerebrospinal fluid 255, 279
CGRP 128
CGS-93129 395

channels 75
ChAT levels 408, 412
chemoreceptor trigger zone (CTZ) 160, 187
chemotherapeutic agents 165
chemotherapy 165, 172, 187, 188
chemotherapy- and radiotherapy-induced nausea and vomiting 172
chemotherapy/radiation-induced sickness 166
chlordiazepoxide 395
chlorpromazine 190
cholera toxin 56
choline acetyl-transferase 69
cholinergic 160
cholinergic agents 407
cholinergic deficits 407
cholinergic neurones 65
cholinomimetic agents 408
cholinomimetics 407
choroid plexus 69
chronic alcoholics 420
chronic cocaine treatment 417
chronotropic effects 137
CIE 191
cinanserin 419
circadian rhythms 240
cisapride 53, 60
cisplatin 160, 161, 162, 165, 166, 188, 189, 190
cisplatin-induced emesis 187
cisplatinum 171, 172, 173, 174, 175, 176
citalopram 370
clomipramine 370
clozapine 290, 369
CNS 5-HT activity 353, 355
co-trimoxazole 192
cocaine 13, 171, 361
cocaine abuse 421
cocaine analogs 359, 361
cocaine intake 421
cocaine metabolite 359
complete loop (aa 216–350) 65
compulsion to drink 421
Con-A stimulated T cell blastogenesis 355
conceptual disorganization 272
copper sulphate 166
coronary arteriography 321
coronary artery disease 322
coronary artery stenosis 307
cortical 407
cortical pyramidal cells 59

corticosteroids 162, 188
cortisol 239, 273
CP-93129 394
CP-94253 394
cranial circulation 128
5-CT (5-carboxamidotryptamine) 25, 60, 332
cyanopindolol 394
cyclic AMP 107
cyclic AMP-dependent protein kinase 107
cycloheximide 166
cyclophosphamide 160, 162, 166, 171, 172, 174, 175, 176, 190
cyclophosphamide/anthracyclin 187
cyclosporine 332
cyclosporine A 355
cyproheptadine 347, 354, 394
cytokine production 354
cytokines 333, 356
cytolytic activity 349
cytotoxic drug 180
cytotoxic-induced emesis (CIE) 187, 188

d-fenfluramine (d-FEN) 349, 354, 355
d-norfenfluramine (d-NF) 393, 395
D2 dopamine 52
DA antagonists 394
DA-dependent gnawing 394
DA release 368
DA_2 271
dacarbazine 171, 172, 174, 176, 188
DAU 6285 110
deep slow wave sleep (SWS) 417
dehydration 188, 191
delayed emesis 188
delayed recall 415
dentate gyrus 68, 332
depolarization 301
depression 61, 211
depressive disorders 58
desensitization 103
dexamethasone 161, 162, 171, 172, 174, 176, 187, 188, 190
diarrhea 161
diazepam 412
dihydroergotamine 127
5,7-dihydroxytryptamine (5,7-DHT) 69, 329
2,5-dimethoxyphenylisopropylamine (2,5-DMA) 51
dilantin 408

diphenhydramine 162, 190
distribution 107
diurnal activity rhythm 255
DOB 53, 419
DOI 53, 419
dopamine (DA) release 367
dopamine-2 271
dopamine 289
dopamine activity 370
dopamine antagonists 188
dopaminergic (D_2) receptors 20, 160
dorsal horn 69
dorsal raphe cells 395
dorsal raphe neurons 55
dorsal raphe nucleus 68, 359, 375
dorsal subiculum 69
doxorubicin 162, 166
dronabinol 188
droperidol 188, 191
Duncan's multiple rank test 173
Dunnett's test 410, 411, 412, 413, 414
dura mater 129
dysthymic patients 420
dystonic reactions 191

eating disorders 383
eighth cranial nerve 160
electroconvulsive shocks 58
electrolesions 408
electrolyte imbalance 188
emesis 159, 165, 167, 171, 173, 179
emetic reflex 167
emetic reflex arc 159
emetic response 171
emetogen 165
emetogenecity 175, 176
endothelial injury 307
endothelin 301
enterochromaffin 176
enterochromaffin cells 166
enteroendocrine cells 176
entorhinal cortex 69
equivalent gene 22
ergolines 50, 51
ergometrine 297, 298, 299, 300, 302
ergometrine maleate 298
ergonovine 297
ergot alkaloids 33
ergotamine 127

escape latency 413
estrogen 397
ethanol abuse 368
ethanol consumption 367
ethanol-induced dopamine release 368
ethanol intoxication 367
ethanol-related behaviors 367
ethanol withdrawal 367
extrapyramidal effects 188
extrapyramidal side-effects 271
extravasation of plasma 128

F344 rats 354
facial motor nucleus 59
feeding behavior 384
fenfluramine (d-F) 393
fenfluramine 337
fentanyl 421
ferret 165, 172, 176
ferret ileum 166
first dose (peak level) 163
flesinoxan 56, 154, 206
flourouracil 162
5-fluorouracil 160
5-HIAA 173, 174, 175, 176, 355, 357
fluoxetine 370, 384
food reinforced alternation 409
forskolin 331, 355
fourth afferent pathway 160
fourth ventricle 160
frontal and entorhinal cortex 68
full agonists 55
fusion protein 67

γ-carbolines 53
G-protein 56, 73, 367
G-protein-coupled receptors 120
G protein-coupled superfamily 65
$G_{i/o}$ proteins 55, 58
G_S 107
G_S proteins 56
gag 159
gastrointestinal tract 176
Gellerman schedule 412
genetic subtype 22
gepirone 55, 56, 369, 401, 409
glutamate 59
glutathione-S-transferase 65
GR 113808 110

granisetron 135, 165, 166, 187, 190, 408
granulocytes 345
guinea pig 166
gynecological laparoscopy 192
gynecological surgery 191

H_2O_2 356
habituation 408, 409
habituation profiles 410, 411
hallucinatory behavior 272
haloperidol 271, 395
headache 38, 161
hemicholinium-3 (HC-3) 412
heterogeneity of the 5-HT_3 receptor 101
^{3}H-paroxetine 198, 199
$5HT_{1D}$ 19
5-HT 51, 69, 167, 171, 172, 173, 175, 176, 297, 288, 299, 301, 302, 329, 330, 333, 353, 354, 355, 356, 357
5-HT autoreceptors 16, 60
5-HT behavioral syndrome 152
5-HT creatinine sulphate 298
5-HT neurons 55
5-HT receptor agonists 302
5-HT receptors 51, 55, 353, 354
5-HT release 174, 175, 177, 354
5-HT-releasing properties 176
5-HT reuptake blockers 205
5-HT reuptake system 60
5-HT syndrome 151
5-HT transporter (5-HTT) 9
5-HT turnover 166
5-HT uptake carrier 55, 60
5-HT_1 331
5-HT_1 receptors 23, 41, 332, 384
5-HT_1/5-HT_2 antagonists 354
5-HT_1-like receptors 73, 85, 297, 301, 302, 326
5-HT_{1A} 21, 25, 50, 51, 52, 53, 65, 331, 332
5-HT_{1A} agonists 55, 141, 241, 393
5-HT_{1A} binding 55
5-HT_{1A} ligands 409, 412, 415
5-HT_{1A} partial agonists 224
5-HT_{1A} receptor 21, 50, 65, 67, 73, 151, 354, 356, 359, 407, 417, 418
5-HT_{1A} receptor agonists 205, 206
5-HT_{1A} receptor antagonists 206
5-HT_{1A} receptor antibodies 70
5-HT_{1A} receptor binding sites 69
5-HT_{1A} receptor ligands 70

5-HT$_{1A}$ receptor protein 70
5-HT$_{1A}$ receptor subtype 408
5-HT$_{1A}$ receptors 55, 179, 330, 332, 355, 375, 401
5-HT$_{1A}$ specific agonists 354
5-HT$_{1B/D}$ 332
5-HT$_{1B/D}$ heteroreceptors 33
5-HT$_{1B/1D}$ receptors 355
5-HT$_{1B}$ 21, 25, 52, 68, 69, 85, 331, 332, 393
5-HT$_{1B}$/5-HT$_{1D}$ receptors 86
5-HT$_{1B}$ receptors 58, 59, 74
5-HT$_{1C/2}$ antagonist ketanserin 58
5-HT$_{1C/2}$ antagonist ritanserin 59
5-HT$_{1C/2}$ antagonists 205, 226
5-HT$_{1C/2}$ receptor antagonists 208
5-HT$_{1C/2}$ receptors 368
5-HT$_{1C}$ 51, 52, 68, 332, 393
5-HT$_{1C}$ receptors 53, 59, 69, 75, 417
5-HT$_{1D\alpha}$ 21, 85
5-HT$_{1D\alpha}$ gene 25
5-HT$_{1D\beta}$ 21, 85
5-HT$_{1D\beta}$ receptor 394
5-HT$_{1D}$ receptor clone 23
5-HT$_{1D}$ 23, 85, 332
5-HT$_{1D}$ ligands 394
5-HT$_{1D}$ receptors 26, 52, 59, 76, 331, 332
5-HT$_{1D}$ specific binding 332
5-HT$_{1E}$ 21
5-HT$_{1E\beta}$ 25
5-HT$_{1E}$ 21, 25, 332
5-HT$_{1E}$ receptor 25
5-HT$_{1F}$ 21, 25, 85
5-HT$_{2/1C}$ receptors 51
5-HT$_2$ 52, 53, 68, 271, 301, 332, 417
5-HT$_2$/5-HT$_{1C}$ agonists 242
5-HT$_2$ antagonists 263, 354
5-HT$_2$ receptors 25, 41, 50, 59, 179, 212, 278, 297, 331, 355
5-HT$_3$ 51, 53, 176
5-HT$_3$ agonists 100
5-HT$_3$ antagonist 159, 171, 189, 192, 205, 354, 378
5-HT$_3$ receptor antagonist 409, 165, 166, 168, 179, 407, 408, 412, 413, 414, 415
5-HT$_3$ receptors 41, 50, 52, 59, 68, 99, 161, 165, 167, 171, 172, 175, 177, 330, 368
5-HT$_4$ 53
5-HT$_4$-receptor agonist BMIU8 [10] 301
5-HT$_4$ receptors 41, 50, 51, 60, 107, 135, 184

5-HT$_5$ 85
5-HT$_5$ receptor 25, 41
5-HT$_6$ 25, 51
5-HTP 329
5-HTT mRNA 9
high-affinity binding 330
high-dose radiotherapy 187
higher cortical centers 187
hippocampus 55, 59, 68, 101, 330, 332, 355
hippocampus pyramidal neurons 56
histamine 166, 302
histaminergic 160
homeostasis 367
HPLC 173
hu 5-HT$_{1D\beta}$ 26
human gene 86
human isolated small coronary arteries 297
human lymphocytes 330
human touncils 330
humoral pathway 160
hydergine 408
hydrogen peroxide 349
hydroxy indole acetic acid 330
5-hydroxyindoleacetic acid (5-HIAA) 171, 172, 255, 278, 330
5-hydroxytryptamine 329
hyoscine 192
hyperphagia 384, 393
hyperpolarization 55, 60, 367
hypolocomotion 395
hypophagia 384
hypothalamic 393
hypothalamus 55, 395

ibotenic acid 412
ICI 170,809 419
^{125}I-Cyanopindolol 332
ICS 205-930 420
ICV 330
IFN-gamma 346
ifosfamide 162
IL-1 receptors 333
IL-1 secretion 356
ileum 102
imipramine 330
immune response 330
immune system (IS) 329, 333, 337, 353, 354
immunocompetent cells 329, 330, 353, 354
immunostaining 68

indole 53
indolealkylamines 50, 51
indoles 110
inferior and superior colliculi 69
inotropic effect 138
insomnia 418
insomniacs 420
interferons 346
interleukin-1 (IL-1) 330, 355, 357
interleukin-2 (IL-2) 330, 346, 355, 356, 357
interneurons 59
interpeduncular nucleus 69
intracellular calcium 302
ipsapirone 401, 409, 418, 420
irindalone 53
ischemic heart disease 321

K_d 60
K^+ 302
K^+ conductance 73
ketanserin 52, 53, 298, 301, 313, 326, 347, 354
keto compounds 53
keto indoles 53
KSS 298, 299, 301

lacunosum 331
large granular lymphocytes 337
lateral septum 55
lesopitron 407, 408, 409, 412
ligand binding assays 166
ligand-gated ion channel 102
light/dark cycle 385
lightheadedness 161
limbic areas 65
lipopolysaccharide (LPS) 355
Lister Hooded rats 409, 413
lisuride 166
localization of 5-HT_{1B} and 5-HT_{1D}
locomotor performance 413
locus coeruleus 59
lorazepam 188
low central serotonin syndrome 257
LSD psychosis 277
LY 297524 110
LY 297582 110
lymphocytes 329, 354
lymphokine production 345

m-chlorophenylpiperazine 282

macrophages 329
major depression 420
major histocompatibility (MHC) products 345
Mantel-Haenszel test 173
MAO promotor activities 7
marmosets 408, 414, 415
mCPP 52, 58, 369, 394, 418
MDL 72222 165, 166, 369, 420
mechloroethamine 166
medial prefrontal cortex 60
median raphe nucleus 69
median septum 65, 69
melphalan 162
membrane potential (E_m) 298
membrane potential 297
memory impairments 407, 410, 411, 415
mental disorders 417
mesocortical systems 407
mesolimbic DA system 375
meta-chlorophenylpiperazine (MCPP) 273
methiothepin 58, 301
methoxyphenethylamines DOM 419
methylprednisolone 187, 188
5-methoxytryptamine 60
5-methyl-urapidil 154
2-methyl-5-HT 59, 60
metoclopramide 159, 160, 161, 162, 166, 171,
    172, 174, 175, 176, 187, 188, 189, 190, 191
metocopramide-dexamethasone 175
mianserin 369, 394
mice thymus 330
microtubule-associated protein MAP-2 69
midbrain dopamine (DA) system 263
migraine 127
MK 212 394
modulation of transmitter release 103
molecular biology of molecular modelling 119
monoamine oxidase (MAO) A and B genes 1
monocytes 345
morphine 166
Morris water maze 414
Morris water maze task 413
motion sickness 160, 166, 187, 192
motor activity 231
mouse habituation model 407
mRNA 59, 331, 354
muramyl dipeptide (MDP) 332
murine thymocytes 354
muscarinic receptors 302

mustine 166

N-alkylpiperidines 52
N4-substituted arylpiperazines 50
Na+/Cl−-dependent serotonin
Na+/K+ ATPase 198
nabilone 188
name-face association 415
1-naphthylpiperazine (1-NP) 50, 52
narcoleptic 420
natural killer (NK) arm 353
natural killer (NK) cell 337, 345, 353, 354, 355
nausea 159, 161, 165, 173, 174, 176, 187
negative feedback 58
neural cell adhesion molecule 345
neuroendocrinologic 273
neuroimmune interactions 353
neuron-specific enolase 69
neurophysiology 231
neuroprotection 401
neurotensin 289
neurotransmitter systems 353
nigro-striatal system 58
NK-cell responsiveness 330
NK-cell surface antigen (CD 16, C56) 356
NK-cell surface proteins 354
N,N,N,-trimethyl quaternary amine analog of 5-HT 51
nomenclature 41
non-5-HT$_3$ receptor mechanism 171
nonadaptive immunity 345
noncisplatin chemotherapy-induced emesis 190
nootropic agents 408
noradrenaline 301, 302
norfenfluramine 384
nucleus accumbens 59, 101, 368, 370, 375
nucleus basalis lesions 409, 410, 411
nucleus basalis magnocellularis 407, 408, 412
nucleus raphe dorsalis 418
nucleus tractus solitarius 101, 165, 166, 330

obesity 201
object discrimination 415
object discrimination and reversal learning tasks 414
object reversal discrimination task 408
obsessive compulsive disorders 231, 249, 284
octretoide 175
8-OH-DPAT 51, 55, 141, 154, 331, 347, 354, 355, 393, 401, 418, 419
8-OH-DPAT-induced hypothermia 394
old age 407, 408, 410, 411
ondansetron 161, 162, 165, 166, 167, 168, 171, 172, 174, 175, 177, 187, 188, 189, 190, 192, 347, 354, 369, 370, 407, 408, 409, 410, 412, 413, 414, 415, 420
ophthalmic surgery 191
opioid mechanisms 164
optimum antiemetic cocktail 190
osmotic minipumps 60
oxygen metabolites 345

paired t test 415
parachlorophenylalalnine 172
paraventricular nucleus (PVN) 330, 395
paroxetine 60
partial agonists 55
perforin 346
peripheral 160
peripheral blood cells 342
pertussis toxin 58
pGEX-KG plasmid 66
pharmacological subtype 20
pharmacology 107
phenothiazines 160
phenylalkylamines 53
phenylbiguanide 60
phospholipase C 75
phosphorylation sites 16
piglet atrium 136
pilocarpine 166
pindolol 52, 243, 347, 395
piracetam-type drugs 408
piriform cortex 59
pituitary hormones 212, 240
placebo 187, 192
plasma 5-HT 166
platelet 5-HT 176
platelet activating factor 307
platelet aggregation 307
platelets 212
pneumocystics 192
polymerase chain reaction (PCR) 66
postoperative nausea and vomiting (PONV) 167, 191, 192
postsynaptic 5-HT$_{1A}$ receptors 57, 393
postsynaptic neurons 55
postsynaptic receptors 367

potassium channels 55
prazosin 154
pregnancy 187
prejunctional 5-HT_1 receptors 38
primate object discrimination and reversal task 407
probenecid 279
prochlorperazine 188, 190, 191
prolactin 239, 273
proliferation 345
propranolol 52
protein intake 385
protein kinases 18
protoveratrine A 166
protracted nausea 192
pseudorandum 412
pseudorandum schedule 414
psychiatric disorders 99
psychological depression 188
psychoneuroimmunology 329
psychotropic drugs 239
pyramidal neurons 59

quinolines 110
quisqualic acid 56

rabies virus 332
radiation 166
radiation-induced emesis 165
radioligand binding 55
radiotherapy 187, 188, 190
radiotherapy-induced emesis 162
raphe area 329
raphe nuclei 55
raphe nuclei of the mesencephalon 367
raphe presynaptic 5-HT_{1A} receptors 393
rapid eye movement (REM) 417, 418
rat spleen 330
rat T-maze 407
RDC4 23
reactive oxygen metabolites 349
receptor architecture 125
receptor binding 152
receptor characterization 41
receptor classification 41
receptor desensitization 111
receptors 88, 159
recombinant interleukin 1 (rIL-1) 357
reinforced alternation training 409

relapse prevention during abstinence 421
repolarization 301
reserpine 9
respiration 159
reticularis nucleus 69
reversal learning task 415
reverse benzoates 53
reverse transcriptase-polymerase chain reaction (RT-PCR) 331
reward 370
risperidone 271, 291
ritanserin 224, 263, 271, 291, 370, 417, 419
RP62203 419
RS-23597-190 110
RU 24969 58, 394, 418
ryanodine 302

SB 203186 110, 135
SC-53606 110
schizophrenia 263, 271, 277, 289
schizophrenic psychosis 271
schizophrenicin 271
schizophrenics 420
scopolamine 407, 408, 409, 410, 411, 412, 413
scopolamine-induced deficits 414
SDZ 205, 557 110, 135
sedation 191
sedative 370
sedative drugs 369
seganserin 420
selective 5-HT uptake inhibitors 282
selective serotonin reuptake inhibitors 223, 249
septum 68
serotonergic (5-HT_3) neurotransmitter receptors 159
serotonergic ligands 354
serotonergic receptors 161
serotonin (5-hydroxytryptamine) 297
serotonin 172, 231, 412
serotonin homeostasis 197
serotonin receptor subtypes 407
serotonin receptors 49
serotonin-selective 225
serotonin transport processes 360
serotonin transporter 197
serotonin uptake inhibitors 197, 200
sertindole 289
sex differences 393
sheep-red-blood-cell (SRBC) 329, 355

single-cell recording 55
sleep disturbances 417
sleep efficiency 418, 420
sleep impairments 418
sleep maintenance 420
sleep-promoting effect 420
sleep regulation 417
sleep time 418
sleep-wakefulness patterns 418
slow excitatory response 60
sodium pentobarbitone 412
somato-dendritic 5-HT autoreceptors 69
somatodendritic 5-HT$_{1A}$ autoreceptors 56, 57
somatodentric autoreceptor 361
somatostatin analog (octreotide) 171
somatostatin analog 175
somatostatin-insensitive mechanism 176
species homologue 22
spinal cord 59
spiperone 53, 56, 59
spiroxatrine 153
spleen 338
splenocytes 342
SRBC 330, 331, 357
stereotaxic techniques 412
stratum Oriens 331
stratum radiatum 331
stress 240
stress-induced hyperthermia 205
striatal neurons 59
striatum 69, 332
subiculum 332
substance P 301
substantia nigra 69, 332
substituted benzamides 160
suicide attempts 259
sumatriptan 25, 33, 127, 297, 298, 399, 300, 301, 302
superfusion 58
superoxide dismutase 349
supraoptic nucleus 330
suspiciousness 272
swallowing 159
sympathetic nerve activity 141
sympatholytic effects 141
synaptosomal preparations 332

T cells 346
T-maze 409

T-maze reinforced alternation task 412
T-maze task 409
tachycardia 138
tail-pinch induced feeding 396
tandospirone 56
tefludazine 53
terminal 5-HT autoreceptor 59
testosterone 255
TFMPP 58, 332, 394
thalamus 69
therapeutic significance 107
third afferent pathway 160
third intracellular loop 66
thromboxane A$_2$ 307
thromboxane mimetic U46619 302
7TM (7 transmembrane domains) 22
tonic activation 58
total body irradiation 190
tractis nucleus solitarius 159
transaminase 161
transporter 11
tricyclic antidepressant drugs 58
trimelalol 166
trimethobenzamide 395
tritiated 5-HT 330
tritiated acetylcholine 415
tropisetron 135, 165, 187, 191, 407, 408, 409, 410, 412, 413, 414, 415
tryptophan 211
tryptophan hydroxylase 213
tryptophan hydroxylase gene 255
5-TT$_{1E}$ receptor 77

ultrasonic distress vocalizations 205
unstable angina 307
unusual thought content 272
urapidil 151
ureas 53
urine 5-HIAA 166

vagal afferent nerve 187
vagotomy 166
vagus nerve 102, 165, 166, 167
vascular endothelium 321
ventricular system 412
vesicular monoamine transporters 15
vestibular apparatus 187
vestibular pathway 160
vigilance states 418

visual 413
voluntary ethanol consumption 370
vomiting 159, 165, 174, 176, 187
vomiting center 159, 166, 167, 187

wakefulness 418
water maze task 408
water maze test 407
WAY 100135 393
Wisconsin general test apparatus 414
wound rupture 191

YAC-1 355

zacopride 60, 166, 407, 408, 409, 410, 412, 413, 414, 415, 420
zydovudine 192
[^{3}H]-5HT 60, 330, 332
[^{3}H]5-HT receptors 332
[^{3}H]8-OH-DPAT 65, 69
[^{3}H]ketanserin 332

Medical Science Symposia Series

1. A. M. Gotto, C. Lenfant, R. Paoletti (eds.) and M. Soma (ass.ed.): *Multiple Risk Factors in Cardiovascular Disease.* 1992   ISBN 0-7923-1938-9
2. A. L. Catapano, A. M. Gotto, Jr., L. C. Smith and R. Paoletti (eds.): *Drugs Affecting Lipid Metabolism.* 1993   ISBN 0-7923-2232-0
3. T. Godfraind, S. Govoni, R. Paoletti and P. M. Vanhoutte (eds.): *Calcium Antagonists. Pharmacology and Clinical Research.* 1993
   ISBN 0-7923-2259-2
4. D. Galmarini, L. R. Fassati, R. Paoletti and S. Sherlock (eds.): *Drugs and the Liver: High Risk Patients and Transplantation.* 1993   ISBN 0-7923-2307-6
5. P. M. Vanhoutte, P. R. Saxena, R. Paoletti, N. Brunello (eds.) and A. S. Jackson (ass.ed.): *Serotonin. From Cell Biology to Pharmacology and Therapeutics.* 1993   ISBN 0-7923-2518-4

KLUWER ACADEMIC PUBLISHERS – DORDRECHT / BOSTON / LONDON